Essentials of Oceanography

Essentials of Oceanography

SECOND EDITION

TOM GARRISON
Orange Coast College
University of Southern California

BROOKS/COLE
™
THOMSON LEARNING

Australia • Canada • Mexico • Singapore • Spain • United Kingdom • United States

BROOKS/COLE

THOMSON LEARNING

Sponsoring Editor: *Nina Horne*
Development Editor: *Mary Arbogast*
Project Development Editor: *Kristin Milotich*
Marketing Team: *Rachel Alvelais, Mandie Houghan,*
  Laura Hubrich, Carla Martin-Falcone
Editorial Assistant: *John-Paul Ramin*
Production Coordinator: *Laurel Jackson*
Production Service: *Joan Keyes, Dovetail Publishing Services*
Manuscript Editor: *Lorraine Anderson*
Permissions Editor: *Catherine Murray Gingras*
Interior Design: *Adriane Bosworth*

Media Editor: *Pat Waldo*
Indexer: *Nancy Ball*
Cover Design: *Roy R. Neuhaus*
Cover Photo: *Douglas Peebles/CORBIS*
Interior Illustration: *Precision Graphics*
Photo Researcher: *Myrna Engler*
Print Buyer: *Vena Dyer*
Typesetting: *G & S Typesetters, Inc.*
Printing and Binding: *R. R. Donnelley & Sons, Inc./*
  Willard Mfg. Division
Cover Printing: *Phoenix Color Corp.*

For more information about this or any other Brooks/Cole product, contact:
BROOKS/COLE
511 Forest Lodge Road
Pacific Grove, CA 93950 USA
www.brookscole.com
1-800-423-0563 (Thomson Learning Academic Resource Center)

Printed in United States of America

10 9 8 7 6 5 4 3 2 1

Library of Congress Cataloging-in-Publication Data

Garrison, Tom 1942–
    Essentials of oceanography / Tom Garrison.—2nd ed.
        p.   cm.
    Includes bibliographical references and index.
    ISBN 0-534-37732-7
    1. Oceanography.   I. Title.
    GC11.2 .G36   2000
    551.46—dc21
                                    00-026039

To my family and my students:
My hope for the future.

Brief Contents

Contents

Preface

This book was written to provide an *interesting,* clear, and relatively brief overview of the marine sciences. It is designed for college and university students who are curious about the Earth's largest feature but who may have little or no formal background in science. Indeed, oceanography—the study of the ocean—is an ideal class for students wishing to fulfill general education requirements. Oceanography is broadly interdisciplinary; students are invited to see the connections between astronomy, economics, physics, chemistry, history, meteorology, geology, and ecology—areas of study they once considered separate. It's no surprise that oceanography courses have become increasingly popular in the last decade.

Students bring a natural enthusiasm to their study of this field. Even the most indifferent reader will perk up when presented with stories of encounters with huge waves, photos of divers and giant squids, tales of exploration under the best and worst of circumstances, evidence that vast chunks of the Earth's surface slowly move, micrographs of glistening diatoms, and data showing the economic importance of seafood and marine materials. If pure spectacle is needed to generate an initial interest in the study of science, oceanography wins hands down!

In the end, however, it is subtlety that triumphs. Studying the ocean reinstills in us the sense of wonder we all felt as children when we first encountered the natural world. The story of the ocean is a story of change and chance; its history is written in the rocks, the water, and the genes of the millions of organisms that have evolved here. My goal is to help the students who use this book to gain an oceanic perspective. "Perspective" is the ability to view things in terms of their relative importance or relationship to one another. An oceanic perspective lets you see this misnamed planet in a new light and helps you plan for its future. You will see that water, continents, seafloors, sunlight, storms, seaweeds, and society are connected in subtle and beautiful ways.

How This Book Is Organized

A broad view of oceanography is presented in 15 chapters, each freestanding (or nearly so) to allow an instructor to assign chapters in any order he or she finds appropriate. The text begins with a brief history of our understanding of oceanography, then continues through the origin of the Earth and ocean, the structure and character of the ocean floor, a discussion of water and its movements and interactions with the more solid Earth, and an ecologically based discussion of marine life. It ends with a look at the character of human relationships with the ocean, especially the way marine resources have been managed and mismanaged.

Each chapter begins with an attractive **vignette**—a short observation, eyewitness account, or description of marine scientists at work. The chapters are written in an engaging style at a level appropriate to students not majoring in science. The number of technical terms is kept to a minimum, and when appropriate to their meanings, the derivations of words are presented. Some of the more complex ideas are initially outlined in broad brushstrokes; then the same concepts are discussed again after you have a clear view of the overall situation. **Measurements** are given in both metric and English systems. At the request of a great many students, we have written out the units (that is, we write "kilometer" rather than "km") to avoid ambiguity and make reading easier.

The **illustration program** is extensive; the photos, charts, graphs, and paintings have been chosen for their utility, clarity, and beauty. **Boxes** in some of the chapters present commentaries of special interest on unique topics or controversies. Chapters begin with a list of **key concepts** and an **outline,** and end with **questions** asked by students (**answers** are provided), a **summary,** and a list of important **terms and concepts** that are defined in an extensive **glossary** at the back of the book. **Study questions** are also included in each chapter. The brief **annotated bibliography** that follows the study questions will be helpful to anyone who wants to know more about a particular topic.

Appendixes will help you master measurements and conversions, geological time, latitude and longitude, and chart projections. In case you'd like to make oceanography your life's work, the last appendix discusses jobs in marine science.

Best of all, the book has been **thoroughly student-tested.** You need not feel intimidated by the concepts presented or the words used to describe them. Students just like you have mastered this material. Read slowly, and go step by step through any parts that give you trouble. Your predecessors have found the ideas presented here to be understandable, useful, inspiring, and applicable to their lives. Best of all, they have found the subject *interesting*!

The Web Site

This book is fully Internet integrated. You'll find its dedicated web site at **http://www.garrisonessentials2.com.** Internet icons at each head and subhead indicate text-specific information available on the site. Questions asked by students, chapter outlines, sample quizzes, flashcards, links, exercises, animations, and tutorial assistance are also accessible at the site. We will periodically update the web site to add features and announce important advances in marine science. *The site is open to anyone without cost or subscription.*

Suggestions for Using This Book

1. *Begin with a preview.* Scout the territory ahead: read the overview that begins the chapter; flip through the assigned pages, reading only the headings and subheadings; look at the figures and read captions that catch your attention.

2. *Keep a pen and paper handy.* Jot down a few questions—any questions—that this quick glance stimulates. Why is the deep ocean cold if the inside of the Earth is so hot? What makes storm conditions like those seen in 1997–1998? Where did sea salt come from? Will global warming actually be a problem? Does anybody still hunt whales? Writing questions will help you focus when you start studying.

3. *Now read in small but concentrated doses.* Each chapter is written in a particluar sequence and tells a story. The logical progression of ideas is going somewhere. Find and follow the organization of the chapter. Stop occasionally to review what you've learned. Flip back and forth to review and preview.

4. *Strive to be actively engaged!* Write notes in the margin, underline occasional passages (underlining whole sections is seldom useful), write more questions, draw on the diagrams, check off subjects as you master them, make flashcards while you read (if you find them helpful). *Use the book!*

5. *Monitor your understanding.* If you start at the beginning of the chapter, you will have little trouble understanding the concepts as they unfold. But if you find yourself at the bottom of the page having only scanned (rather than understood) the material, stop there and start that part again. Look ahead to see where we're going. Remember that other students have been here before, and I have listened to their comments to make the material as clear as I can. *This book was written for you.*

6. *Use the Internet sites.* We have references to more than 1,500 Internet sites throughout the text. If you have access to a computer, fire it up and scan the designated site as you read the associated passage of the book.

7. *Enjoy the journey.* Your instructor will be glad to share his or her understanding and appreciation of marine science with you—you have only to ask. Students, instructors, and authors all work toward a common goal: an appreciation of the beauty and interrelationships a growing understanding of the ocean can provide.

Acknowledgments

Jack Carey at Thomson Learning, the grand master of college textbook publishing, willed the first edition of this book into being. His suggestions have been combined with those of more than 100 peer reviewers and 450 undergraduate students, all of whom gave of their time and knowledge to help me assemble and organize the material. For this edition, I have especially depended on the advice of Steven Benham at Pacific Lutheran University, Donald Buchanan at San Bernadino Valley College, D.L. Clark at University of Wisconsin, Kathleen Flickinger at Maui Community College, Nancy Hinman at University of Montana, Scott D. King at Purdue University, Lawrence Krissek at Ohio State University, Ben leFebvre at City College of San Francisco, Gregory A. Mead at University of Florida, and Rene S. Revuelta at Miami-Dade Community College. As always, I am greatly indebted to my long-suffering departmental colleagues Dennis Kelly, Jay Yett, Robert Profeta, and Joyce Kai-Mott for putting up with me through this book's gestation. A corps of diligent teaching assistants led by Tim Riddle and Matt Bolen worked tirelessly on the book's Internet site. Thanks also to Stanley Johnson, our Dean, and Margaret Gratton, our college president, for supporting and encouraging the faculty to write, engage in community service, and conduct research. And again, gold medals should go to my loyal and supportive family for their patience with my long hours, late nights, crabby moods, and loud Glenn Gould Bach recordings.

Many of the illustrations for this book came from friends and acquaintances, who maintained their cooperative cheer through a blizzard of increasingly anxious e-mails and telephone calls. Bruce Hall, teacher and friend, again contributed photos, as did colleagues Norman Cole, Ron Romanosky, Andreas Rechnitzer, and Don Walsh. Herbert Kauainui Kane again donated his beautiful pictures of Hawaiian topics. Deborah Day and Cindy Clark at Scripps Institution, Jutta Voss-Diestelkamp at the Alfred Wegener Institut in Bremerhaven, Robert Headland at Scott Polar Research Institute, and David Taylor at the Centre for Maritime Research in Greenwich dug through their archives one more time. Don Dixon provided paintings, Dan Burton sent photos, and Andrew Goodwillie printed customized charts. Wim van Egmond contributed striking photomicrographs of diatoms. Bill Haxby at Lamont-Doherty Earth Observatory provided truly beautiful seabed scans. The team at Woods Hole Oceanographic Institution was generous, as always, in providing photographs and diagrams for inclusion. The U.S. Coast Guard, U.S. Navy, Army Corps of Engineers, Breitling-SA, Associated Press, The Smithsonian, NASA, and NOAA were patient and understanding of my needs and deadlines.

The Brooks/Cole team performed the customary miracles. The charge was led by Laurie Jackson, production editor, and, from her redwood forest keep, Joan Keyes, production service and champion of every known means of digital communication. The text was polished by Mary Arbogast, the best developmental editor on the planet, and Lorraine Anderson, a copy editor so excellent she was in constant danger of becoming a co-author. Myrna Engler and Catherine Gingras helped with photo research and permissions. Kristin Milotich continued in her capacity as the standard by which all project development editors must be measured, and editor Nina Horne kept me and my overheated computer and fax machine on the right path. My thanks to all.

A Gift

The ocean's greatest gift to humanity is intellectual—the constant analytical challenge its restless mass presents. Let yourself be swept into this book and the class it accompanies. Ask questions of your instructors, read some of the references, try your hand at the questions at the ends of the chapters. Be optimistic. *Take pleasure in the natural world.* Please write to me when you find errors or if you have comments. Enjoy yourself!

Tom Garrison
tsgarri@attglobal.net

Reviewers

ERNEST E. ANGINO, *University of Kansas*

M. A. ARTHUR, *Pennsylvania State University*

HENRY A. BART, *La Salle University*

STEVEN R. BENHAM, *Pacific Lutheran University*

LATSY BEST, *Palm Beach Community College*

EDWARD BEUTHER, *Franklin and Marshall College*

WILLIAM L. BILODEAU, *California Lutheran University*

JULIE BRIGHAM-GRETTE, *University of Massachusetts at Amherst*

LAURIE BROWN, *University of Massachusetts*

KEITH A. BRUGGER, *University of Minnesota, Morris*

DONALD BUCHANAN, *San Bernadino Valley College*

ZANNA CHASE, *Lamont-Doherty Earth Observatory, Columbia University*

KARL M. CHAUFF, *St. Louis University*

D. L. CLARK, *University of Wisconsin*

WILLIAM COCHLAN, *University of Southern California*

JAMES E. COURT, *City College of San Francisco*

RICHARD DAME, *University of South Carolina, Columbia*

DAVID DARBY, *University of Minnesota at Duluth*

ROBERT J. FELLER, *University of South Carolina, Columbia*

L. KENNETH FINK, JR., *University of Maine*

KATHLEEN FLICKINGER, *Maui Community College*

DIRK FRANKENBURG, *University of North Carolina, Chapel Hill*

ROBERT R. GIVEN, *Marymount College, Rancho Palos Verdes*

WILLIAM GLEN, *U.S. Geological Survey*

KAREN GROVE, *San Francisco State University*

BARRON HALEY, *West Valley College*

JACK C. HALL, *University of North Carolina at Wilmington*

WILLIAM HAMNER, *University of California, Los Angeles*

WILLIAM B. HARRISON III, *Western Michigan University*

DAVID HASTINGS, *University of British Columbia*

TED HERMAN, *West Valley College*

NANCY HINMAN, *University of Montana*

JOSEPH HOLLIDAY, *El Camino College*

ANDREA HUVARD, *California Lutheran University*

JAMES C. INGLE, JR., *Stanford University*

RONALD E. JOHNSON, *Old Dominion University*

SCOTT D. KING, *Purdue University*

JOHN A. KLASIK, *California Polytechnic University, Pomona*

C. ERNEST KNOWLES, *North Carolina State University*

EUGENE KOZLOFF, *Friday Harbor, WA*

LAWRENCE KRISSEK, *Ohio State University*

ALBERT M. KUDO, *University of New Mexico*

FRANK T. KYTE, *University of California, Los Angeles*

LYNTON S. LAND, *University of Texas, Austin*

RICHARD W. LATON, *Western Michigan University*

RUTH LEBOW, *University of California, Los Angeles Extension*

BEN leFEBVRE, *City College of San Francisco*

DOUGLAS R. LEVIN, *Bryant College*

LARRY LEYMAN, *Fullerton College*

TIMOTHY LINCOLN, *Albion College*

DONALD L. LOVEJOY, *Palm Beach Atlantic College*

JAMES MACKIN, *State University of New York, Stony Brook*

DAVID C. MARTIN, *Centralia College*

ELLEN MARTIN, *University of Florida*

JAMES McWHORTER, *Miami-Dade Connnunity College, Kendall Campus*

GREGORY A. MEAD, *University of Florida*

CHRIS METZLER, *Mira Costa College*

RICHARD W. MURRAY, *Boston University*

JOHN MYLROIE, *Mississippi State University*

CONRAD NEWMAN, *University of North Carolina, Chapel Hill*

JAMES G. OGG, *Purdue University*

B. L. OOSTDAM, *Millersville University*

JAN PECHENIK, *Tufts University*

BERNARD PIPKIN, *University of Southern California*

MARK PLUNKETT, *Bellevue Community College*

K. M. POHOPIEN, *Covina, CA*

RENE S. REVUELTA, *Miami-Dade Community College*

RICHARD G. ROSE, *West Valley College*

JUNE R. P. ROSS, *Western Washington University*

WENDY L. RYAN, *Kutztown University*

ROBERT J. SAGER, *Pierce College, Washington*

ROBERT F. SCHMALZ, *Pennsylvania State University*

DON SEAVY, *Olympic College*

SAM SHABB, *Highline Community College*

WILLIAM G. SIESSER, *Vanderbilt University*

RALPH SMITH, *University of California, Berkeley*

SCOTT W. SNYDER, *East Carolina University*

MORRIS L. SOTONOFF, *Chicago State University*

JAMES F. STRATTON, *Eastern Illinois University*

J. COTTER THARIN, *Hope College*

STANLEY ULANSKI, *James Madison University*

J. J. VALENCIC, *Saddleback College*

RAYMOND E. WALDNER, *Palm Beach Atlantic College*

JILL M. WHITMAN, *Pacific Lutheran University*

P. KELLY WILLIAMS, *University of Dayton*

BERT WOODLAND, *Centralia College*

JOHN H. WORMUTH, *Texas A&M University*

RICHARD YURETICH, *University of Massachusetts at Amherst*

MEL ZUCKER, *Skyline College*

History 1

SILENCE

In the winter of 1968 a lonely crew of American astronauts ventured farther from Earth than anyone has traveled before or since. They rode in a spacecraft named for Apollo, the Greek god of light and intelligence. On the afternoon of Christmas Eve the spacecraft turned to orbit its goal, the calm gray moon. Humans had their first close-up view of Earth's lifeless companion, its plaster-of-paris surface dotted with rounded craters, sharp peaks, and dusty plains beneath an airless black sky.

Through the 16 hours they spent in the close vicinity of the moon, the apprehensive astronauts felt their attention increasingly drawn away from the object of their mission to the shining crescent Earth, their watery home. "The vast loneliness is awe-inspiring," radioed James Lovell. "It makes you realize just what you have back there on Earth." The pictures sent down that day were seen by half a billion people; the beauty of the distant ocean world was obvious to all. The faraway Earth was "small and blue and beautiful in that eternal silence." Three days later the spacecraft and its relieved occupants shot into Earth's atmosphere at a speed of 40,000 kilometers (25,000 miles) per hour, slowed, and came safely to rest in the warm, welcoming ocean south of Hawaii. For most of the people of Earth, the concept that our planet is a fragile, rare, lovely object dates from that winter. For the first time we saw ourselves as we truly are.   1-1

Suddenly from behind the rim of the moon, in long, slow-motion moments of immense majesty, there emerges a sparkling blue and white jewel, a light, delicate sky-blue sphere laced with slowly swirling veils of white, rising gradually like a small pearl in a thick sea of black mystery. It takes more than a moment to fully realize this is Earth . . . home.
EDGAR MITCHELL, *APOLLO 14,* JANUARY 1971

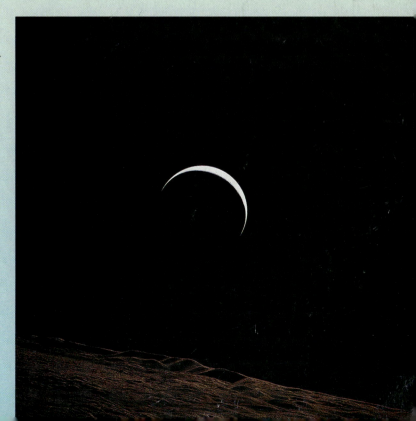

The crescent Earth from beyond the moon, photographed by *Apollo 15* astronauts in 1971. Our water planet shines blue-white in the darkness of space, its surface largely hidden by a turbulent layer of clouds.

Ours is not a particularly large planet, and it is not unusual in overall composition. What *is* extraordinary is the ocean of water dominating its cloud-shrouded surface. This brilliant blue ocean affects and moderates temperature and dramatically influences weather. Its creatures directly provide at least 2% of humanity's food. From beneath its floor is pumped about one-third of the world's petroleum and natural gas. The ocean borders most of Earth's largest cities—nearly half of the planet's 6 billion human inhabitants live within 150 kilometers (240 miles) of a coastline.[1] It is a primary shipping and communication route and a major recreational resource. The dry land on which nearly all of human history has unfolded is hardly visible from space—nearly three-quarters of the planet is covered by water.

More than 97% of the water on or near Earth's surface is contained in the ocean; less than 3% is held in land ice, groundwater, and all the freshwater lakes and rivers (see **Figure 1.1**). The **ocean**[2] can be defined as the vast body of saline water that occupies the depressions of Earth's surface. Traditionally, we have divided the ocean into artificial compartments called *oceans* and *seas* using the boundaries of continents and imaginary lines such as the equator. In fact there are few dependable natural divisions, only one great mass of water. The Pacific and Atlantic Oceans, the Mediterranean and Baltic Seas, so named for our convenience, are in reality only temporary features of a single **world ocean.** In this book we refer to the ocean *as a single entity*, with subtly different characteristics at different locations but with very few natural partitions. Marine scientists have descended to its greatest depths, have explored a few of its peaks, and have photographed and sampled some of its floor, but we know more about the topography of the far side of the moon than we know about the 70.78% of Earth covered by water. Despite our years of research, there is much to be learned. The world ocean is a key to Earth's past and present, and the prime link to its future.

On a *human* scale, the ocean is impressively large—it covers 361 million square kilometers (139 million square miles) of Earth's surface (see **Figure 1.2**). The average depth of the ocean is about 3,796 meters (12,451 feet), the volume of seawater is 1.37 billion cubic kilometers (329 million cubic miles), the average temperature a cool 3.9°C (39°F). Its mass is a staggering 141 billion billion metric tons. If Earth's contours were leveled to a smooth ball, the ocean would cover it to a depth of 2,686 meters (8,810 feet). Average land elevation is only 840 meters (2,772 feet), but average ocean depth is 4½ times greater!

On a *planetary* scale, however, the ocean itself is insignificant. Its average depth is a tiny fraction of Earth's radius—the blue ink representing the ocean on an 8-inch paper globe

---

[1] Throughout this book, metric measurements precede American measurements. For a quick review of metric (SI) units and their abbreviations, please turn to Appendix I.

[2] When an important new term is introduced and defined, it is printed in **boldface type.** These terms are listed at the end of the chapter and defined in the glossary.

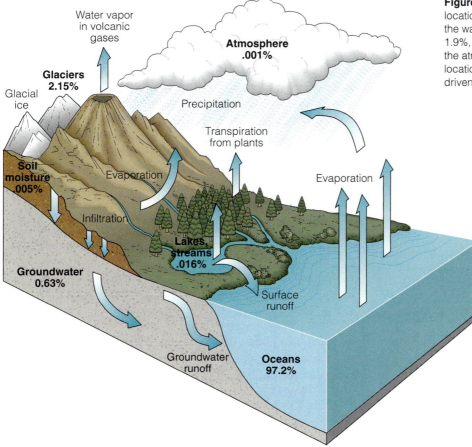

Figure 1.1 The relative amount of water in various locations on or near Earth's surface. More than 97% of the water lies in the ocean. Ice on land contains about 1.9%, groundwater 0.6%, rivers and lakes 0.02%, and the atmosphere 0.001%. The cycling of water among locations is shown in this diagram. The movement is driven by radiant energy from the Sun.

Figure 1.2 (**a**) The proportion of sea versus land, shown on an equal-area projection of Earth. An equal-area projection is a map drawn to represent areas in their correct relative proportions. The Northern Hemisphere is 60.7% sea and 39.3% land, the Southern Hemisphere is 80.9% sea and 19.1% land. Note the extent of the Pacific Ocean, Earth's most prominent single feature. (**b**) Light, air, wind, and water—a typical scene on Earth. Most of the planet's surface is covered by an ocean of water that averages 3,796 meters (12,451 feet) in depth.

Table 1.1 Characteristics of the World Ocean		
Area: 361,100,000 square kilometers (139,400,000 square miles)		
Volume: 1,370,000,000 cubic kilometers (329,000,000 cubic miles)		
Average depth: 3,796 meters (12,451 feet)		
Average temperature: 3.9°C (39.0°F)		
Average salinity: 34.482 grams per kilogram (0.56 ounce per pound), 3.4%		
Most abundant elements (by mass):	Oxygen	(86%)
	Hydrogen	(11%)
	Chlorine	(1.9%)
	Sodium	(1.1%)
	Magnesium	(0.1%)
Age: About 4 billion years		
Future: Uncertain		

is proportionally thicker. The ocean accounts for only slightly more than 0.02% of Earth's mass, or 0.13% of its volume. There is much more water trapped within Earth's hot interior than there is in its ocean and atmosphere. Some characteristics of the world ocean are summarized in **Table 1.1.**

VOYAGING AND DISCOVERY   1-2

It has taken a long time for humans to appreciate the nature of the world, but we're a restless and inquisitive lot, and despite the ocean's great size, we have populated nearly every inhabitable place. This fact was aptly illustrated when the European explorers set out to "discover" the world only to be met by native peoples at almost every landfall! Clearly the oceans did not prevent the spread of humanity. The early history of marine science is closely associated with the history of voyaging. Voyaging had a practical aim: to facilitate travel, trade, and warfare.

Voyaging Begins   1-3

Ocean transportation offers people the benefits of mobility and greater access to food supplies. Any coastal culture skilled at raft building or small boat navigation would have economic and nutritional advantages over their less adept competitors. From the earliest period of human history, then, understanding and appreciation of the ocean and its life forms benefited those patient enough to learn.

The first direct evidence we have of **voyaging,** traveling on the ocean for a specific purpose, comes from records of trade in the Mediterranean Sea. The Egyptians organized shipborne commerce on the Nile River, but the first regular ocean traders were probably the Cretans, or the Phoenicians who inherited maritime supremacy in the Mediterranean after the Cretan civilizations were destroyed by earthquakes and political instability around 1200 B.C. Skilled sailors, the Phoenicians carried their wares through the Straits of Gibral-

tar to markets as distant as Britain and the west coast of Africa. Considering the simple ships they used, this was quite an achievement.

The Greeks began to explore outside the Mediterranean into the Atlantic Ocean around 900–700 B.C. Early Greek seafarers noticed a current running from north to south beyond Gibraltar. Believing only rivers had currents, they decided that this great mass of water, too wide to see across, was part of an immense flowing river. The Greek name for this river was *okeanos.* Our word *ocean* is derived from **oceanus,** a Latin variant of that root. Phoenician sailors were also very much at home in this "river," but like the Greeks, they rarely ventured out of sight of land.

As they went about their business, early mariners began to record information to make their voyages easier and safer—the location of rocks in a harbor, landmarks and the sailing times between them, the direction of currents. These first **cartographers** (chart makers) were probably Mediterranean traders who made routine journeys from producing areas to markets. Their first charts (drawn about 800 B.C.) were merely notes to jog their memory for obvious features along the route. Today's **charts** are graphic representations that depict primarily water and water-related information. (*Maps* primarily represent land.)

In this early time other cultures also traveled on the ocean. The Chinese began to engineer an extensive system of inland waterways, some of which connected with the Pacific Ocean to make long-distance transport of goods more convenient. The Polynesian peoples had been moving easily among islands off the coasts of Southeast Asia and Indonesia since 3000 B.C. and were beginning to settle the mid-Pacific islands. Though none of these civilizations had contact with the others, each developed methods of charting and navigation. All these early travelers were skilled at telling direction by the stars and by the position of the rising or setting sun.

Curiosity and commerce encouraged adventurous people to undertake ever more ambitious voyages. But these voyages were possible only with the coordination of astronomical direction finding and knowledge of the shape and size of Earth, advanced shipbuilding technology, accurate graphic charts (not just written descriptions), and, perhaps most important, a growing understanding of the ocean itself. Marine science, the organized study of the ocean, had its origin in the technical studies of voyagers.

Science for Voyaging   1-4

Progress in applied marine science began at the **Library of Alexandria** in Egypt. Founded in the third century B.C. by Alexander the Great, the library constituted history's greatest accumulation of ancient writings and could be considered the world's first university. Written knowledge of all kinds regarding the characteristics of nations, trade, natural wonders, artistic achievements, tourist sights, investment opportunities, and other items of interest to seafarers was warehoused around its leafy courtyards. Traders quickly realized the competitive benefit of this information, and librarians welcomed their in-

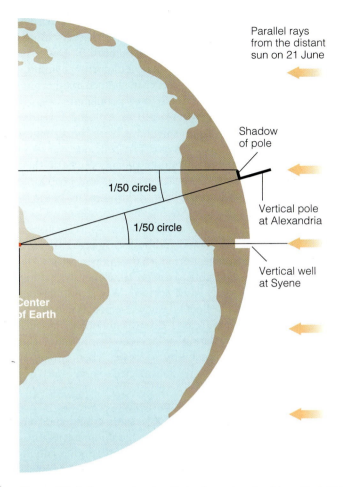

Figure 1.3 A diagram showing Eratosthenes' method for calculating the circumference of Earth. As described in the text, he used simple geometric reasoning based on the assumptions that Earth is spherical and that the sun is very far away. Using this method, he was able to discover the circumference of Earth to within about 8% of its true value. Thus, this knowledge was available more than 1,700 years before Columbus began his voyages. (The diagram is not drawn to scale.)

Parallel rays from the distant sun on 21 June

Shadow of pole

1/50 circle

1/50 circle

Vertical pole at Alexandria

Vertical well at Syene

Center of Earth

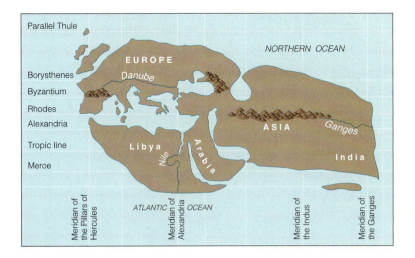

Parallel Thule

NORTHERN OCEAN

EUROPE

Borysthenes
Byzantium
Rhodes
Alexandria
Tropic line
Meroe

Danube

ASIA

Libya
Arabia
Nile

Ganges

India

ATLANTIC OCEAN

Meridian of the Pillars of Hercules

Meridian of Alexandria

Meridian of the Indus

Meridian of the Ganges

Figure 1.4 The world, according to a chart from the third century B.C. Eratosthenes drew latitude and longitude lines through important places rather than spacing them at regular intervals as we do today. The Alexandrian perception of the world is reflected in the size of the continents and the central position of Alexandria at the mouth of the Nile.

terest in return for even more information. Here perhaps was the first instance of cooperation between a university and the commercial community, a partnership that has paid dividends for science and business ever since.

The second librarian at Alexandria (from 235 B.C. until 192 B.C.) was the Greek astronomer, philosopher, and poet **Eratosthenes of Cyrene.** This remarkable man was the first to calculate the circumference of Earth. The Greek Pythagoreans had realized Earth was spherical by the sixth century B.C., but Eratosthenes was the first to estimate its true size.

Eratosthenes had heard from travelers returning from Syene (now Aswan, site of the great Nile dam) that at noon on the longest day of the year the sun shone directly onto the waters of a deep vertical well. In Alexandria, he noticed that a vertical pole cast a slight shadow on that day. He measured the shadow angle and found it to be a bit more than 7°, about one-fiftieth of a circle. He correctly assumed that the sun is a great distance from Earth, so the sun's rays would approach Syene and Alexandria in parallel lines. If the sun were directly over-

head at Syene but not directly overhead at Alexandria, then the surface of Earth must be curved. But what was the *circumference* of Earth?

By studying the reports of camel caravan traders, he estimated the distance from Alexandria to Syene at about 785 kilometers (491 miles). Eratosthenes now had the two pieces of information needed to derive the circumference of Earth by geometry. **Figure 1.3** shows his solution. The size of the units of length (stadia) Eratosthenes used is thought to have been 555 meters (607 yards), and historians estimate his calculation, made in about 230 B.C., was accurate to within about 8% of the true value. Within a few hundred years most people in the West who had contact with the library or its scholars knew Earth's approximate size.

Cartography flourished. The first workable charts representing a spherical surface on a flat sheet were developed by Alexandrian scholars. Latitude and longitude, systems of imaginary lines dividing the surface of Earth, were invented by Eratosthenes. **Latitude lines** were drawn parallel to the equator, and **longitude lines** ran from pole to pole. He drew the lines through prominent landmarks and important places, creating a convenient, though irregular, grid (see **Figure 1.4**). Our present regular grid of latitude and longitude lines was invented by Hipparchus (c. 165–c. 127 B.C.), a librarian who divided the surface of Earth into 360°.[3] A later Egyptian-Greek, Claudius Ptolemy (A.D. 90–168), "oriented" charts by placing east to the right and north at the top. Ptolemy's division of degrees into minutes and seconds of arc is still used by navigators.

Ptolemy also attempted to improve on Eratosthenes' surprisingly accurate estimate of Earth's circumference, but he wrongly calculated a degree as about 50 miles instead of the more correct 70 miles. This error, coupled with his mistake of overestimating the size of Asia, greatly reduced the apparent

---

[3] For more information on latitude and longitude, please see Appendix III.

width of the unknown part of the world between the Orient and Europe. Unfortunately for generations of navigators, Eratosthenes' estimate was forgotten while Ptolemy's persisted.

Though it weathered the dissolution of Alexander's empire, the Alexandrian library did not survive the subsequent period of Roman rule. The last librarian was Hypatia, the first notable woman mathematician, philosopher, and scientist. In Alexandria she was a symbol of science and knowledge, concepts the early Christians identified with pagan practices. The mission of the library, as personified by the last librarian, antagonized the governors and citizens of the city of Alexandria. After years of rising tensions, in A.D. 415 a mob brutally murdered her and burned the library with all its contents. Most of the community of scholars dispersed and Alexandria ceased to be a center of learning in the ancient world. The academic loss was incalculable, and trade suffered because ship owners no longer had a clearinghouse for updating the nautical charts and information upon which they had come to depend. All that remains of the library today is a remnant of an underground storage room. We shall never know the true extent and influence of its collection of more than 700,000 irreplaceable scrolls.

Western intellectual development slackened during the so-called Dark Ages that followed the fall of the Roman Empire in A.D. 476. For almost a thousand years, until the European Renaissance, much of the progress in medicine, astronomy, philosophy, mathematics, and other vital fields of human endeavor was made by the Arabs or imported by them from Asia. For example, the Arabs used the Chinese-invented compass for navigating caravans over seas of sand. During this time the Vikings raided and explored to the south and west, and the Polynesians continued some of the most extraordinary voyages in history.

Voyages of the Oceanian Peoples   1-5

In the history of human migration, no voyaging saga is more inspiring than that of the Polynesian colonizations, the peopling of the central and eastern Pacific islands. A profound knowledge of the sea was required for these voyages, and the story of the Polynesians is a high point in our chronology of marine science applied to travel by sea.

The Polynesians are one of four cultures inhabiting some 10,000 islands scattered across nearly 26 million square kilometers (10 million square miles) of open Pacific ocean (**Figure 1.5**). The Southeast Asian or Indonesian ancestors of the Oceanian peoples, as these cultures are collectively called, spread eastward in the distant past. Although experts vary in their estimates, there is some consensus that by 30,000 years ago New Guinea was populated by these wanderers, and by 20,000 years ago the Philippines were occupied. By around 500 B.C. the so-called cradle of **Polynesia**—Tonga, Samoa, the Marquesas, and the Society Islands—was settled.

For a long and evidently prosperous period, the Polynesians spread from island to island until the easily accessible islands had been colonized. Eventually, however, overpopulation and depletion of resources became a problem. Politics, intertribal tensions, and religious strife shook society. Groups of people scattered in all directions from some of the "cradle" is-

lands during a period of explosive dispersion. Between A.D. 300 and 600, Polynesians successfully colonized nearly every inhabitable island within a vast triangular area (Figure 1.5). Easter Island was found against prevailing winds and currents, and the remote islands of Hawaii were discovered and occupied. These were among the last places on Earth to be populated.

How did these risky voyages into unexplored territory come about? Religious warfare may have been the strongest stimulus to colonization. If the losers of a religious war were banished from the home islands under penalty of death, their only hope for survival was to reach a distant and hospitable new land. Seafaring had been a long tradition in the home islands, but these trips called for radical new technology. Great dual-hulled sailing ships, some capable of transporting up to a hundred people, were designed and built (**Figure 1.6**). New navigation techniques were perfected that depended on the positions of stars barely visible to the north. New ways of storing food, water, and seeds were devised. Whole populations left their home islands in fleets designed especially for long-distance discovery. In some cases, fire was nurtured on board in case of landfall on an island that lacked volcanic flame. But a new island was only a possibility, a dream. Their gods may have promised the voyagers safe deliverance to new lands, but how many fleets set out from the troubled homelands only to fall victim to storms, thirst, or other dangers?

Yet in that anxious time the Polynesians honed and perfected their seafaring knowledge. To a skilled navigator a change in the rhythmic set of waves against the hull could indicate an island out of sight over the horizon. The flight tracks of birds at dusk could suggest the direction of land. The positions of the stars told stories, as did the distant clouds over an unseen island. The smell of the water—or its temperature or salinity or color—conveyed information, as did the direction of the wind relative to the sun, and the type of marine life clustering near the boat. The colors of sunrise and sunset, the hue of the moon—every nuance had meaning, every detail had been passed in ritual from father to son. The greatest Polynesian minds were navigators, and reaching Hawaii was their greatest achievement.

Of all islands colonized by the Polynesians, Hawaii is farthest away, across an ocean whose guide stars were completely unknown to the southern navigators. The Hawaiian Islands are isolated in the northern Pacific. There are no islands of any significance for more than two thousand miles to the south. Moreover, Hawaii lies beyond the equatorial doldrums, a hot and often windless stretch across which these pioneers must somehow have paddled. And yet some fortunate and knowledgeable people colonized Hawaii sometime between A.D. 450 and 600. Try to imagine their feelings of relief and justification upon reaching a promised paradise under a new night sky. Think of that first approach to the high islands of Hawaii, the first unlimited drink of fresh water, the first solid earth after months of uncertainty.

Within a hundred years of their first arrival, Hawaiian navigators were routinely piloting vessels on regular return trips to the Marquesas and the Society Islands (Tahiti and others). Some of the trips were undertaken to import needed food

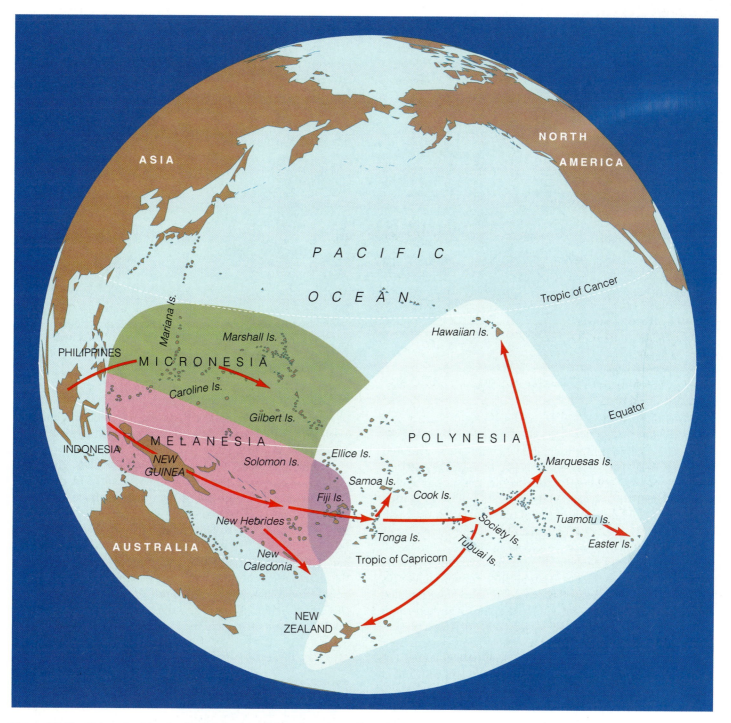

Figure 1.5 The Polynesian triangle. Ancestors of the Polynesians spread from Southeast Asia or Indonesia to New Guinea and the Philippines by about 20,000 years ago. The mid-Pacific islands have been colonized for about 2,500 years, but the explosive dispersion that led to the settlement of Hawaii occurred about A.D. 450–600. Arrows show a possible direction and order of settlement.

species to the newly found islands, but others were made to recruit new citizens and leaders to "green-clad Hawaii."

At a time when seafarers of other civilizations sailed beside the comforting bulk of a charted coast, Polynesians looked to the open sea for sustenance, deliverance, and hope. Their great knowledge of the ocean protected them.

The Age of European Discovery 1-6

Half a world away from their Polynesian counterparts, Renaissance Europeans—having been jolted by internal awakening and external reality—set out to explore the world by sea. They did not undertake exploration for its own sake, however; any voyage had to have a material goal. Trade between East and

Figure 1.6 A voyaging canoe struggles across the Kaiwi Channel toward Molokai. Larger versions carried families and supplies on the long, dangerous route to Hawaii.

West had long been dependent on arduous and insecure desert caravan routes through the central Asian and Arabian deserts. This commerce was cut off in 1453 when the Turks captured Constantinople, and an alternate ocean route was needed.

A European visionary who thought ocean exploration held the key to great wealth and successful trade was **Prince Henry the Navigator,** third son of the royal family of Portugal (**Figure 1.7**). Prince Henry established a center at Sagres for the study of marine science and navigation "... through all the watery roads." Although he personally was not well traveled (he went to sea only twice in his life), captains under his patronage explored from 1451 to 1470, compiling detailed charts wherever they went. Henry's explorers pushed south into the unknown and opened the west coast of Africa to commerce. He sent out small, maneuverable ships designed for voyages of discovery and manned by well-trained crews. For navigation, his mariners used the **compass**—an instrument (invented in China in the fourth century B.C.) that points to magnetic north. Although Arab traders had brought the compass from China in the twelfth century, navigators still considered it a magical tool. They concealed the compass in a special box (predecessor to today's binnacle) and consulted it out of plain view. Henry's students knew Earth was round (but because of the errors of Claudius Ptolemy they were wrong in their estimation of its size).

A master mariner (and skilled salesman), **Christopher Columbus,** "discovered" the New World quite by accident. Native Americans had been living on the continent for about 11,000 years, and the Norwegian Vikings had made about two dozen visits to a functioning colony on the continent 500 years before his noisy arrival; yet Columbus gets the credit. Why? Because his interesting souvenirs, exaggerated stories, inaccu-

rate charts, and promises of vast wealth excited the imagination of royal courts. Columbus made North America a media event without ever sighting it!

Columbus wasn't trying to discover new lands. His intention was to pioneer a sea route to the rich and fabled lands of the East, made famous more than 200 years earlier in the overland travels of Marco Polo. As "Admiral of the Ocean Sea," Columbus was to have a financial interest in the trade routes he blazed. He was familiar with Prince Henry's work and, like all other competent contemporary navigators, knew that Earth was spherical. By sailing west he believed he could come close to his eastern destination, the latitude of which he thought he knew. Because he depended on Ptolemy's data, however, Columbus made the *smallest* estimate of the size of

Figure 1.7 Prince Henry of Portugal, the Navigator. In the mid-1400s, Henry established a center at Sagres for the study of marine science and navigation "... through all the watery roads."

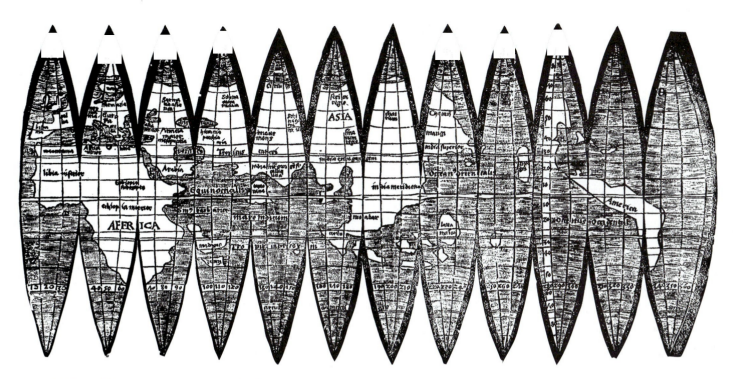

Figure 1.8 The Waldseemüller map, published in 1507—the first map to name America and to show the New World as separate from Asia. The map is gored to form a globe about 10 centimeters (4 inches) in diameter.

Earth by any navigator in modern history; he assumed Earth to be only about half its actual size.

Not surprisingly, Columbus mistook the New World for his goal of India or Japan. He thought that the notable absence of wealthy cities and well-dressed inhabitants resulted from striking the coast too far north or south of his desired latitude. He made three more trips to the New World but died still believing that he had found islands off the coast of Asia. He never saw the mainland of North America and never realized the size and configuration of the continents whose future he had so profoundly changed.

Other explorers quickly followed, and Columbus's error was soon rectified. Charts drawn as early as 1507 included the New World (**Figure 1.8**). Such charts perhaps inspired **Ferdinand Magellan** (**Figure 1.9**), a Portuguese navigator in the service of Spain, to believe that he could open a westerly trade route to the Orient. Unfortunately, the chart makers estimated that the Americas and the Pacific Ocean were much smaller than they actually are. Magellan's expedition set out in September 1519 and returned to Spain three years later. Only 34 of the original crew of 260 survived—Magellan himself was killed by natives in the Philippines in April 1521—but they had proved it was possible to circumnavigate the globe.

The Magellan expedition's return to Spain in 1522 marks the end of the Age of European Discovery. An unpleasant era of exploitation of the human and natural resources of the Americas followed. Native empires were destroyed, and objects of priceless religious and archaeological value were melted into coin to fund European warfare and greed.

Figure 1.9 Ferdinand Magellan, a Portuguese explorer in service to Spain whose expedition was the first to circumnavigate the world.

VOYAGING FOR SCIENCE 1-7

British sea power arose after the Age of European Discovery to compete with the colonial aspirations of France and Spain. Sailing ships require dependable supply and repair stations, especially in remote areas. The great powers sent out expeditions to claim appropriate locations, preferably inhabited by friendly natives eager to help provision ships half a globe from home. The French sent Admiral de Bougainville into the South

Pacific in the mid-1760s. His 1768 claim for France of what is now called French Polynesia opened the area to the powerful European nations. The British followed immediately.

James Cook

1-8

Scientific oceanography begins with the departure from Plymouth Harbor in 1768 of HMS *Endeavour* under the command of **James Cook** of the British Royal Navy (**Figure 1.10**). An intelligent and patient leader, Cook was also a skillful navigator, cartographer, writer, artist, diplomat, sailor, scientist, and dietician. The primary reason for his voyage was to assert the British presence in the South Seas, but the expedition had numerous scientific goals as well. First, Cook conveyed several members of the Royal Society (a scientific research group) to Tahiti to observe the transit of Venus across the disk of the sun. Their measurements verified calculations of planetary orbits made earlier by Edmund Halley (later of comet fame) and others. Then Cook turned south into unknown territory to search for a hypothetical southern continent, which some philosophers believed had to exist to balance the landmass of the Northern Hemisphere. Using the newly invented chronometer to calculate their longitude, Cook and his men found and charted New Zealand, mapped Australia's Great Barrier Reef, marked the positions of tens of small islands, made notes on the natural history and human habitation of these distant places, and initiated friendly relations with many chiefs. Cook survived an epidemic of dysentery contracted by the ship's company while ashore in Batavia (Djakarta) and sailed home to England around the world in 1771. Because of his insistence on cleanliness and ventilation, and because his provisions included cress, sauerkraut, and citrus extracts, his sailors avoided scurvy—a Vitamin-C deficiency disease that for centuries had decimated crews on long voyages.

The Admiralty was deeply impressed. Cook was promoted to the rank of commander and in 1772 was given command of the ships *Resolution* and *Adventure*, in which he embarked on one of the great voyages in scientific history.[4] On this second voyage he charted Tonga and Easter Island, discovered New Caledonia in the Pacific and South Georgia in the Atlantic. He was the first to circumnavigate the world at high latitudes. Though he sailed to 71° south latitude, he never sighted Antarctica. He returned home again in 1775.

Posted to the rank of captain, Cook set off in 1776 on his third, and last, expedition in *Resolution* and *Discovery*. His commission was to find a northwest passage around Canada and Alaska, or a northeast passage above Siberia. He "discovered" the Hawaiian Islands (Hawaiians were there to greet him, of course, as shown in **Figure 1.11**) and charted the west coast of North America. After searching unsuccessfully for a passage across the top of the world, Cook retraced his steps to Hawaii to provision for departure home. On 14 February 1779, after an elaborate farewell dinner with the chief of the island of Hawaii, Cook and his officers prepared to return to *Resolution* anchored in Kealakekua Bay. The Englishmen somehow angered the Hawaiians and were beset by the crowd. Cook, among others, was killed in the fracas.

Cook deserves to be considered a scientist as well as an explorer because of his accuracy and thoroughness, and the completeness of his descriptions. He and the scientists aboard took samples of marine life, land plants and animals, the ocean floor, and geological formations; they also reported their characteristics in their logbooks and journals. His navigation was outstanding, and his charts of the Pacific were accurate enough to be used by the Allies in World War II invasions of the Pacific islands. He drew accurate conclusions, did not exaggerate his findings in his reports, and opened friendly diplomatic relations with many native populations. Cook recorded and successfully interpreted events in natural history, anthropology, and oceanography. Unlike most captains of his day, he cared for his men. He was a thoughtful and clear writer. This first marine scientist peacefully changed the map of the world more than any other explorer or scientist in history.

Figure 1.10 Captain James Cook, Royal Navy, painted in 1776 by Nathaniel Dance, shortly before embarking on his third, and fatal, voyage. Cook is a fully matured, self-confident captain who has twice circled the globe, penetrated into the Antarctic, and charted coastlines from Newfoundland to New Zealand. (Copyright, National Maritime Museum, Greenwich.)

The Sampling Problem

1-9

Marine science advances by the analysis of samples. Sampling of floor sediments or bottom water is not an easy task in the deep ocean. The line used to suspend the sampling device

---

[4] Sailing master in *Resolution* was the 21-year-old William Bligh, later the object of the famous mutiny aboard HMS *Bounty*.

Figure 1.11 On his third Pacific voyage, Captain James Cook, commanding HMS *Resolution,* sailed north from Tahiti with orders to explore the northwestern coast of North America. On 19 January 1778, he became the first European to reach Hawaii. He wrote in his journal, "How do we account for this Nation spreading itself so far over this Vast ocean?"

snakes back and forth as currents strike it, and the weight of the line makes it difficult to tell when the sampler has hit bottom. Deploying and recovering the line is laborious and time-consuming, and sometimes the sampling device does not work properly. Early bottom sampling devices (such as those used by Cook) were simple wax-covered lead weights lowered to shallow bottoms to pick up sediments and test the suitability of anchorages. Later devices took deep-water samples, extracted cores from the sediments, grabbed samples of the bottom, or scooped biological specimens from the ocean floor.

The first researchers to attack the deep sampling problem successfully were British explorers Sir John Ross and his nephew Sir James Clark Ross. During an expedition to scout the Northwest Passage in 1818, Sir John Ross obtained a bottom sample from 1,919 meters (3,296 feet) near Greenland by using a clamping sampler to trap the specimen. Sir James Clark Ross, discoverer of the Ross Sea and the area of Antarctica known as Victoria Land, obtained **soundings** (depth measurements) of 4,433 meters and 4,893 meters (14,545 feet and 16,054 feet) in the South Atlantic.

Sampling techniques improved through the nineteenth century. Using a sounding method perfected in the late 1840s by a U.S. Navy midshipman, American Commodore Matthew Maury used a long lightweight line and lead weight to discover the Mid-Atlantic Ridge, an important hidden range of mountains. Fridtjof Nansen perfected the deep-water sampling bottle bearing his name near the end of the century. Even today, in spite of modern advances, deep sampling remains diffi-cult. As we will see in the next chapters, a new generation of expensive manned and remotely operated vehicles now works at great depths to return samples and pictures to the surface.

SCIENTIFIC EXPEDITIONS 1-10

Great as Cook's contributions undoubtedly were, his three voyages (and those of the Rosses) were not purely scientific expeditions. These men were British naval officers engaged in Crown business, concerned with charting, "foreign relations," and natural phenomena as they applied to Royal Naval matters. The first genuine only-for-science expedition may well have been the *Challenger* expedition of 1872–1876, but the United States got into the act first with a hybrid expedition in 1838.

The United States Exploring Expedition 1-11

After a ten-year argument over its potential merits, the **United States Exploring Expedition** was launched in 1838. It was primarily a naval expedition, but its captain was somewhat more free in maneuvering orders than Cook had been. The work of the scientists aboard the flagship *Vincennes* and the expedition's five other vessels helped to establish the natural sciences as reputable professions in America. Had it not been for the combative and disagreeable personality of its leader, Lieutenant

Charles Wilkes (**Figure 1.12**), this expedition might have become as famous as those of Cook or the later *Challenger* voyage.

The expedition departed on a four-year circumnavigation (**Figure 1.13**). Its goals included showing the flag, whale scouting, mineral gathering, charting, observing, and pure exploration. One unusual goal was to disprove a peculiar theory that Earth was hollow and could be entered through huge holes at either pole.

Wilkes's team explored and charted a large sector of the east Antarctic coast, and it made observations that confirmed the landmass as a continent. Hawaii was thoroughly explored, and Wilkes led an ascent of Mauna Loa, one of the two peaks of Hawaii's tallest volcano. James Dwight Dana, the expedition's brilliant geologist, confirmed Charles Darwin's hypothesis of coral atoll formation (about which more will be found in Chapter 14). The expedition returned with many scientific specimens and artifacts, which formed the nucleus of the collection of the newly established Smithsonian Institution in Washington, D.C. No evidence of polar holes was found.

Upon their return in 1842, Wilkes and his "scientifics" prepared a final report totaling 19 volumes of maps, text, and illustrations. The report is a landmark in the history of American scientific achievement.

Matthew Maury   1-12

At about the time the Wilkes expedition returned, **Matthew Maury** (**Figure 1.14**), a Virginian and U.S. naval officer, was appointed director of the navy's Bureau of Charts and Instru-

Figure 1.12 Lieutenant Charles Wilkes soon after his return from the United States Exploring Expedition.

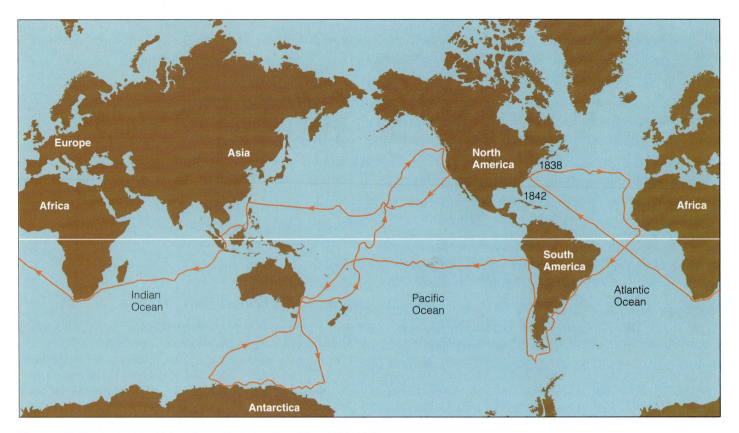

Figure 1.13 The track of the United States Exploring Expedition, 1838–1842.

Figure 1.14 Matthew Fontaine Maury, compiler of winds and currents. Maury was perhaps the first person for whom oceanography was a full-time occupation. This photograph was probably taken in 1853.

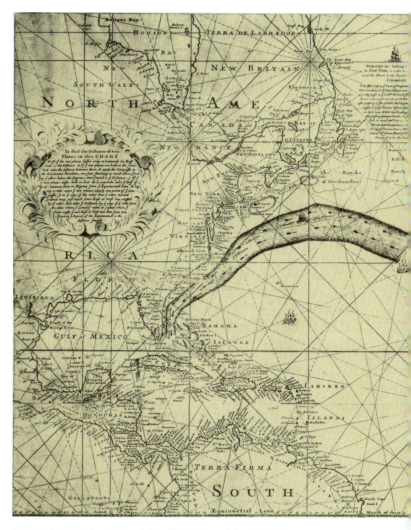

Figure 1.15 Benjamin Franklin's 1769 chart of the Gulf Stream system. His cousin, Timothy Folger, discovered that Yankee whalers had learned to use the Gulf Stream to their advantage. Others, especially English shipowners, were slower to learn. Folger, himself a sea captain, wrote that Nantucket whalers ". . . in crossing it have sometimes met and spoke with those packets who were in the middle of and stemming it. We have informed them that they were stemming a current that was against them to the value of three miles an hour and advised them to cross it, but they were too wise to be counseled by simple American fishermen."

ments. There he studied a huge and neglected treasure trove of ships' logs with their many regular readings of temperature and wind direction. By 1847 Maury had assembled much of this information into coherent wind and current charts. Maury began to issue these charts free to mariners in exchange for logs of their own new voyages.

Slowly a picture of planetary winds and currents began to emerge. Maury himself was a compiler, not a scientist, and he was vitally interested in the promotion of maritime commerce. His understanding of currents was built on the work of Benjamin Franklin, who nearly a hundred years earlier had noticed the peculiar fact that the fastest ships were not always the fastest ships—that is, hull speed did not always correlate with out-and-return time on the European run. Franklin's cousin, a Nantucket merchant named Tim Folger, noted Franklin's puzzlement and provided him with a rough chart of the "Gulph Stream" that he (Folger) had worked out. By staying within the stream on the outbound leg and adding its speed to their own, and by avoiding it on their return, captains could traverse the Atlantic much more quickly. It was Franklin who published, in 1769, the first chart of any current (**Figure 1.15**).

But it was Maury who was the first person to sense the worldwide pattern of surface winds and currents. His work became famous in 1849 when the California gold rush made his sailing directions invaluable for fast trips around Cape Horn. His crowning achievement, *The Physical Geography of the Seas,* a book explaining his discoveries, was published in 1855. Maury, considered by many to be the father of physical oceanography, was perhaps the first man to undertake the systematic study of the ocean as a full-time occupation.

The *Challenger* Expedition 1-13

The first sailing expedition devoted completely to marine science was conceived by a professor of natural history at Scotland's University of Edinburgh, Charles Wyville Thomson, and his Canadian-born student John Murray. Stimulated by their own curiosity and by the inspiration of Charles Darwin's 1831–1836 voyage in HMS *Beagle,* they convinced the Royal Society and the British government to provide a Royal Navy ship and trained crew for a "prolonged and arduous voyage of exploration across the oceans of the world." Thomson and Murray even coined a word for their enterprise: **oceanography.** Though the term literally implies only marking or charting, it has come to mean the science of the ocean. The government and the Royal Society agreed to the endeavor provided a portion of any financial gain from discoveries was handed over to the Crown. This arranged, the scientists made their plans.

HMS *Challenger,* a 2,306-ton steam corvette (**Figure 1.16**), set sail on 21 December 1872 on a four-year voyage that circumnavigated the world and covered 127,600 kilometers (79,300 nautical miles). Although the captain was a Royal Navy officer, the six-man scientific staff directed the course of the voyage. *Challenger's* track is shown in **Figure 1.17.**

One important mission of the ***Challenger* expedition** was to investigate Edinburgh professor Edward Forbes's contention that life below 549 meters (1,800 feet) was impossible because of high pressure and lack of light. The steam winch on board made deep sampling practical, and samples from depths as great as 8,185 meters (26,850 feet) were collected off the Philippines. Through the course of 492 deep soundings with mechanical grabs and nets at 362 stations (including 133 dredgings), Forbes was proven resoundingly wrong. With each hoist, animals new to science were strewn on the deck; in all, staff biologists discovered 4,717 new species! **Figure 1.18** shows one of the large trawl nets used in making some of these discoveries.

The scientists also took salinity, temperature, and water density measurements during these soundings. Each reading contributed to a growing picture of the physical structure of the deep ocean. They completed at least 151 open water trawls and stored 77 samples of seawater for detailed analysis ashore. The expedition collected new information on ocean currents, meteorology, and the distribution of sediments; the locations and profiles of coral reefs were charted. Thousands of pounds of specimens were brought to British museums for study. Manganese nodules, brown lumps of mineral-rich sediments, were discovered on the seabed, sparking interest in deep-sea mining.

Figure 1.16 Lieutenant Pelham Aldrich, first lieutenant of HMS *Challenger,* kept a detailed journal of the *Challenger* expedition. With accuracy and humor he kept this record in good weather and bad, and had the patience and skill to include watercolors of the most exciting events. This is part of the first page of his journal.

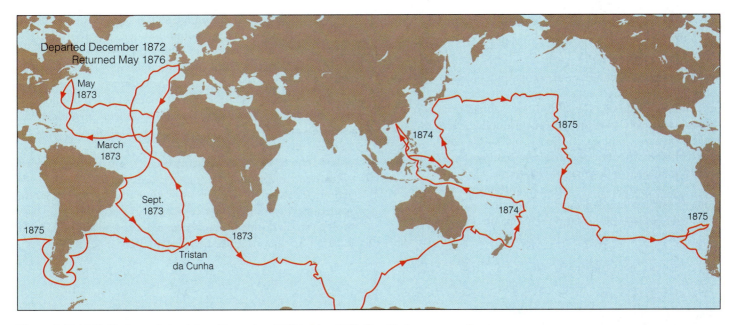

Figure 1.17 HMS *Challenger*'s track from December 1872 to May 1876. The *Challenger* expedition remains the longest continuous oceanographic survey on record.

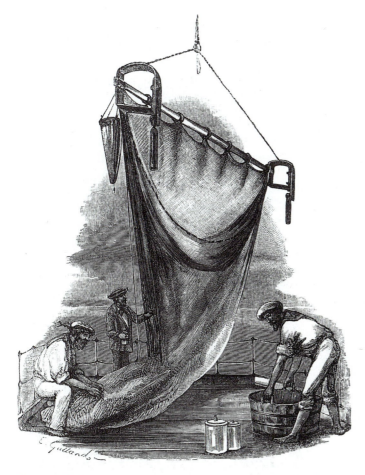

Figure 1.18 Emptying a trawl net, an engraving from the *Challenger Report*.

This first purely oceanographic investigation was an unqualified success. The discovery of life in the depths of the oceans stimulated the new science of marine biology. The scope, accuracy, thoroughness, and attractive presentation of the researchers' written reports made this expedition a high point in scientific publication. The *Challenger Report*, the record of the expedition, was published between 1880 and 1895 by Sir John Murray in a well-written and magnificently illustrated 50-volume set; it is still used today. Indeed, it was the 50-volume *Report*, rather than the cruise, that provided the foundation for the new science of oceanography. The expedition's many financial spin-offs indicated that pure research was a good investment, and the British government realized quick profits from the exploitation of newly discovered mineral deposits on islands. The *Challenger* expedition remains history's longest continuous scientific oceanographic expedition.

With successes like these, the pace of exploration accelerated. American naturalist Alexander Agassiz, sailing in 1877 on the U. S. Coast and Geodetic Survey ship *Blake,* collected data corroborating the *Challenger* material at 355 deep sea stations. The distribution of manganese nodules was found to be widespread. Further work by Agassiz and his students around the turn of this century on the survey ship *Albatross* helped train a generation of influential American marine biologists. In 1886 the Russians entered the field of marine exploration with the three-year cruise of *Vitiaz* under the leadership of S. O. Makarov; their main contribution was a careful analysis of salinity and temperature of North Pacific water.

TWENTIETH-CENTURY VOYAGING FOR SCIENCE

1-14

In the twentieth century, oceanographic voyages became more technically ambitious and expensive. Scientist–explorers sought out and investigated places that had once been too difficult to reach. New electronic and optical devices aided navigation and sampling. In the last half of the century, high-speed shipboard computers made it possible for marine scientists to analyze data while still at sea.

Polar oceanography made dramatic advances early in the century. Newly designed ships and new methods of food storage made polar exploration possible in the last years of the nineteenth century. In 1893 Fridtjof Nansen began studying the north polar ocean in *Fram,* a ship designed specifically to withstand the crushing pressure of sea ice (**Figure 1.19**). In the next 20 years Nansen and others probed the polar ocean depths. Researchers confirmed the feeding relationships between whales and plankton and collected much data about the whale population of the southern ocean—not out of scientific curiosity, but because whales were a source of oil and baleen ("whalebone").

In 1925 the German ***Meteor* expedition,** which crisscrossed the South Atlantic for two years, introduced modern optical and electronic equipment to oceanographic investigation. Its most important innovation was to use an **echo sounder,** a device that bounces sound waves off the ocean bottom, to study the depth and contour of the seafloor (**Figure 1.20**). The echo sounder revealed to *Meteor* scientists a var-

a

b

Figure 1.19 (**a**) Fridtjof Nansen, pioneering Norwegian oceanographer and polar explorer, looking every inch the Viking. In 1908 Nansen became the first professor of oceanography, a post created for him at Christiania University. (**b**) Nansen's 123-foot schooner *Fram* ("forward"). With 13 men, *Fram* sailed on 22 June 1893 to the high Arctic with the specific purpose of being frozen into the ice. *Fram* was designed to slip up and out of the frozen ocean, and drifted with the pack ice to within about 4° of the North Pole. The whole harrowing adventure took nearly four years. The ship's 1,650-kilometer (1,025-mile) drift proved that no Arctic continent existed beneath the ice. Living conditions aboard can be sensed from this recently discovered photograph.

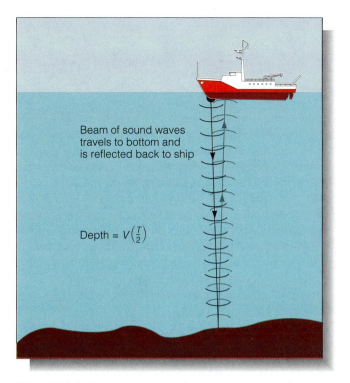

Figure 1.20 Echo sounders sense the contour of the seafloor by beaming sound waves to the bottom and measuring the time required for the sound waves to bounce back to the ship. If the round-trip travel time and wave velocity are known, distance to the bottom can be calculated. This technique was first used on a large scale by the German research vessel *Meteor* in the 1920s.

In the diagram:

Beam of sound waves travels to bottom and is reflected back to ship

$$\text{Depth} = V\left(\frac{T}{2}\right)$$

Figure 1.21 *Glomar Challenger,* a research ship conceived and built in the early 1960s by an international consortium of oceanographic institutions and the U.S. National Science Foundation as part of the Deep Sea Drilling Project. Its goal: to test the then-radical hypothesis that continents are moved across Earth's surface by seafloor spreading. Operated by the Deep Sea Drilling Project from 1968–1983, the 122-meter (400-foot) ship used computers to maintain her position with the precision needed to complete the drilling of cores up to a mile long.

ied and often extremely rugged bottom profile rather than the flat floor they had anticipated.

Atlantis, launched in 1931, was the first U.S. research ship built specifically for ocean studies. Investigations by her scientists confirmed Matthew Maury's findings of a mid-Atlantic ridge and helped to discover its extent. The 32-meter (104-foot) schooner *E. W. Scripps,* under the direction of Harald Sverdrup, began a wide-ranging program of chemical, biological, and geophysical exploration off the coast of southern California in 1937. These voyages led to publication in 1942 of *The Oceans,* the first modern reference work on all phases of marine science.

In October 1951 a new HMS *Challenger* began a two-year voyage that would make precise depth measurements in the Atlantic, Pacific, and Indian Oceans and in the Mediterranean Sea. With echo sounders, measurements that would have taken the crew of the first *Challenger* nearly four hours to complete could be made in seconds. *Challenger II's* scientists discovered the deepest part of the ocean's deepest trench, naming it Challenger Deep in honor of their famous predecessor. In 1960 U. S. Navy Lieutenant Don Walsh and Jacques Piccard descended into Challenger Deep in *Trieste,* a Swiss-designed, blimplike bathyscaphe (see Chapter 4 opener).

In 1968 the drilling ship *Glomar Challenger* (**Figure 1.21**) set out to test a controversial hypothesis about the his-

tory of the ocean floor. It was capable of drilling into the ocean bottom beneath more than 6,000 meters (20,000 feet) of water and recovering samples of seafloor sediments. These long and revealing plugs of seabed provided confirming evidence for seafloor spreading and plate tectonics. (The details will be found in Chapter 3.) In 1985 deep-sea drilling duties were taken over by the much larger and more technologically advanced ship *JOIDES Resolution.*[5] The new ship contains equipment capable of drilling in water 8,100 meters (27,000 feet) deep and houses the most completely equipped geological laboratories ever put to sea (see Figure 5.15).

Modern voyages are usually made in well-equipped surface ships, but a few recent expeditions have been accomplished in vessels that would have astonished earlier explorers. As can be seen in **Box 1.1,** a very fast, immensely strong, silent nuclear submarine makes an ideal platform for oceanographic research. For an idea of how far the technology of exploration has progressed in this century, compare conditions aboard *Fram* (in Figure 1.19).

---

[5] JOIDES stands for Joint Oceanographic Institutions for Deep Earth Sampling.

The SCICEX Expeditions   BOX 1.1

Modern technology has eased the burden of high-latitude travel. In 1958, under the command of Captain William Anderson, the U.S. nuclear submarine *Nautilus* cruised beneath the North Pole during a submerged transit from Point Barrow, Alaska, to the Norwegian Sea. A warm, strong, stable nuclear submarine is a nearly ideal platform for conducting oceanographic research at high latitudes.

At the request of oceanographic researchers, the U.S. Navy more recently initiated a series of research cruises in the northern polar ocean. Called SCICEX (for Scientific Ice Exercise) and funded by the government, the cruises began in 1993 and are scheduled to continue through the year 2000. **Figure a** shows USS *Pargo,* a fast and powerful *Sturgeon*-class submarine hardened for surfacing through ice, during 1993 operations in the Arctic. A staff of seven scientists spent 45 days operating in the Arctic Ocean during a 1996 research cruise aboard USS *Pogy,* another *Sturgeon*-class submarine. More than 12,800 kilometers (9,000 miles) of underway data—including bathymetry, gravity anomaly, temperature, salinity, ice draft, and images of the underside of the ice—were collected; more than

1,500 water samples were taken for biological and chemical analysis; and four long-term observation buoys were placed. The 1998 and 1999 expeditions aboard USS *Hawkbill* (**Figure b**) made use of a sub-bottom profiler to provide the first images of the shallow strata of the Arctic Ocean floor. The navy has agreed to host annual 45- to 60-day research cruises aboard *Sturgeon*-class submarines through the year 2000. If the $200 million in conversion costs (and $10 million in annual upkeep) were forthcoming, the navy might be persuaded to keep at least one *Sturgeon*-class submarine in oceanographic service past the vessel's decommissioning date. Polar science would be immeasurably richer for it.   1-15

a USS *Pargo,* a *Sturgeon*-class submarine strengthened for operation beneath the northern ice cap, surfaced at the edge of the Arctic ice pack during the 38-day 1993 SCICEX cruise. A scientific staff of seven was embarked along with a full operating crew; 39 shoreside researchers participated in the analysis of samples and data taken during the cruise. A strong, fast, quiet, powerful (and pleasingly warm) nuclear submarine makes an ideal platform for oceanographic research at high latitudes. USS *Pargo* is 89 meters (292 feet) long and 9.75 meters (32 feet) wide. Her speed and depth capability are classified.

b USS *Hawkbill* (SSN-666) at the North Pole during the 1999 SCICEX expedition. A crew member stands ready to offload instruments for oceanographic analysis. Compare working conditions aboard *Hawkbill* to those aboard *Fram* in Figure 1.19b!

THE RISE OF OCEANOGRAPHIC INSTITUTIONS

1-16

The demands of scientific oceanography have become greater than any single voyage can meet. Oceanographic institutions, agencies, and consortia evolved in part to insure continuity of effort. The first of these coordinating bodies was founded by Prince Albert I of Monaco, who endowed his country's oceanographic laboratory and museum in 1906. The most famous alumnus of Albert's Musée Oceánographique is Jacques-Yves Cousteau, co-inventor in 1943 of the scuba underwater breathing system. Monaco also became the site of the International Hydrographic Bureau, founded in 1921 as an association of maritime nations. This bureau published one of the first general charts of the ocean showing bottom contours.

Figure 1.22 *Kaiko,* the deepest-diving vehicle presently in operation, descended to a measured depth of 10,914 meters (35,798 feet) near the bottom of the Challenger Deep on 24 March 1995. The small remotely operated vehicle (ROV) sends information back to operators in the mother ship by fiber-optic cable. *Kaiko* is operated by JAMSTEC, a Japanese marine science consortium.

A consortium of Japanese industries and governmental agencies established the Japan Marine Science and Technology Center (JAMSTEC) in 1971. In 1989 JAMSTEC launched *Shinkai 6500,* now the deepest-diving manned submersible. *Kaiko,* a remotely controlled robot sister that became fully operational in 1995 (**Figure 1.22**), is the deepest-diving vehicle presently in service.

In the United States, the three preeminent oceanographic institutions are the Woods Hole Oceanographic Institution on Cape Cod, founded in 1930 (and associated with the Massachusetts Institute of Technology and the neighboring Marine Biological Laboratory, founded in 1888); the Scripps Institution of Oceanography, founded in La Jolla, California, in 1892 and affiliated with the University of California in 1912 (**Figure 1.23**); and the Lamont–Doherty Earth Observatory of Columbia University, founded in 1949.[6] These and other U.S. institutions engaged in marine research are funded largely by the National Science Foundation and the U.S. Office of Sea Grant.

The U.S. government has been active in oceanographic research. Within the Department of the Navy are the Office of Naval Research, the Office of the Oceanographer of the Navy, the Naval Oceanic and Atmospheric Research Laboratory, and the Naval Ocean Systems Command. These agencies are responsible for oceanographic research related to national defense. The National Oceanic and Atmospheric Administration (**NOAA**), founded within the Department of Commerce in 1970, includes the National Ocean Service, the National Weather Service, the National Marine Fisheries Service, and the Office of Sea Grant. NOAA seeks to facilitate commercial uses of the ocean.

SATELLITE OCEANOGRAPHY

1-17

The National Aeronautics and Space Administration (NASA), organized in 1958, has become an important institutional contributor to marine science. For four months in 1978, NASA's *Seasat,* the first oceanographic satellite, beamed oceanographic data to Earth. More recent contributions have been made by satellites beaming radar signals off the sea surface to determine wave height, variations in sea-surface contour and temperature, and other information of interest to marine scientists.

The first of a new generation of oceanographic satellites was launched in 1992 as a joint effort of NASA and the Centre National d'Études Spatiales (the French space agency). The centerpiece of **TOPEX/Poseidon,** as the project is known, is a satellite orbiting 1,336 kilometers (835 miles) above Earth in an orbit that allows coverage of 95% of the ice-free ocean every ten days. The *TOPEX/Poseidon* satellite

---

[6] There are, of course, other prominent institutions involved in studying the ocean. A list of some of them, and some thoughts on how a student might enter the field, appear in Appendix V, "Working in Marine Science."

a

b

Figure 1.23 (**a**) The Woods Hole Oceanographic Institution, Woods Hole, Massachusetts. Marine science has been an important part of this small Cape Cod fishing community since Spencer Fullerton Baird, then assistant secretary of the Smithsonian Institution, established the U.S. Commission of Fish and Fisheries there in 1871. The Marine Biological Laboratory was founded in 1888; the Oceanographic Institution in 1930. (**b**) The Scripps Institution of Oceanography, La Jolla, California. Begun in 1892 as a portable laboratory-in-a-tent, Scripps was founded by William Ritter, a biologist at the University of California, San Diego. Its first permanent buildings were erected in 1905 on a site purchased with funds donated by philanthropic newspaper owner E. W. Scripps and his sister, Ellen.

(**Figure 1.24**) is supplied with a positioning device that allows researchers to determine its position to within one centimeter (½ inch) of Earth's center. The radar transmitters aboard can then determine the height of the sea surface with unprecedented accuracy. Other experiments in this five-year program include sensing water vapor over the ocean, determining the precise location of ocean currents, and determining wind speed and direction. Satellite oceanography is an important frontier, and many discoveries made by satellites are discussed in later chapters.

Figure 1.24 The joint U.S.-French *TOPEX/Poseidon* satellite orbits 1,336 kilometers (835 miles) above Earth in an orbit that allows coverage of 95% of the ice-free ocean every ten days. The satellite was launched in 1992 and is supplied with a positioning device that allows researchers to determine its position to within 1 centimeter (½ inch) of Earth's center. Such accuracy makes possible very accurate determination of sea-surface height by radar transmitters on board.

1. How did this wet planet get the name "Earth"?

We are terrestrial organisms, so the name is understandable. The modern English word *Earth* is derived from the Anglo-Saxon word *eorthe,* which itself comes from the ancient German word for Earth, *Erde.* The earliest German tribes were landlocked people who had no knowledge of the true extent of the world ocean. We seem to be stuck with Earth.

2. OK, then where did the word *ocean* come from?

Ocean derives from the Greek word *okeanos,* meaning "outer sea" (in contrast to the Greeks' "inner sea," the Mediterranean). The term connotes a large moving river. Okeanos was also a mythical Greek titan who was god of the sea before Poseidon and father of the oceanides, or ocean nymphs. The later Latin name for the ocean was *oceanus.* The Latin word evolved into the Middle English term *occean,* in which the double *c* was pronounced as a *k.* This later became the English word *ocean* (which, until this century, was pronounced with a soft *c* as in the word *celery*). Perhaps the best name for Earth would be Oceanus, but so far I've not had much luck in getting the name changed!

3. You noted that the Chinese invented the compass and built an extensive canal system for trade. Did they make contributions to oceanic understanding?

Yes, indeed, and the story has a surprising end. The Chinese appear to have invented the compass around A.D. 1000 and were the first to use it in long-distance navigation. Around the year 1100, the Chinese began to engineer an extensive system of inland waterways, some of which connected with the Pacific Ocean to make long-distance transport of goods more convenient.

As the Dark Ages distracted Europeans, Chinese navigators became more skilled, and their vessels grew larger and more seaworthy. Then they set out to explore their world. Between 1405 and 1433, Admiral Zheng He commanded the greatest fleet the world had ever known. At least 317 ships and 37,000 men undertook seven missions to explore the Indian Ocean, Indonesia, and around the tip of Africa into the Atlantic. Their aim: to display the wealth and power of the young Ming dynasty and to "show kindness to people of distant places." The largest ship in the fleet, with nine masts and a length of 134 meters (440 feet), was a huge treasure ship carrying objects of the finest materials and craftsmanship. The mission of the fleet was not to accumulate such treasure, but to give it away! Indeed, the primary purpose of these expeditions was to convince all nations with which the fleet had contact that China was the only truly civilized state and beyond any imaginable need for knowledge or assistance.

Many technical innovations had been required to make such an ambitious undertaking possible. In addition to the compass, the Chinese invented the central rudder, watertight compartments, and sophisticated sails on multiple masts, all of which were critically important for the successful operation of large sailing vessels.

Despite having these technical advances, the Chinese intentionally abandoned oceanic exploration in 1433. The political winds had changed, and the cost of the "reverse tribute" system was judged too great. In all, until the twentieth century, the Chinese made very few contributions to our understanding of the ocean. Still, their voyaging technology filtered into the West and made subsequent discoveries possible.

4. Since Columbus was unsuccessful in his attempt to sail around the world, and Magellan died without completing his circumnavigation, who was the first captain to complete the trip?

Sir Francis Drake, of England, was the first captain to sail his own ship around the world. His expedition, begun in 1577, lasted three years. In the eastern Pacific he raided Spanish merchants and captured a fortune in gold, silver, coins, and precious stones. He was the first European to sight the west coast of what is now Canada and claimed California for Queen Elizabeth. His transit of the Pacific lasted 68 days, and on his return to England in September 1580 he concluded spice trade negotiations with various heads of state; some of the agreements remain in effect to this day. A wealthy man after the voyage, he was knighted by the Queen and went on in 1588 to help defeat the Spanish Armada.

Little was known of Drake's explorations until comparatively recently. The trade and geographical information he brought home to England was considered so valuable that it was given the highest security classification—thus few people saw it or benefited from it.

5. How do modern navigators find their position at sea?

Very dull story. They push a few buttons on a small box and read their latitude and longitude directly on a screen. This is accomplished by analysis of radio transmissions from satellites. The satellite system is rapidly becoming affordable: For about $100, you can now buy a small, hand-held portable device capable of receiving Global Positioning System (GPS) satellite signals. The GPS system is accurate to about 20 meters and can even tell you which way to move to get home (or anywhere else you want to go). None of these methods is nearly as much fun as the old-fashioned sextant-and-chronometer method, but I suspect that any of the explorers mentioned in this chapter would be very impressed by our new tools.

CHAPTER SUMMARY

Earth is a water planet, possibly one of few in the galaxy. An ocean covering 71% of its surface has greatly influenced its rocky crust and atmosphere. The ocean dominates Earth, and

---

[7] Each chapter ends with a few questions students have asked me after a lecture or reading assignment. These questions and their answers may be interesting to you, too.

the average depth of the ocean is about 4½ times the average height of the continents above sea level. Life on Earth almost certainly evolved in the ocean; the cells of all life forms are still bathed in salty fluids.

The early history of marine science is closely associated with the history of voyaging. The first marine studies had a practical aim: to facilitate travel, trade, and warfare. Later the search for new knowledge became a goal in itself. The first part of this chapter focused on *marine science for voyaging*— it looked at some of the voyagers and their voyages, the inventions that made their adventures possible, and some of the discoveries they made. The second part discussed *voyaging for marine science*, including the British *Challenger* expedition, the first wholly-for-science oceanographic research voyage. The contributions of a few of the founders of modern marine science were summarized, and the rise of oceanographic institutions and satellite oceanography was outlined.

STUDY QUESTIONS

1. Which hemisphere contains the greatest percentage of ocean? Is most of Earth's water in the ocean?

2. Which is greater: the average depth of the ocean floor or the average height of the continents above sea level?

3. How did the library at Alexandria contribute to the development of marine science? What happened to most of the information accumulated there? Why do you suppose the residents of Alexandria became hostile to the librarians and the many achievements of the library?

4. What were the stimuli to Polynesian colonization? How were the long voyages accomplished? How were Polynesian voyages different from (and similar to) those of the Vikings?

5. What were the main stimuli to European voyages of exploration during the Age of Discovery? Why did it end?

6. What were the contributions of Captain James Cook? Does he deserve to be remembered more as an explorer or as a marine scientist?

7. What was the first purely scientific oceanographic expedition, and what were some of its accomplishments?

8. Who was probably the first person to undertake the systematic study of the ocean as a full-time occupation? Are his contributions considered important today?

9. Sketch briefly the major developments in marine science since 1900. Do individuals, separate voyages, or institutions figure most prominently in this history?

10. In your opinion, where does the future of marine science lie?

FOR FURTHER STUDY

The World Ocean

Benchley, P., and J. Gradwohl, eds. 1995. *Ocean Planet: Writings and Images of the Sea.* New York: Abrams. This companion volume to a major Smithsonian Institution exhibit contains photographs and artwork and writings by Jacques-Yves Cousteau, Rachel Carson, John McPhee, Peter Matthiessen, Benjamin Franklin, and others. Highly recommended.

Carson, R. 1951. *The Sea Around Us.* New York: Houghton Mifflin. The book that introduced many of us to the wonders of the ocean when we were young.

Kennish, M. J., ed. 1994. *Practical Handbook of Marine Science. 2d ed.* Boca Raton, FL: CRC Press, Inc. An excellent summary of physical and biological oceanographic data, primarily in the form of graphs and tables. A thorough bibliography follows the introduction.

Scientific American Presents, Fall 1998, is devoted to the topic of "The Oceans."

Slocum, J. 1899. *Sailing Alone Around the World.* Reprint. New York: Sheridan House, 1972. Inspiring autobiographical account of the first single-handed circumnavigation.

The History of Marine Science

Bellwood, P. S. 1991. "The Austronesian Dispersal and the Origin of Languages." *Scientific American,* July, 88–93. The Polynesian voyages are reflected in their languages.

Boorstin, D. 1983. *The Discoverers.* New York: Random House. Elegantly written, very informative sections on exploration and on the instruments invented to make the discoveries possible.

Coakley, B. J. 1998. "Forty Days in the Belly of the Beast." *Scientific American*, Fall, 36–7. A firsthand account of a SCICEX expedition (see **Box 1.1**).

Deacon, M. 1971. *Scientists and the Sea, 1650–1900: A Study of Marine Science*. London: Academic Press. A comprehensive history.

Hale, J. R. 1998. "The Viking Longship." *Scientific American*, February, 56–63. Long, narrow ships packed with dedicated warriors helped to make the Vikings the dominant power in Europe for three centuries, beginning in about A.D. 800.

Hough, R. 1994. *Captain James Cook*. New York: Norton. Hough provides a readable biography with more detail than usual about the deterioration of Cook's psychological outlook during the third voyage. Still, "Cook stood out like a diamond amidst junk jewelry."

Kane, H. *Voyagers*. 1991. Bellevue, WA: Whalesong. Deftly written and stirringly illustrated book on the Hawaiian colonization and other Polynesian topics. Perhaps hard to find but not to be missed!

Linklater, E. 1972. *Voyage of the* Challenger. Garden City, NY: Doubleday. Nicely written account, beautifully illustrated with contemporary photographs and excerpts from the logs.

Matkin, J. 1992. *At Sea with the Scientifics: The* Challenger *Letters of Joseph Matkin*. P. F. Rehbock, ed. Honolulu: University of Hawaii Press. A detailed account of the *Challenger* expedition from a member of the below-decks crew.

Schlee, S. 1973. *The Edge of an Unfamiliar World—A History of Oceanography*. New York: Dutton. Complete and well written.

Sobel, Dava. 1995. *Longitude—The True Story of a Lone Genius Who Solved the Greatest Scientific Problem of His Time*. New York: Walker & Co. A delightful short book that tells the story of Harrison's chronometer with style and sympathy.

Viola, H. J., and C. Margolis, eds. 1987. *Magnificent Voyagers: The United States Exploring Expedition 1838–1842*. Washington, DC: Smithsonian Institution Press.

Wilford, J. N. 1981. *The Mapmakers*. New York: Knopf. An entertaining history of cartography from Eratosthenes to satellite mapping.

For additional readings, go to InfoTrac College Edition, your online research library at:

http://infotrac.thomsonlearning.com/

Origins 2

SOLITUDE

The ocean has always challenged the human spirit. Meeting that challenge *alone* is a supreme triumph of humanity over the uncontrollable and unpredictable forces of nature. Whatever the reasons for their voyages, all who travel alone by sea have experienced profound solitude, loneliness, helplessness, and—if things went well—the exaltation of success.

The first man to sail alone around the world was Joshua Slocum, a Massachusetts sailor who went to sea when he was 14. He was 51 when he began his solitary voyage. He started from Boston on 24 April 1895 in *Spray*, an 11-meter (37-foot) sloop he had rebuilt from a derelict hulk. More than three years and 74,000 kilometers (46,000 miles) later, on 3 July 1898, he returned to Boston, then sailed *Spray* to Newport, Rhode Island, where he tied the boat to the cedar spike driven into the bank that had held her when she was first launched. "I could bring her no nearer home."

In his book, *Sailing Alone Around the World,* Slocum wrote about the peace and heightened awareness that intimate contact with the ocean can bring.

Joshua Slocum, shown here rounding the tip of Cape Horn in his 37-foot sloop, *Spray,* was the first person to sail around the world alone. After a difficult voyage of more than three years and 46,000 miles, he returned to Newport, Rhode Island, and tied *Spray* to the same cedar spike driven into the bank that had held her when she was first launched. "I could bring her no nearer home."

The fog lifted just before night, and I was afforded a look at the sun just as it was touching the sea. I watched it go down and out of sight. Then I turned my face eastward, and there, apparently at the very end of the bowsprit, was the smiling full moon rising out of the sea. Neptune himself coming over the bows could not have startled me more. "Good evening, sir," I cried; "I'm glad to see you." Many a long talk since then I have had with the man in the moon; he had my confidence on the voyage.

About midnight the fog shut down again denser than ever before. One could almost "stand on it." It continued so for a number of days, the wind increasing to a gale. The waves rose high, but I had a good ship. Still, in the dismal fog I felt myself drifting into loneliness, an insect on a straw in the midst of the elements. I lashed the helm, and my vessel held her course, and while she sailed I slept.

During these days a feeling of awe crept over me. My memory worked with startling power. The ominous, the insignificant, the great, the small, the wonderful, the commonplace—all appeared before my mental vision in magical succession. Pages of my history were recalled which had been so long forgotten that they seemed to belong to a previous existence. I heard all the voices of the past laughing, crying, telling what I had heard them tell in many corners of the Earth. The loneliness of my state wore off when the gale was high and I found much work to do. When fine weather returned, then came the sense of solitude, which I could not shake off. . . .

The ocean is still as vast, as quiet, as furious, as inspiring as it was in Slocum's day, but our understanding of it has grown immensely. I think he would have enjoyed the insights that nearly a century of progress in marine science has added to the things he saw, heard, and felt.   2-1

Source: Slocum, J. 1899. *Sailing Alone Around the World*. Reprint. New York: Sheridan House, 1972.

KEY CONCEPTS

1. Marine science—oceanography—is the process of discovering unifying principles in data obtained from the ocean.

2. Science is the process of asking questions about the observable world and then testing the answers to those questions. The external world, not internal conviction, is the testing ground for scientific beliefs.

3. The objects comprising our solar system condensed about 5 billion years ago from a thin cloud that had been enriched by heavy elements made in exploding stars.

4. Earth is density stratified. During its formation, heavy materials fell inward to form the core, and lighter substances rose to become the outer layers. Ocean and atmosphere are the least dense of these layers.

5. Earth first had a solid surface about 4.6 billion years ago. The ocean formed from clouds of steam and water vapor when Earth's surface became cool enough to allow liquid water to rest on the surface.

6. Life probably arose on Earth shortly after its formation, about 4 billion years ago. Life may have arisen in the deep ocean.

7. Other planets may have—or may have had—oceans.

CHAPTER AT A GLANCE

Marine Science, Oceanography, and the Nature of Science

Origins
Galaxies and Stars
Earth and Ocean
The Origin of Life

Ocean Worlds

MARINE SCIENCE, OCEANOGRAPHY, AND THE NATURE OF SCIENCE 2-2

Marine science (or **oceanography**) is the process of discovering unifying principles in data obtained from the ocean, its associated life forms, and the bordering lands. It draws on several disciplines, integrating the fields of geology, physics, biology, chemistry, and engineering as they apply to the ocean and its surroundings. *Marine geologists* focus on questions such as the composition of the inner Earth, the mobility of the crust, the characteristics of seafloor sediments, and the history of Earth's climate. Some of their work touches on areas of intense scientific and public concern, including earthquake prediction and the distribution of valuable resources. *Physical oceanographers* study and observe wave dynamics, currents, and ocean–atmosphere interaction. Their predictions of long-term climate trends are becoming increasingly important as pollutants change Earth's atmosphere. *Marine biologists* work with the nature and distribution of marine organisms, the impact of oceanic and atmospheric pollutants on the organisms, the isolation of disease-fighting drugs from marine species, and the yields of fisheries. *Chemical oceanographers* study the ocean's dissolved solids and gases, and the relationships of these components to the geology and biology of the ocean as a whole. *Marine engineers* design and build oil platforms, ships, harbors, and other structures that enable us to use the ocean wisely. Other marine specialists study weather forecasting, ways to increase the safety of navigation, methods to generate electricity, and much more. Virtually all marine scientists specialize in one area of research, but they also must be familiar with related specialties and appreciate the linkages among them. **Figure 2.1** shows marine scientists in action.

Marine scientists today are asking some critical questions about the origin of the ocean, the age of its basins, and the nature of the life forms it has nurtured. We are fortunate to live at a time when scientific study may be able to answer some of those questions. **Science** is a systematic *process* of asking questions about the observable world and then testing the answers to those questions. Scientists gather and study information (data), but the information itself is not science. Science interprets raw information by constructing a general explanation with which the information is compatible.

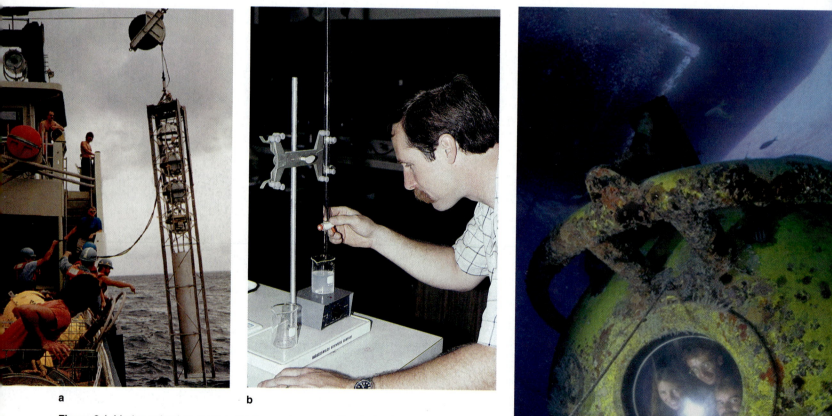

Figure 2.1 Marine scientists at work. (**a**) A sound research float being launched from the Woods Hole Oceanographic Institution's research ship *Oceanus*. The probe will drop to a depth of 3,500 meters (11,500 feet) and produce a low-frequency tone once each day for tracking. (**b**) A marine chemist tests a water sample to find the concentrations of dissolved elements. (**c**) Three researchers peer from a viewing port in the *Aquarius Habitat* off Key Largo, Florida. *Aquarius* rests on the seabed 15 meters (49 feet) below the surface, and serves as a base from which divers can study the area.

a   b   c

Curiosity	A question arises about an event or situation: Why and how does this happen? Why are things this way?
Observations, measurements	Our senses are brought to bear: What is happening? Under what circumstances? When? How does it operate? Does there appear to be a dependable cause-and-effect relationship at work?
Hypothesis	Educated guesses are made. Controlled experiments are planned to prove or disprove potential cause-and-effect relationships. A good hypothesis can predict future occurrences under similar circumstances.
Experiments	Tests are undertaken in nature or in the laboratory. These tests permit manipulating and controlling the conditions under which observations are made.
Theory	Patterns emerge. If one or more of the relationships hold, the hypothesis becomes a theory, an explanation accepted by most researchers.
Law	Theories can evolve into larger constructs: laws. Laws explain events in nature that occur with unvarying uniformity under identical conditions.

Figure 2.2 An outline of the scientific method, a systematic process of asking questions about the observable world and testing the answers to those questions. There is not a single "scientific method." Science depends on a critical attitude about being shown rather than being told, and taking logical approaches to problem solving. Application of a scientific method leads us to truth based on the observations and measurements that have been made—a work in progress, never completed. The external world, not internal conviction, must be the testing ground for scientific beliefs.

Scientists start with a question—a desire to understand something they have observed or measured. They then make an informed guess called a **hypothesis,** a speculation about the natural world that can be tested and verified or disproved by further observations and controlled experiments. (An **experiment** is a test that simplifies observation in nature or in the laboratory by manipulating or controlling the conditions under which observations are made.) Hypotheses consistently supported by observation or experiment are advanced to the status of **theory,** a statement of relationship accepted by most scientists. The largest constructs, known as **laws,** are principles explaining events in nature that have been observed to occur with unvarying uniformity under the same conditions.

Theories and laws in science do not arise fully formed or all at once. Scientific thought progresses as a continuing chain of questioning, testing, and matching theories to observations. A theory is strengthened if new facts support it. If not, the theory is modified or a new explanation is sought. The power of science lies in the ability of the process to operate *in reverse*;

that is, in the use of a theory or law to make predictions and anticipate new facts to be observed.

This procedure, often called the **scientific method,** is an orderly process by which theories are verified or rejected. It is based on the assumption that nature "plays fair"—that the rules governing natural phenomena do not change capriciously as our powers of questioning and observing improve. We believe that the answers to our questions about nature are *ultimately knowable.*

There is no one "scientific method." Some researchers observe, describe, and report on some subject and leave it to others to hypothesize. Scientists don't have one single method in common—the general method they employ is a critical attitude about being *shown* rather than being *told,* and taking a logical approach to problem solving. **Figure 2.2** summarizes the main points.

Nothing is ever proven absolutely true by the scientific method. Theories may change as our knowledge and powers of observation change; thus all scientific understanding is tentative. The conclusions about the natural world that we reach by the process of science may not always be popular or immediately embraced, but if those conclusions consistently match observations, they may be considered true.

This book shows some of the results of the scientific method as it has been applied to the world ocean. It presents facts, interpretations of facts, examples, stories, and some of the crucial discoveries that have led to our present understanding of the ocean and the world on which it formed. As the results of science change, so will the ideas and interpretations presented in books like this one.

ORIGINS 2-3

We have always wondered about our origins—how Earth was formed, how the ocean arose, and how life came to be. In the last 50 years researchers using the scientific method have determined a tentative age for the ocean, Earth, and the universe. They have developed hypotheses about how matter is assembled, how stars and planets are formed, and even how life may have arisen. Many of the details are still sketchy, of course, but their hypotheses have predicted some important recent discoveries in subatomic physics and molecular biology. Perhaps the most dramatic recent discoveries in natural science have been those dealing with the origin and history of the universe.

The universe apparently had a beginning. The **big bang,** as that event is modestly named, probably occurred about 13 billion years ago. All of the mass and energy of the universe is thought to have been concentrated at one geometric point at the beginning of time, the moment when the expansion of the universe began. We don't know what initiated the expansion, but it continues today and will probably continue for billions of years, perhaps forever.

The very early universe was unimaginably hot, but as it expanded it cooled. About a million years after the big bang,

Figure 2.3 This spiral galaxy in the constellation Coma Berenices is very similar in size and structure to our own Milky Way galaxy. The galaxy, photographed in 1999 by the Hubble Space Telescope, is 62 million light years away and about 56,000 light years in diameter. The stars in the foreground are in our galaxy. Note the clouds of dust and gas that obscure our view of some parts of this galaxy's spiral arms. Planets (and probably oceans) are made of such material.

temperatures fell enough to permit the formation of atoms from the energy and particles that had predominated up to that time. Most of these atoms were hydrogen, then as now the most abundant form of matter in the universe. About a billion years after the big bang, this matter began to congeal into the first galaxies and stars.

Galaxies and Stars

 2-4

A **galaxy** is a huge rotating aggregation of stars, dust, gas, and other debris held together by gravity. There are perhaps 50 billion galaxies in the universe and 50 billion stars in each galaxy. Our galaxy is named the **Milky Way** (*galaktos* = milk).[1] A galaxy very similar to our own is shown in **Figure 2.3.**

The **stars** that make up a galaxy are massive spheres of incandescent gases. They are usually intermingled with diffuse clouds of gas and debris. In spiral galaxies like the Milky Way, the stars are arrayed in spiral arms radiating from the galactic center. Other galaxies are elliptical or irregular in shape. Our part of the Milky Way is populated with many stars, but distances within a galaxy are so huge that the star nearest the sun is about 42 trillion kilometers (26 trillion miles) away.

Our sun is a typical star. The sun and its family of planets, called the **solar system,** is located about three-fourths of the way out from the galaxy's center in a spiral arm called the Orion arm. We orbit the galaxy's brilliant core, taking about

230 million years to make one orbit even though we are moving at about 280 kilometers per second (half a million miles an hour). Earth has made about 20 circuits of the galaxy since the ocean formed.

Our planet and our sun probably had a common origin. The members of the solar system are thought to have condensed from a thin cloud—the **solar nebula**—which had been enriched with heavy elements created and released by the explosive deaths of nearby stars. By about 5 billion years ago, the solar nebula was a rotating disk-shaped mass of about 75% hydrogen, 23% helium, and 2% other material including heavier elements, gases, dust, and ice (**Figure 2.4**). Like a spinning skater bringing in her arms, the nebula spun faster as it condensed. Material concentrated near its center became the **protosun.** Much of the outer material eventually became **planets,** the smaller bodies that orbit a star and do not shine by their own light.

The new planets formed in the disk of dust and debris surrounding the young sun through a process known as **accretion**—the clumping of small particles into large masses (**Figure 2.5**). Bigger clumps with stronger gravity pulled in most of the condensing matter. Near the protosun, where temperatures were highest, the first materials to solidify were substances with high boiling points, mainly metals and certain rocky minerals. The planet Mercury, closest to the sun, is mostly iron because iron is a solid at high temperatures. Somewhat farther out, in the cooler regions, magnesium, silicon, water, and oxygen condensed. Methane and ammonia accumulated in the frigid outer zones. Earth's mix of water, silicon–oxygen compounds, and metals results from its position within that

---

[1] Because they can be useful and interesting, the derivations of words are sometimes included.

a

b

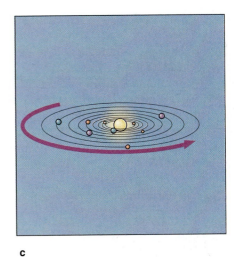

c

Figure 2.4 The formation of the solar system from the solar nebula. Because the nebula was rotating (**a**), it contracted into a disk (**b**). The protosun condensed at the center (**c**), and when the planets accreted, their orbits all lay in nearly the same plane. In (**c**), Earth is shown in blue. Note that this figure is not drawn to scale—the nebula in (**a**) is much larger than the diameter of planetary orbits in (**c**).

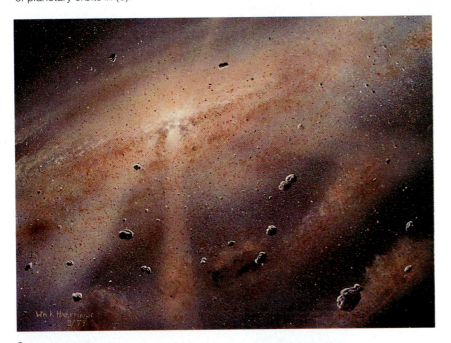

a

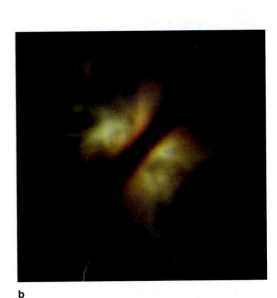

b

Figure 2.5 (**a**) In this artist's conception, our sun begins to shine at the center of a spinning disk of dust, ice, gases, and rock. The accretion phase is in progress—the bits will condense into planets and other bodies. (**b**) A solar nebula about the size of the one that formed our solar system is seen in this 1998 image from the Hubble Space Telescope. The new star is hidden from direct view by a thick, dark disk crossing the center of the image. (The same sort of disk is represented in Figures 2.4b and 2.5a.) The disk—seen here edge-on—is about 130 billion kilometers (80 billion miles) in diameter, about 15 times the diameter of Neptune's orbit around our sun. Dark clouds and bright wisps above and below the disk suggest it is still growing from infalling dust and gas. (**c**) The first image of a possible planet outside our solar system. The bright object lies at the end of a filament of dust and gas extending from a newly formed sun.

accreting cloud. The planets of the outer solar system—Jupiter, Saturn, Uranus, and Neptune—are composed mostly of methane and ammonia ices, because those gases can only congeal at cold temperatures.

The period of accretion lasted perhaps 50 to 70 million years. The protosun became a star when hydrogen fusion temperature was reached. The violence of its nuclear reactions sent radiation sweeping past the inner planets, clearing the area of excess particles and ending the period of rapid accretion. Gases like those we now see on the giant outer planets may once have surrounded the inner planets, but this rush of solar energy and particles stripped them away.

Earth and Ocean

2-5

The young Earth, formed by the accretion of cold particles, was probably homogeneous throughout. Then, in the midst of the accretion phase, Earth's surface was heated by the impact of meteors and other falling debris. This heat, combined with gravitational compression and heat from the decay of radioactive elements accumulating within the newly assembled planet, caused Earth to partially melt. Gravity pulled most of the iron inward to form the planet's core. The friction of the sinking iron heated Earth even more. At the same time, a slush of lighter minerals—silicon, magnesium, aluminum, and oxygen-bonded compounds—rose toward the surface, forming Earth's crust (**Figure 2.6**). This important process, called **density stratification**, lasted perhaps 100 million years.

Then Earth began to cool. Its first surface is thought to have formed about 4.6 billion years ago. That surface did not remain undisturbed for long. A planetary body somewhat larger than Mars smashed into the young Earth. The rocky mantle of the impactor was ejected to form a ring of debris around Earth, and its metallic core fell into Earth's core and joined with it. Could a similar cataclysm happen today? The issue is addressed in Box 12.1.

Radiation from the energetic young sun then stripped away our planet's outermost layer of gases, its first atmosphere; but soon gases that had been trapped inside the forming planet burped to the surface to form a second atmosphere. This volcanic venting of volatile substances—including water vapor—is called **outgassing** (**Figure 2.7a**). As the hot vapors rose, they condensed into clouds in the cool upper atmosphere. Though most of Earth's water was present in the solar nebula during the accretion phase, recent research suggests that a barrage of icy comets colliding with Earth may also have contributed a portion of the accumulating mass of water, this ocean-to-be (**Figure 2.7b**).

Earth's surface was so hot that no water could settle there, and no sunlight could penetrate the thick clouds. After millions of years the upper clouds cooled enough for some of the outgassed water to form droplets. Hot rains fell toward Earth, only to boil back into the clouds again. As the surface became cooler, water collected in basins and began to dissolve minerals from the rocks. Some of the water evaporated, cooled, and fell again. The world ocean was gradually accumulating.

a

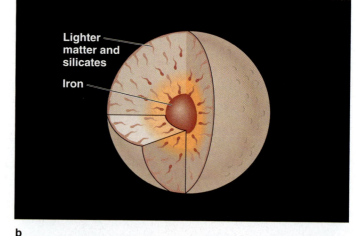

Lighter matter and silicates

Iron

b

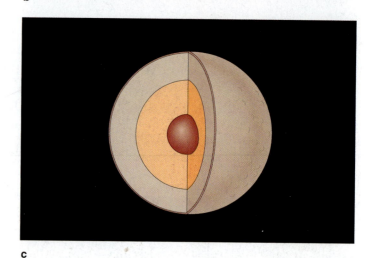

c

Figure 2.6 A representation of the formation of Earth. (**a**) The planet grew by the aggregation of particles. Meteors and asteroids bombarded the surface, heating the new planet and adding to its growing mass. At the time, Earth was composed of a homogeneous mixture of materials. (**b**) Earth lost volume because of gravitational compression. High temperatures in the interior turned the inner Earth into a semisolid mass; dense iron (red drops) fell toward the center to form the core, while less dense silicates move outward. Friction generated by this movement heated Earth even more. (**c**) The result of density stratification is evident in the formation of an inner and outer core, a mantle, and the crust.

a

These heavy rains may have lasted for about 25 million years. Large amounts of water vapor and other gases continued to escape through volcanic vents during that time and for millions of years thereafter. The ocean grew deeper. Recent studies suggest that water may have covered Earth's entire surface for some 200 million years before the continents emerged. Although most of the ocean was in place about 4 billion years ago, ocean formation continues very slowly even today: About 0.1 cubic kilometer (0.025 cubic mile) of new water is added to the ocean each year, mostly as steam flowing from volcanic vents and in the form of small cometary fragments.

The composition of that early atmosphere (often called the primitive atmosphere) was much different from today's. Geochemists believe it may have been rich in carbon dioxide and water vapor, with lesser amounts of carbon monoxide, nitrogen, hydrogen, and hydrogen chloride, and traces of ammonia and methane. Beginning about 3.5 billion years ago, this mixture began a gradual alteration to its present composition, mostly nitrogen and oxygen. At first this change was brought about by carbon dioxide dissolving in seawater to form

Figure 2.7 Sources of the ocean. (**a**) Outgassing. Volcanic gases emitted by fissures add water vapor, carbon dioxide, nitrogen, and other gases to the atmosphere. Volcanism was a major factor in altering Earth's original atmosphere; the action of photosynthetic plants was another. (**b**) Comets may have delivered some of Earth's surface water. Intense bombardment of the early Earth by large bodies—comets and asteroids—probably lasted until about 3.8 billion years ago.

b

carbonic acid, then combining with crustal rocks. The chemical breakup of water vapor by sunlight high in the atmosphere also played a role. About 2 billion years ago, the ancestors of today's green plants greatly accelerated the rate of oxygen production by the action of photosynthesis.

The Origin of Life

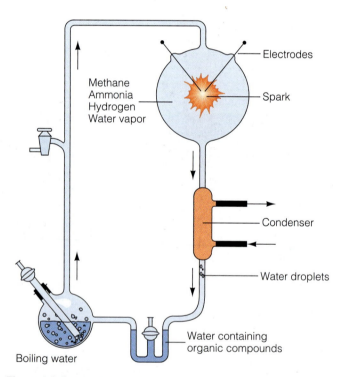 2-6

Life, at least as we know it, would be inconceivable without large quantities of water. Water has the ability to retain heat, moderate temperature, dissolve many chemicals, and suspend nutrients and wastes. These characteristics make it a mobile stage for the intricate biochemical reactions that allowed life to begin and prosper on Earth.

As early as 1929, biologist J.B.S. Haldane proposed that exposing a primitive atmosphere to ultraviolet radiation or lightning might produce some of the same carbon compounds found in living things. Building on this idea in a classic 1953 experiment, Stanley Miller mixed together water vapor, ammonia, methane, and hydrogen—gases that researchers then thought were abundant in the early atmosphere of Earth—

and passed an electric spark through them for a week. In that short time, his apparatus produced a number of different amino acids and other organic compounds (**Figure 2.8**). The electric discharge had provided energy for the formation of these simple molecules.

Recent analysis casts doubt on this hypothesis. The violent collisions now believed to have occurred early in Earth's formation would have prevented the formation of large quantities of methane and ammonia. Where did the candidate carbon compounds come from? There is growing consensus that these organic materials were transported to Earth by the comets, asteroids, meteors, and interplanetary dust particles that crashed into our planet during its birth.

Since Miller's first experiments, other mixtures thought to reflect more accurately the early atmosphere of Earth have been tested, with similar results. When exposed to ultraviolet light, heat, and electrical spark, these mixtures produce simple sugars and most of the biologically important amino acids. They even produce small proteins and nucleotides (components of the molecules that transmit genetic information between generations). The main chemical requirement seems to be the absence (or near absence) of free oxygen, a compound that can disrupt any unprotected large molecule.

Did *life* form in these experiments? No, the compounds that formed are only building blocks of life. But the experiments do tell us something about the commonality and unity of life on Earth. The facts that these crucial compounds can be synthesized so easily and are present in virtually all living forms are probably not coincidental. Those compounds are "permitted" by physical laws and by the chemical composition of this planet. The experiment also underscores the special role of water in life processes. The fact that all life, from a jellyfish to a dusty desert weed, depends on saline water within its cells to dissolve and transport chemicals is certainly significant. It strongly suggests that simple, self-replicating—living—molecules arose somewhere in the early ocean.

The early steps in the evolution of living organisms from simple organic building blocks, a process known as **biosynthesis,** are still speculative. One idea, popularized in the 1950s, suggests that life may have originated in shallow tidal pools at the ocean's edge. Evaporation of water from these pools would have concentrated the amino acid and nucleotide building blocks into a rich organic soup.

Unfortunately for this hypothesis, the sunlight needed to energize the reactions might not have been present; even if it had been, strong sunlight has harmful effects on unprotected large molecules. Planetary scientists now suggest that the sun was faint in its youth. It put out so little heat that the ocean may have been frozen to a depth of around 300 meters (1,000 feet). The ice would have formed a blanket that kept most of the ocean fluid and relatively warm. Periodic fiery impacts by asteroids, comets, and meteor swarms could have thawed the ice, but between batterings it would have reformed. In 1994, chemists Jeffrey Bada and Stanley Miller suggested that organic material may have formed and then been trapped beneath the ice—protected from the atmosphere, which

Figure 2.8 Stanley Miller's 1953 apparatus for producing organic molecules from a mixture of gases thought at the time to have been abundant in Earth's primitive atmosphere. Biochemists have since learned that the early atmosphere was rich in carbon dioxide, water vapor, hydrogen, and nitrogen, with little or no methane and ammonia. But passing an electric spark through either mixture produces simple sugars and most of the biologically important amino acids. The substances that form in these experiments are not alive, but they are the building blocks of life.

Labels in figure: Electrodes · Spark · Methane Ammonia Hydrogen Water vapor · Condenser · Water droplets · Water containing organic compounds · Boiling water

Figure 2.9 Environment for biosynthesis? Weak sunlight and violent conditions at Earth's surface may have made the origin of life more likely near deep-ocean hydrothermal vents similar to the one shown here.

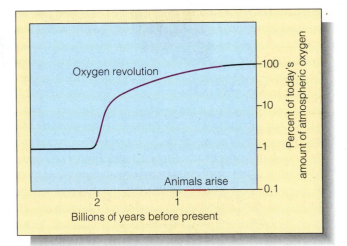

Figure 2.10 The oxygen revolution, a time of radical change in the composition of Earth's atmosphere, occurred between about 2 billion and 400 million years ago. The activity of photosynthetic organisms—mostly photosynthetic bacteria—resulted in a rapid rise in the amount of oxygen in the air, making the evolution of animals possible. Animals are thought to have arisen between 900 and 700 million years ago.

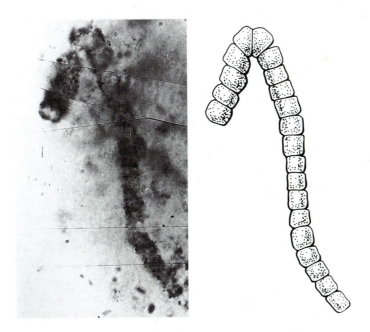

Figure 2.11 Fossil of a bacteria-like organism (with an artist's reconstruction) that photosynthesized and released oxygen into the atmosphere. Among the oldest fossils ever discovered, this microscopic filament from northwestern Australia is about 3.5 billion years old.

contained chemical compounds capable of shattering the complex molecules. The first self-sustaining molecules might have arisen deep below the layers of surface ice, on clays or pyrite crystals near warm hydrothermal vents on the ocean floor (**Figure 2.9**).

A similar biosynthesis cannot occur today. Living things have changed the conditions in the ocean and atmosphere, and those changes are not consistent with any new origin of life. For one thing, green plants have filled the atmosphere with oxygen (**Figure 2.10**). For another, some of this oxygen (as ozone) now blocks much of the ultraviolet radiation from reaching the surface of the ocean. And finally, the many tiny organisms present today would gladly scavenge any large organic molecules as food.

How long ago might life have begun? The oldest fossils yet found, from northwestern Australia, are between 3.4 and 3.5 billion years old (**Figure 2.11**). They are remnants of fairly complex bacteria-like organisms, indicating that life must have originated even earlier, probably only a few hundred million years after a stable ocean formed. Evidence of an even more ancient beginning has been found in the form of carbonaceous residues in some of the oldest rocks on Earth, from Akilia Island near Greenland. These 3.85-billion-year-old specks of carbon bear a chemical fingerprint that researchers believe could only have come from a living organism. Life and Earth have grown old together; each has greatly influenced the other.

OCEAN WORLDS

 2-7

Water planets are probably uncommon in the universe, but water itself is not scarce. In the solar system, for example, Jupiter has hundreds of times more water than Earth does, nearly all of it in the form of ice. Astronomers have even located water molecules drifting in free space. But water in liquid form is unexpected.

Consider the conditions necessary for a large permanent ocean of liquid water to form on a planet. An ocean world

Figure 2.12 A photograph of the icy surface of Europa—a moon of Jupiter—taken by the *Galileo* spacecraft on 20 February 1997. The gravitational pull of Jupiter twists Europa, cracking the ice crust and warming the interior. In some areas the ice has broken into large pieces that have shifted away from one another but fit together like a jigsaw puzzle. This suggests the ice crust is lubricated by slush or water. A global ocean greater in volume than Earth's ocean may lie beneath the movable crust. It may also be salty—salinity has been detected by *Galileo*'s magnetometers. This photo shows an area about 100 kilometers by 140 kilometers (62 miles by 87 miles); the smallest visible objects are about 180 meters (200 yards) across. The image is startlingly similar to the icebergs and ice floes that cover Earth's polar seas in springtime.

must move in a nearly circular orbit around a stable star and the distance of the planet from the star must be just right, to provide a temperature environment in which water remains liquid. Unlike most stars, a water planet's sun must not be a double or multiple star, or the orbital year would have irregular periods of intense heat and cold. The materials that accreted to form the planet must have included both water and substances capable of forming a solid crust. The planet must be large enough so that its gravity will keep the atmosphere and ocean from drifting off into space.

Are these conditions met anywhere else? Is our pleasant blue world the only water planet in the galaxy? Forty-three planets had been discovered outside our solar system by mid-2000. Because they are so far away, we are unable to see details on their surfaces. But two likely nearby candidates have been studied: Jupiter's moon Europa, and the planet Mars.

The spacecraft *Galileo* passed close to Europa, a body about the same size as our moon, in early 1997. Photos sent to Earth revealed a cracked, icy crust covering what appears to be a slushy mix of ice and water (**Figure 2.12**). The jigsaw-puzzle pattern of ice pieces appears to have formed when the ice crust cracked apart, moved slightly, and then froze together again. The underlying ocean is probably kept slushy or liquid by heat escaping Europa's interior, and by gravitational friction of tidal forces generated by Jupiter itself. Though the surface of the ice is as cold as the surface of Jupiter, the liquid interior of the ocean, cradled deep in rocky basins, may be warm enough to sustain life. More missions to Europa are being planned.

Europa may have an icy ocean now, but Mars, a much nearer neighbor, probably had an ocean in the distant past. An ocean may have occupied the low places on the surface of Mars between 3.2 billion and 1.2 billion years ago, when conditions were warmer. In 1998, cameras aboard *Mars Global Surveyor* sent photos from the surface that may show evidence that water once flowed from rock layers below the surface onto the bottom of a crater, leaving sediments that look suspiciously marine (see Figure 5.17). Where is the water now? Mars has become much colder in the past billion years, perhaps because of the loss of greenhouse gases in the atmosphere. If a large quantity of water is present, it probably lies beneath the surface in the form of permafrost (**Figure 2.13**).

And life? Special conditions were necessary for the formation of life on our planet. Earth's gravity is strong enough to retain an ocean but not strong enough to crush the life forms that came from it. Our planet has a magnetic field provided by an iron core to deflect radiation that would otherwise harm the genetic instructions of living things. A single moon provides gentle tides to encourage life forms to leave the ocean and reside on land. The atmosphere is relatively clear—so that sunlight penetrates to the surface—but holds enough moisture to form rains and winds that drive air and ocean currents. Further, the upper air contains ozone, which protects large molecules against the most harmful ultraviolet rays. This

Figure 2.13 Current models suggest that early in its history, Mars had a thick, carbon-dioxide–rich atmosphere, much like the atmosphere of the early Earth. The atmosphere kept Mars warm and allowed water to flow freely, and an ocean may have existed on Mars about 3.2 billion years ago. Over the eons, rocks on the Martian surface absorbed the carbon dioxide and the atmosphere grew thin and cold. The ocean disappeared, most of its water binding to rocks or freezing beneath the planet's surface. In this 1976 photo, morning fog glows blue-gray in the heavily eroded Candor Chasma Valley of Mars. The complex layered terrain may be due to erosion associated with an ancient lake or ocean. The pattern of sediments on the canyon bottom strikes geologists as suspiciously marine, perhaps the remnant of an ancient seabed. (See also Figure 5.17.)

combination may be rare in the galaxy. We have no clear evidence of life elsewhere. We should enjoy and protect the water planet we call home.

QUESTIONS FROM STUDENTS

1. You wrote that "nothing is ever proven absolutely true by the scientific method." What good is it, then? Can't we depend on the process of science?

One philosopher of science has described truth as a liquid—it flows around ideas and is hard to grasp. The progressive improvement in our understanding of nature is subject to the limitations inherent in our observations. As our observations become more accurate, so do our conclusions about the natural world. But because observation (and interpretation of observation) is never perfect, truth can never be absolute. In the 1920s, for example, astronomers assumed that the universe was limited to our own Milky Way galaxy. Observations made with a large new telescope on Mount Wilson in California by Harlow Shapley and Edwin Hubble allowed them to measure more distant objects. Galaxies were discovered in profusion, "like grains of sand on a beach," in Shapley's words. Thus truth changed its shape. In a more directly marine example, one of the *Challenger* expedition's responsibilities was investigating Edinburgh professor Edward Forbes's contention that life below about 550 meters (1,800 feet) was impossible. Consistent sampling below that depth was not practical until the advent of *Challenger*'s steam winch. Sure enough, when observations were made, life was there, and truth changed. We learn as we go. We depend on the underlying assumption that nature "plays fair"—that is, is consistent and does not capriciously change the rules as our powers of observation grow. What we have learned so far is of inestimable practical and aesthetic value, and we have only scratched the surface.

2. How do scientists know how old Earth, the solar system, and the galaxy are? How can they calculate the age of the universe?

The age estimates presented in this chapter are derived from interlocking data obtained by many researchers using different sources. One source is meteorites, chunks of rock and metal formed at about the same time as the sun and planets and out of the same cloud. Many have fallen to Earth in recent times. We know from signs of radiation within these objects how long it has been since they were formed. That information, combined with the rate of radioactive decay of unstable atoms in meteorites, moon rocks, and in the oldest rocks on Earth, allows astronomers to make reasonably accurate estimates of how long ago these objects formed.

As for the age of the universe itself, by June 1999 astronomers had obtained very accurate measurements of its rate of expansion. By calculating backward, they found the universe would have begun its expansion between 12 and 13.4 billion years ago.

3. You mentioned one hypothesis regarding the origin of life. Are there others?

Yes, indeed. Divine creation is one, but it is permanently impossible to test that idea using the scientific method. Another is *panspermia,* which proposes that life formed (or was created) in space and was then distributed throughout the galaxy. Still another involves intelligences "seeding" planets with DNA (or molecules of similar function), later to return and see what useful creatures have resulted. Neither of these ideas can be tested, at least for now. Moreover, both in a sense beg the question; that is, they assume that some extraterrestrial life already existed, but they don't deal with how that life may have arisen.

4. What happened to Joshua Slocum after his single-handed circumnavigation?

He continued to work periodically as a master mariner but spent most of his time writing, lecturing, exhibiting at expositions and fairs, selling curios, farming (a vocation he did not enjoy), and reading. In 1905 he set sail, again in *Spray,* for the West Indies. After two more trips to the same islands, he departed in 1909, at age 65, in his well-traveled boat to explore South America's Orinoco and Amazon Rivers. He disappeared, and no trace of him or his boat was ever found.

CHAPTER SUMMARY

We have learned much about our planet using the scientific method, a systematic process of asking and answering questions about the natural world. Marine science applies the scientific method to the ocean, the planet of which it is a part, and the living organisms dependent on it.

Most of the atoms that make up Earth and its inhabitants were formed within stars. Stars form in the dusty spiral arms of galaxies and spend their lives changing hydrogen and helium to heavier elements. As they die, some stars eject these elements into space by cataclysmic explosions. The sun and the planets, including Earth, probably condensed from a cloud of dust and gas enriched by the recycled remnants of exploded stars. Earth formed by the accretion of cold particles. Heat from infalling debris and radioactive decay partially melted the planet, and density stratification occurred as heavy materials sank to its center and lighter materials migrated toward the surface. Our moon was formed by debris ejected when a planetary body somewhat larger than Mars smashed into Earth.

The ocean formed later, as water vapor trapped in Earth's outer layers escaped to the surface through volcanic activity during the planet's youth. Comets may also have brought some water to Earth. Life originated in the ocean soon after its formation—life and Earth have grown old together. We know of no other planet with a similar ocean, but water is abundant in interstellar clouds and other water planets are not impossible to imagine.

TERMS AND CONCEPTS TO REMEMBER

accretion	oceanography
big bang	outgassing
biosynthesis	planets
density stratification	protosun
experiment	science
galaxy	scientific method
hypothesis	solar nebula
laws	solar system
marine science	stars
Milky Way	theory

STUDY QUESTIONS

1. Can the scientific method be applied to speculations about the natural world that are not subject to test or observation?

2. What are the major specialties within marine science?

3. What is biosynthesis? Where and when do researchers think it might have occurred on our planet? Could it happen again this afternoon?

4. Would you expect ocean worlds to be relatively abundant in the galaxy? Why or why not?

5. Earth has had three distinct atmospheres. Where did each one come from, and what were the major constituents and causes of each?

6. How old is Earth? On what is that estimate based?

7. Where did Earth's heavy elements come from?

8. What is density stratification? What does it have to do with the present structure of Earth?

FOR FURTHER STUDY

The Nature of Science

Carey, S. S. 1998. *A Beginner's Guide to Scientific Method.* 2d ed. Belmont, CA: Wadsworth. A brief, nontechnical introduction to the basic methods underlying all good scientific research.

Sagan, C. 1996. *The Demon-Haunted World: Science as a Candle in the Dark.* New York: Random House. In his last book, Carl Sagan provides an antidote to the obscurantism of those who would seize upon myth and magic as answers to the problems of humankind.

The Origin of Earth and the Ocean

Cloud, Preston. 1988. *Oasis in Space—Earth's History from the Beginning.* New York: Norton.

Consolmango, G. 1994. *Worlds Apart: A Textbook in Planetary Science.* Englewood Cliffs, NJ: Prentice-Hall. A thorough and readable overview of an emerging specialty.

Kasting, J. F. 1998. "The Origins of Water on Earth." *Scientific American,* Fall special issue on "The Oceans," 16–22. Kasting suggests that cometary impacts early in Earth's history contributed most of Earth's surface water.

Levin, H. L. 1996. *The Earth Through Time.* New York: Saunders. A thorough and readable discussion of the geological and biological history of Earth.

Sagan, C. 1980. *Cosmos.* New York: Random House. The most popular scientific book ever published, and with good reason. A magnificently written and illustrated description of the cosmos. A must for anyone wishing to understand Earth's place in the universe.

Trefil, J. 1985. *Space, Time, Infinity.* Washington, DC: Smithsonian Books. Great illustrations, clear and well-written text.

The Origin of Life

Bernstein, M. P., et al. 1999. "Life's Far-Flung Materials." *Scientific American,* July, 42–49. Life may owe its start to complex organic molecules in the heart of an interstellar cloud.

Bjerklie, D., B. Hillenbrand, and J. O. Jackson. 1993. "How Did Life Begin?" *Time,* 11 October, 68–74. Excellent summary article on the most recent speculations.

Brack, André, ed. 1999. *The Molecular Origins of Life.* Cambridge: Cambridge University Press. A compendium of recent articles that illustrates how difficult it can be to draw conclusions in this area of study.

Delaney, J. R. 1998. "Life on the Seafloor and Elsewhere in the Solar System." *Oceanus* 41 (no. 2): 10–13. If volcanoes plus oceans can support life at the vents, why not on other planetary bodies?

Gaidos, E. J., et al. 1999. "Life in Ice-Covered Oceans." *Science* 284 (no. 5420, 4 June): 1631–3. Energetics of aquatic life, and possible prevalence elsewhere.

Horgan, J. 1991. "In the Beginning . . ." *Scientific American,* February, 116–25; Thoughts on the origin of life by a number of researchers.

Trefil, J. 1995. "Life on Earth—Was It Inevitable?" *Smithsonian,* November, 32–41. In a universe filled with prebiotic compounds, it may be only a small step for some of them to hook up in ways that lead directly to life.

Oceans Elsewhere

Head, J. W. III., et al. 1999. "Possible Ancient Oceans on Mars: Evidence from Mars Orbiter Laser Altimeter Data." *Science* 286 (no. 5447, 10 December): 2134–7. High-resolution altimetric data suggest that a lowland-encircling geologic contact represents the ancient shoreline of a large standing body of water in middle Mars history.

Kargel, J. S. 1998. "The Salt of Europa." *Science* 280 (no. 5367, 22 May): 1211–2. Analysis of Europa's magnetic field suggests that water beneath the icy crust is saline.

Kargel, J. S., and R. G. Strom. 1996. "Global Climatic Change on Mars." *Scientific American,* November, 80–88. Excellent overview, wonderfully illustrated.

Kerr, R. 1998. "Geologists Take a Trip to the Red Planet." *Science* 282 (no. 5395, 4 December): 1807–9. An overview of current thinking about the moist past of Mars.

Pappalardo, R. T., et al. 1999. "The Hidden Ocean of Europa." *Scientific American,* October, 54–63. Strong evidence suggests a liquid ocean—the solar system's largest—beneath the icy crust of one of Jupiter's moons.

For additional readings, go to InfoTrac College Edition, your online research library at:

http://infotrac.thomsonlearning.com/

Earth Structure and Plate Tectonics 3

A DANGEROUS BREAKTHROUGH

In October 1987 the scientists and crew of the Scripps Institution research vessel *Melville* experienced a breakthrough of sorts—the eruption of an undersea volcano directly beneath their ship! *Melville* was on an expedition to collect rock and water samples from the MacDonald Seamount, a submerged volcano in French Polynesia, located 1,100 kilometers (700 miles) west of Pitcairn Island. When the research team arrived on station they noticed large patches of greenish-brown water containing fine particles of volcanic ash, which suggested recent volcanic activity. The crew were lowering their gear to the top of the seamount, which rises to within 40 meters (130 feet) of the surface, when huge bubbles of gas and steam suddenly engulfed the ship, making "horrendous clangs and clamors" as they burst against *Melville*'s hull. Chocolate-colored water containing steaming lava balls too hot to hold in bare hands streamed to the surface. The ship's depth recorder and hull-mounted water temperature sensor broke down, but no one was injured. The eruption lasted about five minutes.

The MacDonald Seamount is the last in a chain of submerged and emergent volcanoes stretching 2,000 kilometers (1,200 miles) across the floor of the South Pacific. Here magma (molten rock) rises from deep inside Earth. The composition of gases within the young rock would tell *Melville* researchers about the forces and conditions that cause island chains to form, and something about the history of the Pacific floor itself. Samples they obtained that morning helped to confirm recent theories about oceanic *hot spots*, themselves a verification of the theory of plate tectonics. Seldom do researchers have the good fortune to be in exactly the right place at exactly the right time; scientific breakthroughs are generally less threatening to life and property!

In this chapter we investigate the chemical and physical properties of Earth, the operation of the great forces such as volcanism that have shaped its

Lava meets the ocean on the active island of Hawaii.

surface, and some of the smaller forces that continue to surprise scientists. In the next chapter we describe the geologic features generated by these forces. 3-1

Powerful forces inside our planet are continually forming the major features of its surface. These forces determine the outlines and locations of the continents and ocean floors. They build and destroy mountains and seabeds, raise islands, power volcanoes, form deep trenches, and, through earthquakes, influence the lives of millions of people. Few discoveries in marine science have been as exciting to geologists as the recent advances in our understanding of how these internal forces work. They have pieced together a view of Earth's interior based on measurements of heat leaking from its depths, the chemical composition of volcanic gases, local variations in the pull of gravity, and even the analysis of meteorites thought to have coalesced from the same material as Earth. They have discovered that Earth is *layered*—it looks a bit like the inside of an onion (**Figure 3.1**).

The nature and properties of the layers have been learned largely through the study of shock waves associated with distant earthquakes.

EARTHQUAKES AND EARTHQUAKE WAVES 3-2

Low-frequency pulses of energy generated by the forces that cause earthquakes—**seismic waves**—can spread rapidly through Earth in all directions and then return to the surface. Analysis of seismic waves reveals information about the nature of Earth's interior, much as a tap on a melon can tell the buyer if it's ripe.

On the last Friday of March 1964, one of the largest earthquakes ever recorded struck 144 kilometers (90 miles) east of Anchorage, Alaska (**Figure 3.2**). The release of energy ruptured the surface of Earth for 800 kilometers (500 miles) between the small port of Cordova in the east and Kodiak Island in the west. In some places the vertical movement of the crust was 3.7 meters (12 feet); one small island was lifted 38 feet. Horizontal movement caused the greatest damage: 65,000 square kilometers (25,000 square miles) of land abruptly moved west. In 4½ minutes of violent shaking, Anchorage had moved sideways 2 meters (6.6 feet), the town of

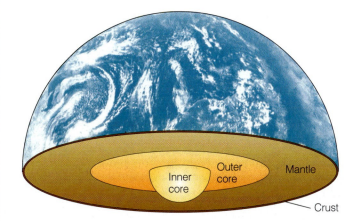

Figure 3.1 Earth in cross section, showing the major internal layers.

Figure 3.2 Alaska earthquake damage, 27 March 1964. The collapse of Fourth Avenue in Anchorage. Violent shaking for 4½ minutes during this earthquake of moment-magnitude 8.9 caused the ground to liquefy and slide, undermining buildings and causing the pavement to sink more than 3 meters (10 feet) in places. Seismic waves associated with the earthquake passed through Earth and emerged at distant locations, carrying information about Earth's interior.

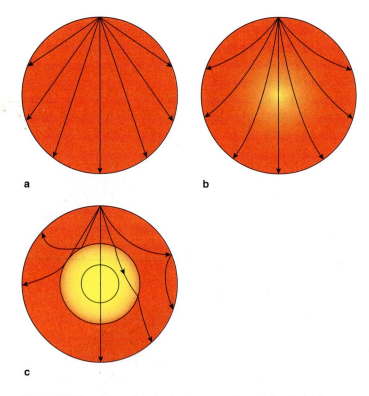

Figure 3.3 Possible paths of seismic waves through Earth. (**a**) If Earth were uniform (homogeneous) throughout, seismic waves would radiate from the site of an earthquake in straight lines. (**b**) If the density, or rigidity, of Earth increased evenly with depth, seismic wave velocity would increase with depth, and the waves would bend smoothly upward toward the surface. (**c**) If Earth were layered inside, some seismic waves would be reflected at the boundaries between layers while others were bent. Seismic evidence shows that Earth is layered.

Seward 14 meters (46 feet)! More than 75% of the state's commerce was disrupted and thousands were homeless. Damage exceeded $750 million, and, considering the violence of the earthquake, it's a wonder that only 115 lives were lost.

Geological research stations over much of the world saw extraordinarily large seismic waves arrive from Alaska. Though they passed through Earth, the waves were so powerful that many **seismographs** (*seismos* = earthquake + *graph* = to write), instruments that sense and record earthquakes, were physically damaged.

You might think that seismic waves would travel at constant speeds from an earthquake, and their paths through the interior would be straight lines (as in **Figure 3.3a**). This would be true if Earth were of uniform (homogeneous) composition. In fact, the speed of seismic waves increases with depth because the density of the materials they encounter increases in the interior of Earth. Waves descending at shallow angles tend to bend up toward the surface as they travel into regions of faster velocity (**Figure 3.3b**).[1] The sensitive seismographs in service in the 1960s enabled researchers to study the ways seismic waves bounce off abrupt transitions between Earth's inner layers (**Figure 3.3c**). Technological advances have continued, and recent computer-assisted analysis of seismic waves has contributed to an understanding of the composition, thickness, and structure of the layers inside of Earth. The shaken citizens of Anchorage could not have known at the time, but their earthquake helped to solidify our present model of the interior of Earth.

---

[1] This phenomenon, known as *refraction,* is explained in Figure 6.18.

Each layer of Earth has different chemical and physical characteristics. One classification of Earth's interior emphasizes chemical composition. The outermost layer is the lightweight, brittle, aptly named **crust.** The crust beneath the ocean differs in thickness, composition, and age from the crust of the continents. The thin **oceanic crust** is primarily **basalt,** a heavy, dark-colored rock composed mostly of oxygen, silicon, magnesium, and iron. By contrast, the most common material in the thicker **continental crust** is **granite,** a familiar speckled rock composed mainly of oxygen, silicon, and aluminum. The deepest crust—beneath great mountain ranges—extends only 70 kilometers (44 miles) below Earth's surface. The **mantle,** the layer beneath the crust, is thought to consist mainly of oxygen, iron, magnesium, and silicon. Most of Earth is mantle—it accounts for 68% of Earth's mass and 83% of its volume. The outer and inner **cores,** which consist mainly of iron and nickel, lie beneath the mantle at Earth's center.

Chemical make-up is not the only important distinction between layers. Different conditions of temperature and pressure occur at different depths, and these conditions influence the physical properties of the materials. The behavior of a rock is determined by three factors: temperature, pressure, and the rate at which a deforming force (stress) is applied. Geologists have therefore devised another classification of Earth's interior based on *physical* rather than *chemical* properties:

- The **lithosphere** (*lithos* = rock)—Earth's cool, rigid outer layer—averages 100 kilometers (62 miles) in thickness. It is comprised of the continental and oceanic crusts *and* the outermost cool and rigid portion of the mantle.
- The **asthenosphere** (*asthenes* = weak) is the hot, slowly flowing layer of upper mantle below the lithosphere extending to a depth of about 300 kilometers (187 miles).
- The **mantle** extends to the core. The asthenosphere and the mantle below the asthenosphere (the **lower mantle**) have a similar chemical composition, but the lower mantle is hotter, denser, and flows more slowly.
- The **core** is divided into two parts. The outer core is a viscous liquid with a density about four times that of the crust. The inner core is a solid with a maximum density about six times that of crustal material. Both parts are extremely hot, with an average temperature of about 5,500°C (9,900°F). Recent evidence indicates that the inner core may be as hot as 6,600°C (12,000°F) at its center, hotter than the surface of the sun!

Figure 3.4 shows Earth's internal layers. Note that the rigid sandwich of crust and upper mantle—the lithosphere—floats on (and is supported by) the denser deformable asthenosphere. Research has shown that slabs of Earth's relatively cool and solid surface—its lithosphere—float and move independently of one another over the hotter, partially molten asthenosphere layer directly below.

Internal Heat 3-4

The interior of Earth is hot. The main source of that heat is **radioactive decay,** a process that generates heat when unstable forms of elements are transformed into new elements. As you read in Chapter 2, radioactive decay within the newly formed Earth released heat that contributed to the melting of the original mass. Most of the melted iron sank toward the core, generating friction and thus releasing huge amounts of energy. By now almost all of the heat generated by the formation of the core has dissipated, but radioactive elements within Earth still continue to decay and produce new heat. Today most of the radioactive heating takes place in the crust and upper mantle rather than in the deeper layers. Some of this heat journeys toward the surface by **conduction,** a process analogous to the slow migration of heat along a skillet's handle. Some heat also rises by **convection** in the asthenosphere. Convection occurs when a fluid is heated, expands and becomes less dense, and rises. (Convection also causes air to rise over a warm radiator, as shown in Figure 7.4.)

Even after 4.6 billion years, heat continues to flow from within Earth. This heat powers the construction of mountains and volcanoes, causes earthquakes, moves continents, and shapes ocean basins.

Isostatic Equilibrium 3-5

Why do large regions of continental crust stand high above sea level? If the asthenosphere is nonrigid and deformable, why don't mountains sink because of their weight and disappear? Another look at **Figure 3.4** will help to explain the situation. The mountainous parts of continents have "roots" extending into the asthenosphere. The continental crust and the rest of the lithosphere "float" on the denser asthenosphere. The situation involves buoyancy, the principle that explains why ships float.

Buoyancy is the ability of an object to float in a fluid by displacing a volume of that fluid equal in mass to the floating object's own mass. A steel ship floats because its shape displaces a volume of water equal in weight to its own weight plus the weight of its cargo. Thus, an empty containership displaces a smaller volume of water than the same ship when fully loaded (**Figure 3.5**). The water supporting the ship is not "strong" in the mechanical sense; water does not support a ship in the same way a strong steel bridge supports the weight of your car. Buoyancy, rather than mechanical strength, supports the ship and its cargo.

Any part of a continent that projects above sea level is supported in the same way. Consider the continent containing Mount Everest, highest of Earth's mountains at 8.85 kilometers (29,035 feet) above sea level. Mount Everest and its neighboring peaks are not supported by the *mechanical* strength of the materials within Earth—nothing in our world is that strong. Over a long period, and under the tremendous weight of the overlying crust, the asthenosphere behaves like a dense, viscous, very slowly moving fluid. The mountainous upper surface of the continent floats high above sea level because the

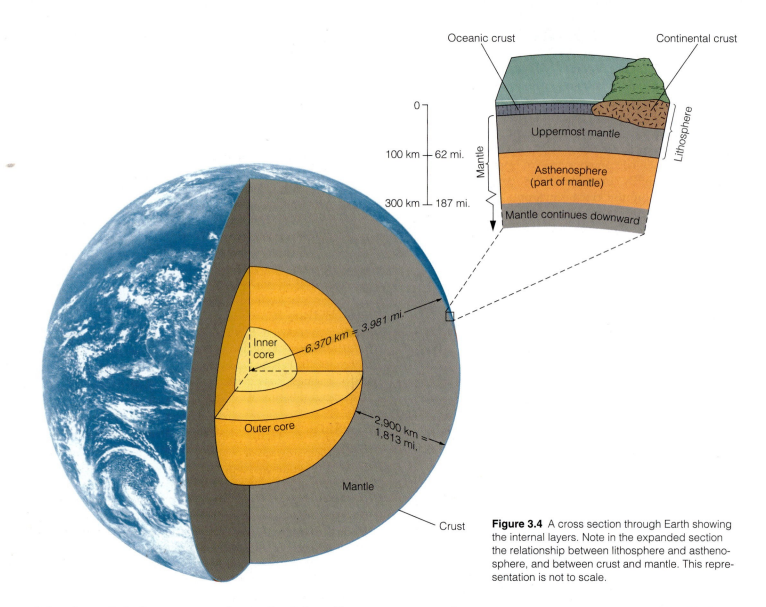

Figure 3.4 A cross section through Earth showing the internal layers. Note in the expanded section the relationship between lithosphere and asthenosphere, and between crust and mantle. This representation is not to scale.

lithosphere of which it is a part sinks into the deformable asthenosphere until it has displaced a volume of asthenosphere equal in mass to its own mass. The continent's mountains rest at great height in balance with their subterranean underpinnings, susceptible to rising or falling as erosion or crustal

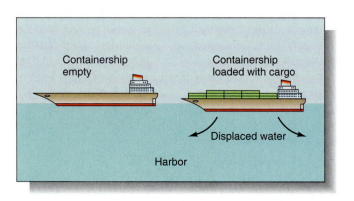

Figure 3.5 The principle of buoyancy. A ship sinks until it displaces a volume of water equal in weight to the weight of the ship and its cargo. If the cargo is removed, the ship will rise.

stresses dictate. Lower regions are supported by shallower roots. Like a ship floating in water, the entire continent stands in **isostatic equilibrium** (*isos* = equal + *stasis* = standing).

What happens when a mountain erodes? In much the same way a ship rises when cargo is removed, Earth's crust will rise in response to the reduced load. Ancient mountains that have undergone millions of years of erosion often expose rocks that were once embedded deep within their roots. This kind of isostatic readjustment results in the thinning of the continental crust beneath the mountains, and subsidence beneath areas of deposited sediments. This process is shown in **Figure 3.6.**

Unlike the asthenosphere on which the lithosphere floats, crustal rock does not flow at normal surface temperatures. A ship reacts to any small change in weight with a corresponding small change in vertical position in the water, but an area of continent or ocean floor cannot react to every small weight change because its edges are mechanically bound to adjacent crustal masses. When the force of uplift or downbending exceeds the mechanical strength of the adjacent rock, the rock will fracture along a plane of weakness—a **fault.** The adjacent

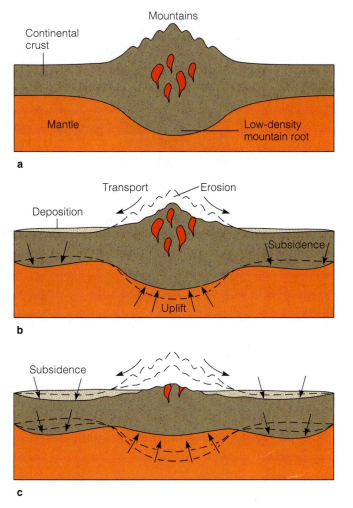

Figure 3.6 Erosion and isostatic readjustment can cause continental crust to become thinner in mountainous regions. As mountains are eroded, isostatic uplift causes their roots to rise, as happens when a ship is unloaded. Further erosion exposes rocks that were once embedded deep within the peaks. Deposition of sediments away from the mountains often causes nearby crust to sink.

crustal fragments will move vertically in relation to each other. This sudden adjustment of the crust to isostatic forces by fracturing, or faulting, is one cause of earthquakes.

Curious Coincidences?   3-6

Look for the site of the 1964 Alaska earthquake in **Figure 3.7.** Many thousands of significant earthquakes have occurred on Earth in recent times, and the Anchorage quake occurred in an area of pronounced seismic activity. Notice that earthquakes *do not occur at random places.* Seismic activity appears concentrated in lines and arcs across Earth's surface. Sometimes these lines coincide with the edge of a continent (as along the western edge of the Americas), and sometimes the lines trace the middle or sides of the ocean basins.

Other curious coincidences present themselves. As Leonardo da Vinci noticed on early charts, in some regions the

continents look as if they would fit together like jigsaw puzzle pieces if the intervening ocean were removed. In 1620 Francis Bacon also wrote of a "certain correspondence" between shorelines on either side of the South Atlantic. In 1885 Edward Suess, a respected German scientist, suggested that the Southern Hemisphere's continents might once have been a single large land mass. He based his belief in part on the similarities of fossils found on these continents, especially fossils of the fern *Glossopteris.* Suess was not taken seriously by his colleagues because he could not explain how the continents had moved.

As they probed the submerged edges of the continents, researchers found that the ocean bottom nearly always sloped gradually out to sea for some distance and then dropped steeply to the deep ocean floor. They realized that these shelflike continental edges were extensions of the continents themselves. In the few locations where they had measurements, researchers found that the "fit" between South America and Africa, impressive at the shoreline, was even better along the submerged edges of the continents.

Though so accurate a fit almost certainly could not have occurred by chance, no one had yet proposed a mechanism that could separate whole continents into moving pieces.

Continental Drift   3-7

Into the fray stepped **Alfred Wegener,** a busy German meteorologist and polar explorer (**Figure 3.8**). In a lecture in 1912 he proposed the startling and original theory of **continental drift.** Wegener suggested that all Earth's land had once been joined into a single supercontinent surrounded by an ocean. He called the land mass **Pangaea** (*Pan* = all + *gaea* = Earth) and the surrounding ocean **Panthalassa** (*pan* = all + *thalassa* = ocean). Wegener thought Pangaea had broken into pieces about 200 million years ago. Since then, he said, the pieces had moved to their present positions and were still moving.

The greatest block to the acceptance of continental drift lay in geology's view of Earth's mantle. The available evidence seemed to suggest that a deep, solid mantle supported the crust and mountains mechanically (not isostatically) from below. Drift would be impossible with this kind of subterranean construction. A few perceptive seismic researchers then noticed that the upper mantle reacted to earthquake waves as if it were a deformable mass, not a rigid solid. Perhaps such a layer would resemble a slug of iron heated in a blacksmith's forge; it would deform with pressure and even flow slowly. Established geologists dismissed this interpretation, saying that the mountains would simply fall over or sink without rigid underpinnings. By 1926 the "drifters" were in full retreat. When Wegener died on an expedition across Greenland in 1930, his theory was already in eclipse.

The Idea Transformed   3-8

The concept of continental drift refused to die, however; those neatly fitted continents provided a haunting reminder of Wegener to anyone looking at an Atlantic chart. In 1935 a

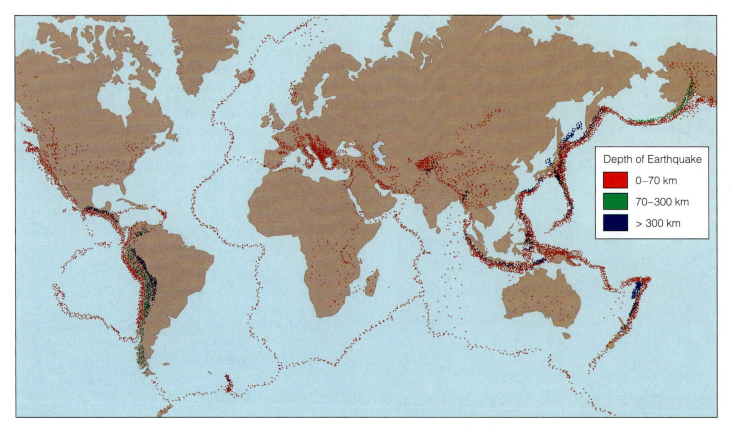

Figure 3.7 Seismic events worldwide, January 1977 through December 1986. The locations of about 10,000 earthquakes are colored red, green, and blue to represent event depths of 0–70 kilometers, 70–300 kilometers, and below 300 kilometers, respectively.

Depth of Earthquake

- 0–70 km
- 70–300 km
- > 300 km

Japanese scientist, Kiyoo Wadati, speculated that earthquakes and volcanoes near Japan might be associated with continental drift. In 1940, seismologist Hugo Benioff plotted the locations of deep earthquakes at the edges of the Pacific. His charts revealed the true extent of the **Pacific Ring of Fire,** a circle of violent geologic activity surrounding much of the Pacific Ocean. Seismographs were now beginning to reveal a worldwide pattern of earthquakes and volcanoes. Deep earthquakes did not occur randomly over Earth's surface but were concentrated in zones that extended in lines along Earth's crust. Benioff, Wadati, and others wondered what could cause such an orderly pattern of deep earthquakes. Figure 3.7 is based on a plot of about 30,000 earthquakes. Notice the odd pattern they form—almost as if Earth's lithosphere is divided into sections! Benioff's sensitive seismographs also began to gather strong evidence for a partially molten, nonrigid layer in the upper mantle. Could the continents somehow be sliding on that?

Other seemingly unrelated bits of information were accumulating. **Radiometric dating** of sediments and rocks was perfected after World War II. This technique is based on the discovery that unstable, naturally radioactive elements lose particles from their nuclei and change into new stable elements. The radioactive decay occurs at a constant rate, and measuring the ratio of radioactive to stable atoms in a sample provides its age. To the surprise of many geologists, the maximum age of the ocean floor and its overlying sediments was ra-

diometrically dated at less than 200 million years, only about 4% of the age of Earth.[2] The centers of the continents are much older—some parts of the continental crust are more than 3.9 billion years old, about 85% of the age of Earth. *Why was oceanic crust so young?*

Attention quickly turned to the deep ocean floors, the complex profiles of which were now being revealed by **echo sounders,** devices that measure depth by bouncing high-frequency sound waves off the bottom (see Figure 1.21). The overall shape of the Mid-Atlantic Ridge was slowly revealed. The ridge's conformance to shorelines on either side of the Atlantic (**Figure 3.9**) raised many eyebrows.

Research ships compiled more complete, accurate charts of the submerged edges of continents. At England's Cambridge University, Sir Edward Bullard used an early computer to process these data to achieve the best possible fit of the continental jigsaw puzzle pieces around the Atlantic. The fit was astonishingly good (**Figure 3.10**).

Mantle studies were keeping pace. The first links in the Worldwide Standardized Seismograph Network, begun during the International Geophysical Year in 1957, were beginning to report data from seismic waves reflected and refracted through the planet's inner layers. This information verified the existence of a layer in the upper mantle that caused a decrease

---

[2] For a discussion of geological time, please see Appendix II.

Figure 3.8 Alfred Lothar Wegener writes in his journal at the beginning of his last expedition in Greenland in 1930. His remarkable book *The Origin of Continents and Oceans* was published in 1915. In it, he outlined interdisciplinary evidence for his theory of continental drift.

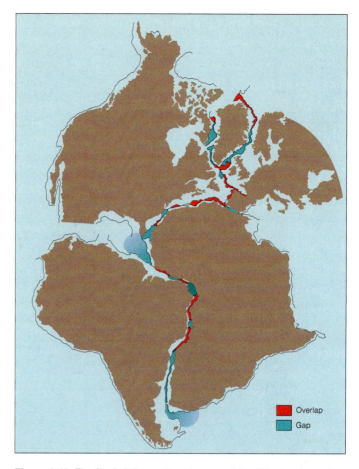

Figure 3.10 The fit of all the continents around the Atlantic at a water depth of about 137 meters (450 feet), as calculated by Sir Edward Bullard at the University of Cambridge in the early 1960s.

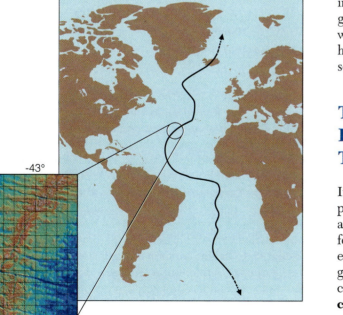

Figure 3.9 The Mid-Atlantic Ridge, showing its conformance to the coastlines of the adjacent continents. The inset shows a detail of the ridge—the central rift is clearly visible.

in the velocity of seismic waves. This finding strongly suggested that the layer was not solid. Perhaps the lithosphere was isostatically balanced in this deformable layer, and perhaps continents could move around in it *if* a suitable power source existed.

THE BREAKTHROUGH: FROM SEAFLOOR SPREADING TO PLATE TECTONICS 3-9

In 1960 Professor Harry Hess of Princeton University proposed a radical idea to explain the features of the ocean floor and the "fit" of the continents. He suggested that new seafloor forms at the Mid-Atlantic Ridge (and the other newly discovered ocean ridges) and spreads outward from this line of origin. Continents would be pushed aside by the same forces that cause the ocean to grow. Perhaps this motion is powered by **convection currents,** slow-flowing circuits of material within the asthenosphere.

Seafloor spreading, as the new theory was called, pulled many loose ends together. If the mid-ocean ridges were **spreading centers** and sources of new ocean floor rising from

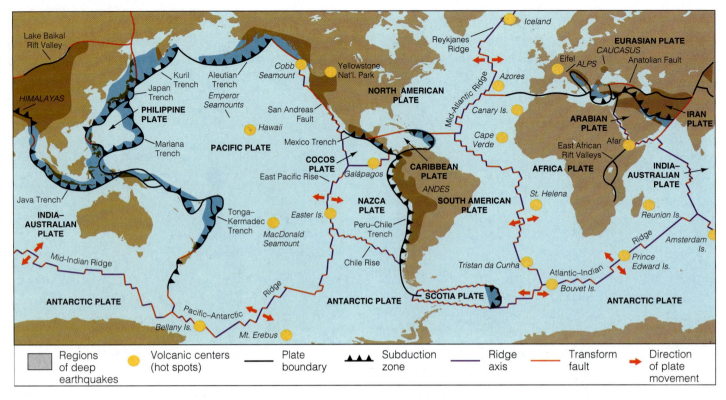

Figure 3.11 The major lithospheric plates, showing their directions of relative movement and the location of the principal hot spots. Compare this figure to Figure 3.7 and note the correspondence of plate boundaries and earthquake locations.

the asthenosphere, they should be hot. They were. If the new oceanic crust cooled as it moved from the spreading center, it should shrink in volume and become more dense, and the ocean should be deeper farther from the spreading center. It was. Sediments at the edges of the ocean basin should be thicker than those near the spreading centers. They were, and they were also older.

Did this mean that Earth was continuously expanding? Since there was no evidence for a growing Earth, the creation of new crust at spreading centers would have to be balanced by the destruction of crust somewhere else. Then researchers discovered that the crust plunges down into the mantle along the periphery of the Pacific. The process is known as **subduction,** and these areas are called **subduction zones** (also known as *Wadati-Benioff zones*). The zones of concentrated earthquakes (see again Figure 3.7) were found in regions of crustal formation (spreading centers) and crustal destruction (subduction zones).

In 1965 the ideas of continental drift and seafloor spreading were integrated into the overriding concept of **plate tectonics** (*tekton* = builder), primarily by the work of **John Tuzo Wilson,** a geophysicist at the University of Toronto. In this theory, Earth's outer layer consists of about a dozen separate major lithospheric **plates,** each about 70 to 100 kilometers (43 to 65 miles) thick, floating on the asthenosphere. When heated from below, the deformable asthenosphere expands, becomes less dense, and rises. It turns aside when it reaches the lithosphere, and drags the plates laterally until turning under again

to complete the circuit. The large plates include both continental and oceanic crust. The major plates, which jostle about like huge flats of ice on a warming lake, are shown and named in **Figure 3.11.** Plate movement is slow in human terms, averaging about 5 centimeters (2 inches) a year—about as fast as your fingernails grow. The plates interact at converging, diverging, or slipping boundaries, sometimes sliding below the surface or wrinkling into mountains. Most of the million or so earthquakes and volcanic events each year occur along plate boundaries.

We now know that *through the great expanse of geologic time, this slow movement remakes the surface of Earth, expands and splits continents, forms and destroys ocean basins.* The light, ancient granitic continents ride high in the lithospheric plates, rafting on the moving asthenosphere below. Plate movement is caused in part by friction against the plate from convection currents flowing in the asthenosphere. In subduction, gravity drags heavy basaltic ocean floor (and its overlying layer of sediment) into the mantle at a subduction zone. Partially melted material from the subducted plate may separate into lighter components and rise, solidifying to form the heart of a continent or reemerging through a volcano. Because the ocean floor itself acts as a vast "conveyor belt" transporting accumulated sediment to subduction zones where the seafloor sinks into the asthenosphere, no marine sediments are of great age. This process, diagrammed in **Figure 3.12** has progressed since Earth's crust first solidified and cooled. Literally and figuratively, it all fits. This recent understanding

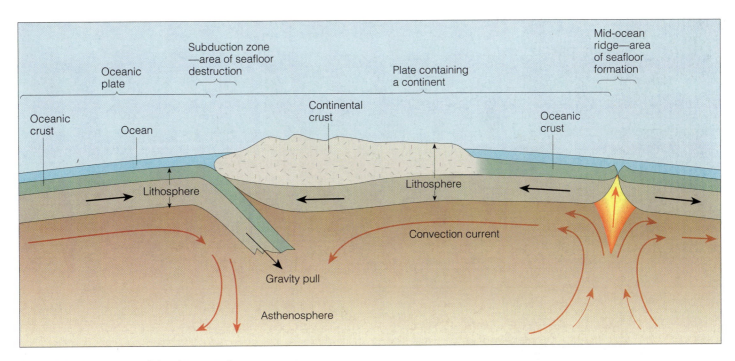

Figure 3.12 An overview of the plate tectonics process.

of the ever-changing nature of Earth has given fresh meaning to historian Will Durant's warning: "Civilization exists by geological consent, subject to change without notice."

PLATE BOUNDARIES 🔲 3-10

The lithospheric plates shown in Figure 3.11 float on a dense, deformable asthenosphere and are free to move relative to each other. Plates interact with neighboring plates along their mutual boundaries. In **Figure 3.13,** movement of Plate A to the left (west) requires it to slide along its north and south margins. An overlap is produced in front (to the west), and a gap is created behind (to the east). Different places on the margins of Plate A experience separation and extension, convergence and compression, and transverse movement (shear).

The three types of plate boundaries that result from these interactions are called *divergent, convergent,* and *transform boundaries,* depending on their sense of movement. **Figure 3.14** provides an overview of the plate boundary interactions we will study next.

Divergent Plate Boundaries— Forming Ocean Basins 🔲 3-11

The spreading center at the Mid-Atlantic Ridge is a **divergent plate boundary,** a line along which two plates are moving apart. Oceanic crust forms along divergent plate boundaries. The formation of the Atlantic is shown in **Figure 3.15.** Heat from the lower crust and mantle accumulated beneath the continents. This heat caused the asthenosphere to expand and rise, thus lifting and fracturing the lighter, solid lithosphere above. The plate and its embedded continent split in two, and a new

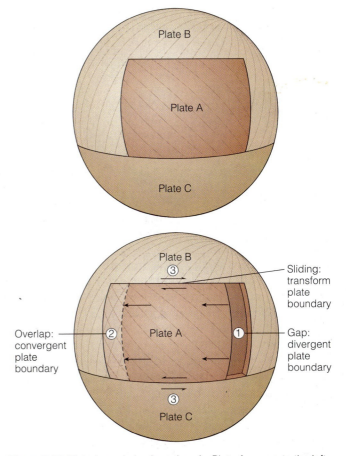

Figure 3.13 Plate boundaries in action. As Plate A moves to the left (west), a gap forms behind it ① and an overlap with Plate B forms in front ②. Sliding occurs along the top and bottom sides ③. The margins of Plate A experiene the three types of interactions: At ①, extension characteristic of divergent boundaries; at ②, compression characteristic of convergent boundaries; and at ③, the shear characteristic of transform plate boundaries. Figure 3.14 shows these kinds of boundaries in more detail.

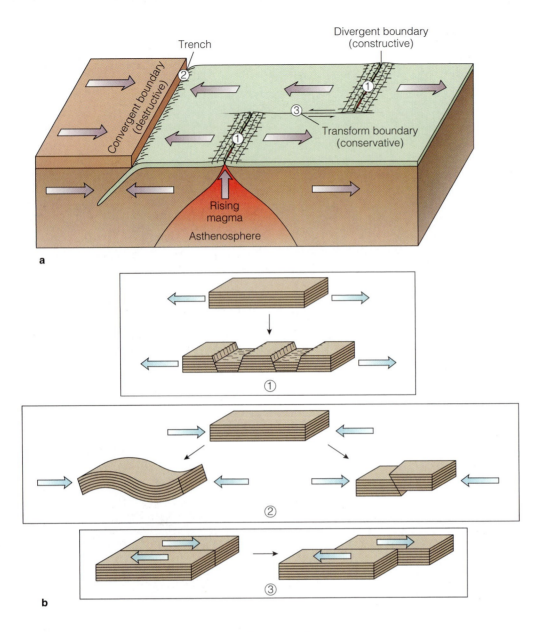

a

b

Figure 3.14 (a) Plate boundaries on the surface of a sphere. The divergent ①, convergent ②, and transform ③ margins match those of Plate A in Figure 3.13. (b) Extension of divergent boundaries causes splitting and rifting ①, compression at convergent boundaries produces buckling and shortening ②, and translation at transform boundaries causes shear ③.

plate boundary—a rift valley—was formed between the pieces. As the broken plate separated at this new spreading center, molten rock called **magma** rose into the crustal fractures. (Magma is called *lava* when found aboveground.) Some of the magma solidified in the fractures; some erupted from volcanoes on the seafloor. Together these processes produced new oceanic crust. The South Atlantic, a large new ocean basin formed between the diverging plates, is shown in **Figure 3.15e.** A long mid-ocean ridge divided by a central rift valley traverses the ocean floor roughly equidistant from the shorelines in both the North and South Atlantic, terminating north of Iceland.

Plate divergence is not confined to the Atlantic, nor has it been limited to the last 200 million years. As may also be seen in Figure 3.11, the Mid-Atlantic Ridge has counterparts in the Pacific and Indian Oceans. The Pacific floor, for example, diverges along the East Pacific Rise and the Pacific Antarctic Ridge, spreading centers that form the eastern and southern boundaries of the great Pacific Plate. In East Africa, rift valleys have formed relatively recently as plate divergence begins to

separate another continent. As happened in the Red Sea (**Figure 3.15d**), the ocean will invade when the rift becomes deep enough. **Figure 3.16** shows how divergence has formed other ocean basins. About 20 cubic kilometers (4.8 cubic miles) of new ocean crust forms each year.

Convergent Plate Boundaries— Recycling Crust, Building Island Arcs and Continents

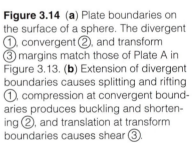 3-12

Since Earth is not getting larger, divergence in one place must be offset by convergence in another. Crust is destroyed at **convergent plate boundaries,** regions of violent geologic activity where plates are pushing together. South America, embedded in the westward-moving South American Plate, encounters the Pacific's Nazca Plate as it moves eastward. The relatively thick and light continental lithosphere of South America rides up and over the heavier oceanic lithosphere of the Nazca Plate, which is subducted along the deep trench

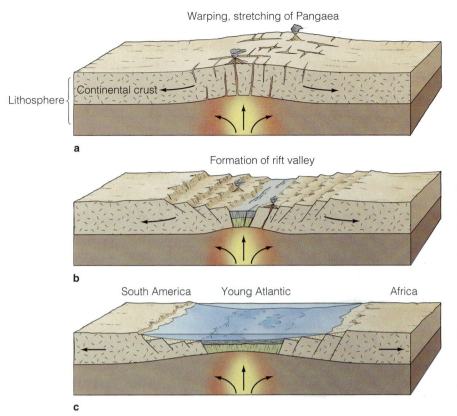

a Warping, stretching of Pangaea

Lithosphere { Continental crust

b Formation of rift valley

c South America   Young Atlantic   Africa

Figure 3.15. A model for the formation of a new plate boundary: the breakup of Pangaea and the formation of the Atlantic. (**a**) As the lithosphere began to crack, a rift formed beneath the continent, and molten basalt from the asthenosphere began to rise. (**b**) As the rift continued to open, the two new continents were separated by a growing ocean basin. Volcanoes and earthquakes occur along the active rift area, which is the mid-ocean ridge. The East African Rift Valley currently resembles this stage. (**c**) A new ocean basin (shown in green) forms beneath a new ocean. The Red Sea (**d**) currently resembles this stage. Note the remarkable sawtooth configuration of the peaks on the horizon, and their similarity to the diagram. (**e**) The South Atlantic in cross section, showing the mid-ocean ridge in the middle of the growing basin.

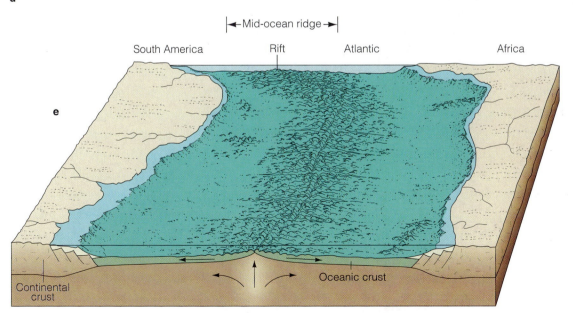

d

|←— Mid-ocean ridge —→|

South America   Rift   Atlantic   Africa

e

Continental crust   Oceanic crust

49

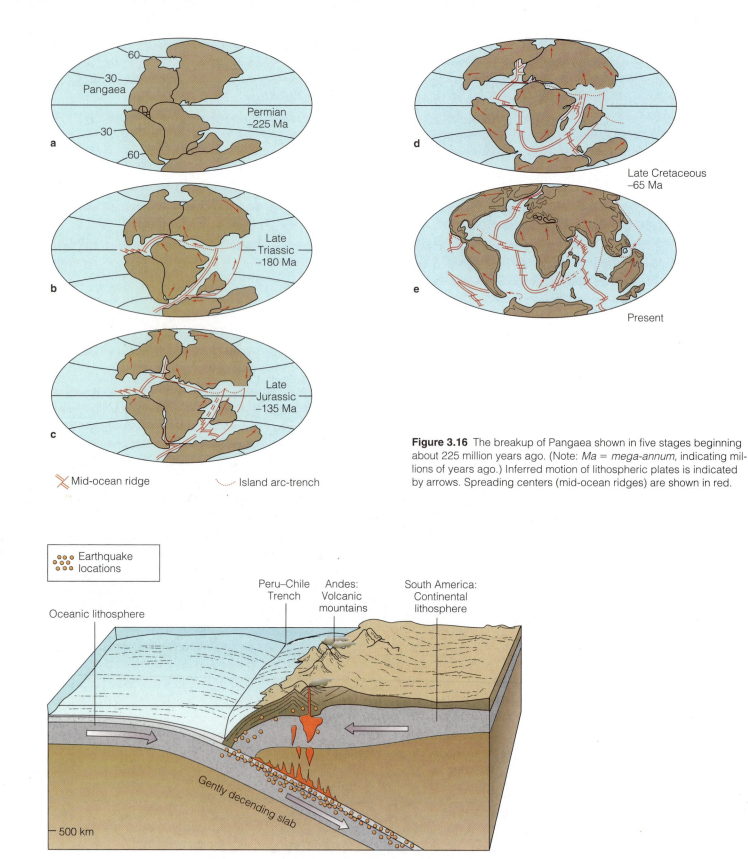

Figure 3.16 The breakup of Pangaea shown in five stages beginning about 225 million years ago. (Note: *Ma = mega-annum,* indicating millions of years ago.) Inferred motion of lithospheric plates is indicated by arrows. Spreading centers (mid-ocean ridges) are shown in red.

✕ Mid-ocean ridge   ⋯ Island arc-trench

Figure 3.17 A cross section through the west coast of South America, showing the convergence of a continental plate and an oceanic plate. The subducting oceanic plate becomes more dense as it descends, its downward slide propelled by gravity. At a depth of about 80 kilometers (50 miles), heat drives water and other volatile components from the subducted sediments into the overlying mantle, lowering its melting point. Masses of the melted material, rich in water and carbon dioxide, rise to power Andean volcanoes.

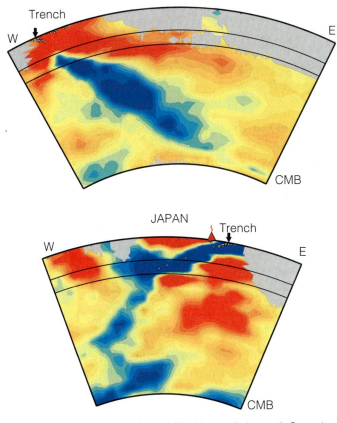

CENTRAL AMERICA

Trench

W

E

CMB

JAPAN

Trench

W

E

CMB

Figure 3.18 Vertical slices through Earth's mantle beneath Central America and Japan. Colder material is shown in blue, warmer in red. The distribution of colder material suggests that the subducting slabs beneath both areas have penetrated to the core–mantle boundary (CMB) at a depth of 2,800 kilometers (1,700 miles).

that parallels the west coast of South America. **Figure 3.17** is a cross section through these plates.

The subducting plate's periodic downward lurches cause earthquakes. Some of the oceanic crust and its sediments will melt as the plate plunges downward, forming a magma rich in water and carbon dioxide. In places this magma rises through overlying layers to the surface and causes volcanic eruptions. The active volcanoes of Central America and South America's Andes Mountains are a product of this activity, as are the area's numerous earthquakes. The North American Cascade volcanoes, including Mount St. Helens, result from similar processes. Most of the subducted crust mixes with the mantle. As shown in **Figure 3.18,** some of it continues downward through the mantle, eventually reaching the core–mantle boundary 2,800 kilometers (1,700 miles) beneath the surface! Subduction at converging oceanic plates was responsible for the great Alaska earthquake of 1964. Plate convergence (and divergence) is faster in the Pacific than in the Atlantic, in a few places reaching a rate of 18 centimeters (7 inches) a year. You can now clearly see the source of the Pacific Ring of Fire.

In the previous example, continental crust met oceanic crust. What happens when two *oceanic* plates converge? One of the colliding plates will usually be older, and therefore cooler and denser, than the other. Pulled by gravity, this heav-

ier plate will slip steeply below the lighter one into the asthenosphere. The ocean bottom is distorted in these areas to form deep trenches, the ocean's greatest depths. Water and carbon dioxide trapped with the melting rock of the subducting plate rise into the overlying mantle, lowering its melting point. This fluid mix of magma and subducted material forms a relatively light magma that powers vigorous volcanoes, but the volcanoes emerge from the seafloor rather than from a continent. These volcanoes appear in patterns of curves on the overriding oceanic crust; when they emerge above sea level they form curving arcs of islands (**Figure 3.19**).

Convergent margins are vast "continent factories" where materials from the surface descend and are heated, compressed, partially liquefied, separated, mixed with surrounding materials, and recycled to the surface. Relatively light continental crust is the main product, and it is produced at a rate of about 1 cubic kilometer (0.24 cubic miles) per year. Some geophysicists believe all of Earth's continental crust may have originated from granitic rock produced in this way. The island arcs may have coalesced to form larger and larger continental masses.

Two plates bearing continental crust can also converge. The most spectacular example of such a collision, between the India-Australian and Eurasian Plates some 45 million years ago, formed the Himalayas. Neither plate edge is being subducted; instead, both are compressed, folded, and uplifted, as **Figure 3.20** shows. The lofty top of Mount Everest is made of rock formed from sediments deposited long ago in a shallow sea!

Transform Plate Boundaries—Fracturing Crust
3-13

In some places, crustal plates shear laterally past one another. These areas are called **transform plate boundaries.** Crust is neither produced nor destroyed at this type of boundary, but the potential for earthquakes can be great as the plate edges slip past one another. The eastern boundary of the Pacific Plate is a long transform fault system. California's San Andreas Fault (**Figure 3.21**) is merely the most famous of the many faults marking the junction between the Pacific and North American Plates. The Pacific Plate moves steadily, but its movement is stored elastically at the North American Plate boundary until friction is overcome. Then the Pacific Plate lurches in abrupt jerks to the northwest along much of its shared border with the North American Plate, an area that includes the major population centers of California. These jerks cause California's famous earthquakes. Because of this movement, western California is gradually sliding north along the rest of North America; some 50 million years from now, it will encounter the Aleutian Trench.

Plate Interactions—A Summary
3-14

There are, then, two kinds of plate divergences: divergent oceanic crust (such as that in the mid-Atlantic) and divergent continental crust (as in the Rift Valley of East Africa). And there are three kinds of plate convergences: oceanic crust toward continental crust (west coast of South America), oceanic crust

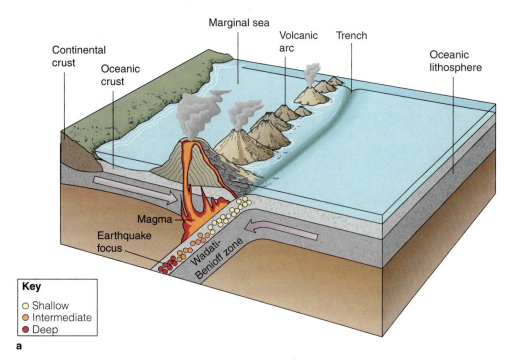

Key
- ○ Shallow
- ○ Intermediate
- ● Deep

a

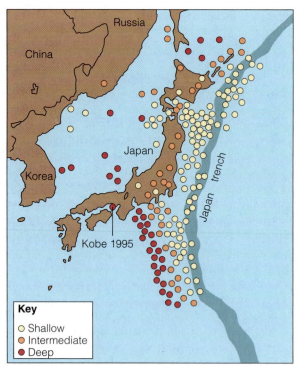

Key
- ○ Shallow
- ○ Intermediate
- ● Deep

b

Figure 3.19 (**a**) The formation of an island arc along a trench as two oceanic plates converge. The volcanic islands form as masses of magma reach the seafloor. The Japanese islands were formed in this way. (**b**) The distribution of shallow, intermediate, and deep earthquakes for part of the Pacific Ring of Fire in the vicinity of the Japan trench. Note that earthquakes occur only on one side of the trench, the side on which the plate subducts. The site of the catastrophic 1995 Kobe subduction earthquake is marked.

toward oceanic crust (northern Pacific), and continental crust toward continental crust (Himalayan Mountains). Transform boundaries mark locations at which crustal plates move past one another (San Andreas Fault). Each of these movements produces a distinct topography, and each zone contains potential dangers for its human inhabitants.

THE CONFIRMATION OF PLATE TECTONICS

3-15

The theory of plate tectonics has had the same effect on geology that the theory of evolution has had on biology. In each case a catalog of seemingly unrelated facts was unified by a powerful central idea. Many discoveries contributed to our present understanding of plate tectonics. Some of the compelling evidence for plate tectonics is outlined next.

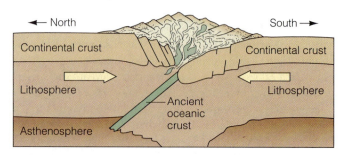

Figure 3.20 A cross section through southern China, showing the convergence of two continental plates. Neither plate is dense enough to subduct; instead, their compression and folding uplift the plate edges to form the Himalayas. Notice the massive supporting "root" beneath the emergent mountain needed for isostatic equilibrium.

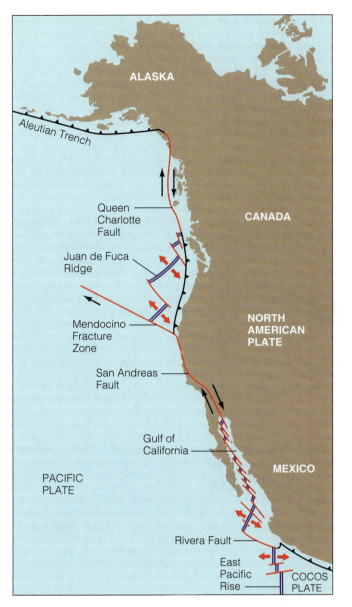

Figure 3.21 A long transform plate boundary, which includes California's San Andreas Fault.

Hot Spots 3-16

Hot spots are stationary sources of heat in the mantle. Hot spots are not always located at plate boundaries—indeed, their origin lies far below the realm of plate tectonics in plumes of magma rising from the core–mantle boundary. As lithospheric plates slide over these fixed locations, they are weakened from below by rising heat and magma. A volcano can form over the hot spot, but because the plate is moving, the volcano is carried away from its source of magma after a few million years and becomes inactive. It is replaced at the hot spot by a new volcano a short distance away. A chain of volcanoes and volcanic islands results.

Figure 3.22 shows the most famous of these "assembly line" chains, which extends from the old eroded volcanoes of the Emperor Seamounts to the still-growing island of Hawaii. In fact, the abrupt bend in the chain was caused by a change in direction of the Pacific Plate from a largely northward to a more westward movement about 40 million years ago. The next Hawaiian island—already named Loihi—is building on the ocean floor to the southeast. Now about 1,000 meters (3,200 feet) beneath the surface, Loihi will break the surface about 30,000 years from now.

There are other hot spots in the Pacific. The island chains formed by their activity "jog" in the Hawaiian pattern, indicating that they are positioned on the same lithospheric plate. Chains of undersea volcanoes in the Atlantic, centered on the Mid-Atlantic Ridge, suggest a similar process is at work there. Hot spots can exist beneath continental crust as well; Yellowstone National Park is believed to be over a hot spot beneath the westward-moving North American Plate.

Atolls and Guyots 3-17

Atolls are ring-shaped islands of coral reefs and reef-derived sediment centered over submerged inactive volcanoes. The coral animals that build atolls can live only in the upper sunlit layer of seawater. How, then, did their skeletal remains end up at a depth of 1,280 meters (4,222 feet) within Eniwetok Atoll? This surprising discovery was made in 1954 when bore holes were being drilled in preparation for the first hydrogen bomb tests. Plate tectonics suggests an answer. Coral animals can build atop the skeletons of their dead predecessors at a rate of about 1 centimeter (½ inch) each year. Coral animals living on a volcano's flanks can grow upward as the crust beneath the volcano slowly cools and contracts (sinks) during its movement away from the warm spreading center where it formed (see Figure 14.16). A deep column of coral skeletons can accumulate if the rate of sinking is less than about 1 centimeter per year. Thus, the coral record traces plate subsidence for millions of years into the past, supporting the plate tectonics theory.

Guyots were discovered by Harry Hess during his service as commander of a U. S. Navy transport in the Pacific during the Second World War. With the ship's echo sounder he found chains of odd flat-topped submerged volcanic mountains and named them after Princeton's first professor of geology, Arnold Guyot (pronounced ghee-*oh*). More than 500 guyots have been located, and plate tectonics neatly explains the formation of most of them. Like atolls, they were once volcanic peaks standing above sea level. As the plate on which they are riding cooled, contracted, and was carried away from the spreading center, they became inactive and stopped growing. They were "shaved" flat by wave action as they sank beneath the ocean surface. Evidence of ancient beaches along their rims suggests that this hypothesis is correct. **Figure 3.23** shows a sinking progression of guyots.

Terranes 3-18

Buoyant continental and oceanic plateaus, fragments of granitic rock and sediments, can be rafted along with a plate and scraped off onto a continent when the plate is subducted. This process is similar to what happens when a sharp knife is scraped across a table top to remove pieces of cool candle wax. The

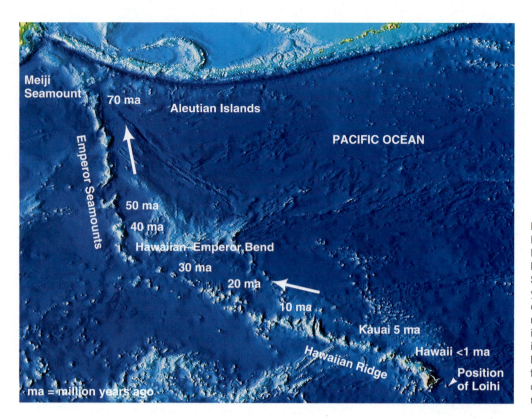

Figure 3.22 The Hawaiian chain, islands formed one by one as the Pacific Plate slid over a hot spot. The oldest known member of the chain, the Meiji Seamount, formed about 70 million years ago (70 Ma), and the bend in the chain shows that the plate changed direction about 40 million years ago. The island of Hawaii still has active volcanism, and the next island in the chain, Loihi, has begun building on the ocean floor. The upper arrow shows the initial direction of plate motion, the lower arrow shows the present direction.

wax accumulates and wrinkles on the knife blade in the same way land masses and ocean sediments accumulate against the face of a continent as the lithosphere in which they are embedded reaches a plate boundary. Plateaus, isolated segments of seafloor, ocean ridges, ancient island arcs, and parts of continental crust that collect on the face of a continent are called **terranes.** The thickness and low density of terranes prevents their subduction. A simplified account of terrane accumulation is diagrammed in **Figure 3.24.**

Terranes are surprisingly common. New England, much of North America west of the Rocky Mountains, and all of Alaska appear to be composed of this sort of crazy-quilt assem-

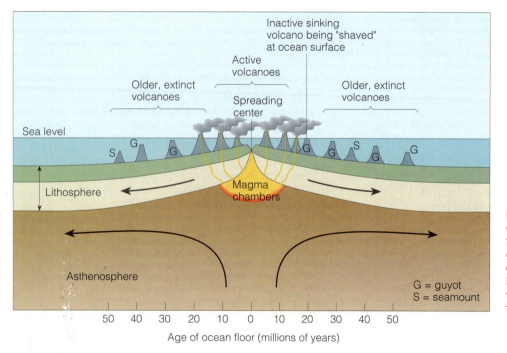

Figure 3.23 The process by which guyots (G) and seamounts (S) form. Guyots have flat tops because they have grown tall enough to be eroded by waves at the ocean's surface. Seamounts have a similar origin but retain their more pointed volcano shape because they never reach the surface.

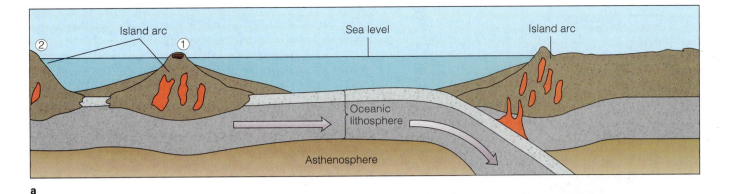

a

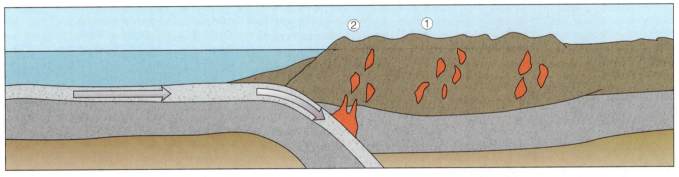

b

c

Figure 3.24 A model for the accumulation of island arcs along the edge of a continental margin. (**a**) Two island arcs ① and ② lie offshore from a continental margin. (**b**) The nearest arc is "scraped off" onto the continent as the intervening ocean floor is subducted. (**c**) Eventually the second island arc complex is accreted as well. These collections are known as terranes. Continents are thought to grow by collecting terranes. Most of North America west of the Rocky Mountains is thought to consist of accumulated island arc complexes.

blage of material, some of which has evidently arrived from thousands of miles away in the Southern Hemisphere!

Paleomagnetism

 3-19

The young ocean basins hold the most convincing evidence for plate tectonics. This evidence was obtained through the magnetic analysis of rocks formed over the last 200 million years.

A compass needle points toward the magnetic North Pole because it aligns with Earth's magnetic field. Tiny particles of iron-bearing magnetic minerals are found in basaltic magma. These minerals act like miniature compass needles; as they cool, their magnetic fields align with Earth's magnetic field. Thus the orientation of Earth's magnetic field becomes frozen in the rock as it solidifies. Any later change in the strength or direction of Earth's magnetic field will not significantly change the characteristics of the field trapped within the solid rocks

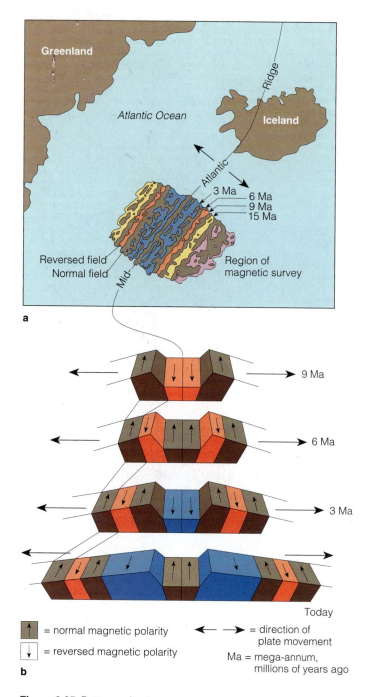

a

b

= normal magnetic polarity

= reversed magnetic polarity

= direction of plate movement

Ma = mega-annum, millions of years ago

Figure 3.25 Patterns of paleomagnetism and their explanation by plate tectonic theory. (**a**) When scientists conducted a magnetic survey of a spreading center, the Mid-Atlantic Ridge, they found bands of weaker and stronger magnetic fields frozen in the rocks. (**b**) The molten rocks forming at the spreading center take on the polarity of the planet when they are cooling and then move slowly in both directions from the center. When Earth's magnetic field reverses, the polarity of new-formed rocks changes, creating symmetrical bands of opposite polarity.

The "fossil," or remanent, magnetic field of a rock is known as **paleomagnetism** (*paleos* = ancient).

A **magnetometer** measures the amount and direction of residual magnetism in a rock sample. In the late 1950s geophysicists towed sensitive magnetometers just above the ocean floor to detect the weak magnetism frozen in the rocks. When plotted on charts, the data revealed an odd pattern of symmetrical magnetic "stripes" or bands on both sides of a spreading center (**Figure 3.25a**). The tiny compass needles contained in the rocks of some bands join with Earth's present magnetic orientation to enhance the strength of the local magnetic field, while the needles in rocks in adjacent bands weaken it. What could cause such a pattern?

In 1963 geologists Drummond Matthews and Frederick Vine proposed a clever interpretation. They knew that Earth's magnetic field reverses at irregular intervals of a few hundred thousand years. (Reversals may be caused by small differences in rotation rates between Earth's inner and outer cores.) In a time of reversal a compass needle would point south instead of north, and any particles of magnetic material in fresh seafloor basalt at a spreading center would be imprinted with the reversed field. The alternating magnetic stripes represent rocks with alternating magnetic polarity—one band having normal polarity (magnetized in the same direction as today's magnetic field direction), the next band having reversed polarity (opposite from today's direction). These researchers realized that the pattern of alternating weak and strong magnetic fields was symmetrical because freshly magnetized rocks born at the ridge are spread apart and carried away from the ridge by plate movement (**Figure 3.25b**). Similar magnetic patterns have been found on land and independently dated by other means.

By 1974, scientists had compiled charts showing the paleomagnetic orientation—and the age—of the seafloors of the Eastern Pacific and the Atlantic (**Figure 3.26**). Plate tectonics beautifully explains these patterns, and the patterns themselves are among the most compelling of all arguments for the theory.

Paleomagnetic data have recently been used to measure spreading rates, to calibrate the geologic time scale, to reconstruct continents, and to understand the movement of terranes. Paleomagnetism has been among the most productive specialties in geology for the past two decades.

Though there is clearly much to learn, plate tectonics is already an especially powerful predictive theory. Discoveries and insights made by the researchers mentioned in this chapter, and hundreds of others, have borne out the intuition of Alfred Wegener. Our understanding of the process will evolve as more data become available, but there seems very little chance that geologists will ever return to the dominant pre-1960 view of a stable and motionless crust. Plate tectonics theory shows us the picture of an actively cycling Earth and an ever-changing surface with *a single world ocean* changing shape and shifting positions as the plates slowly move.

The configuration of the ocean basins—discussed in the next chapter—is the result of plate tectonics activity. Armed with an understanding of the theory, you will find that the variety of features will make more sense to you.

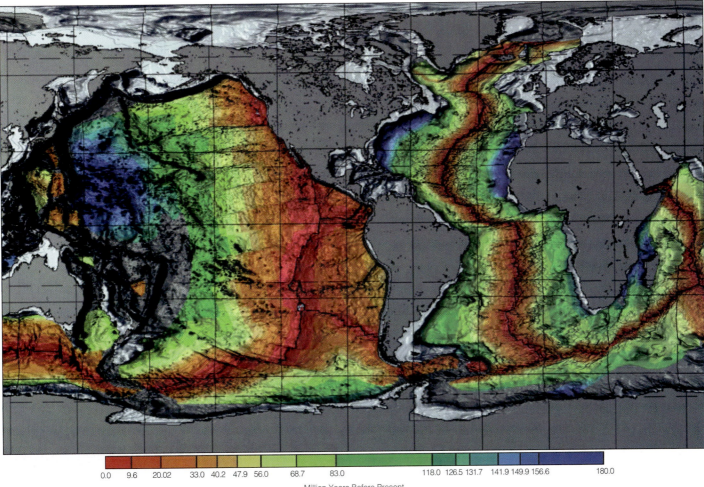

0.0   9.6   20.02   33.0   40.2 47.9 56.0   68.7   83.0   118.0 126.5 131.7   141.9 149.9 156.6   180.0

Million Years Before Present

Figure 3.26 The age of the ocean floors. The colors seen here represent an expression of seafloor spreading over the last 200 million years as revealed by paleomagnetic patterns.

1. What is the difference between crust and lithosphere? Between lithosphere and asthenosphere?

Lithosphere includes crust (oceanic and continental) and the relatively cool, rigid uppermost mantle down to the hotter, deformable asthenosphere. The velocity of seismic waves in the crust is much different from that in the mantle. This suggests differences in chemical composition or crystal structure, or both. The lithosphere and asthenosphere have different physical characteristics: The lithosphere is generally rigid, but the asthenosphere is capable of slow movement. Asthenosphere and lithosphere also transmit seismic waves at different speeds.

2. What are the most abundant elements comprising Earth?

You may be surprised to learn that oxygen accounts for about 46% of the mass of Earth's crust. On an atom-for-atom basis, the proportion is even more impressive: Of every 100 atoms of Earth's crust, 62 are oxygen. Most of this oxygen is not present as the gaseous element but is combined with other atoms into oxides and other compounds. Most of the familiar crustal rocks and minerals are oxides of aluminum, silicon, and iron (rust, for example, is iron oxide).

In Earth as a whole, iron is the most abundant element, making up 35% of the mass of the planet, while oxygen accounts for 30% overall. Remember, most of the mass of the universe is hydrogen gas. Oxygen and iron are abundant on Earth only because nuclear reactions in stars can transform light elements like hydrogen into heavy ones.

3. How common are large earthquakes?

About every two days, somewhere in the world, there's an earthquake of a magnitude of from 6 to 6.9 on the Richter scale—roughly equivalent to the quake that shook Northridge and the rest of southern California in January 1994; or Kobe, Japan, in January 1995. Once or twice a month, on average, there's a 7 to 7.9 quake somewhere. There is about one 8 to 8.9 earthquake—similar in magnitude to the 1964 earthquake in Alaska—each year.

Ocean Basins 4

TRIESTE AT THE BOTTOM

Studying deep ocean basins has always presented daunting obstacles. The most difficult problem is to reach extreme depths, but, amazingly, scientists have visited the bottom of the deepest ocean basin. On 23 January 1960, U.S. Navy Lieutenant Don Walsh and Dr. Jacques Piccard descended to a depth of 11,022 meters (6.85 miles) into the Challenger Deep, an area of the Mariana Trench discovered in 1951 by the British oceanographic research vessel *Challenger II*.[1] The vehicle used in the descent was the **bathyscaphe** *Trieste* (**Figures a** and **b**) a deep-diving submersible designed like a blimp with a very strong and thick (and cramped) steel crew sphere suspended below. A blimp uses helium gas for buoyancy, but a gas would be compressed by water pressure; so gasoline, which is relatively incompressible, was used to provide lift.

a *Trieste* before launching. Note the crew sphere suspended below the gasoline-filled flotation hull.

First, *Trieste* made a series of increasingly deep practice dives; then the big day arrived. Don Walsh wrote of the experience in 1979:

> At about 600 feet we entered a zone of deepening twilight where colors faded into gray. By 1,000 feet the light had gone completely. We turned out the lights in the sphere to watch for the luminescent creatures that are sometimes visible at this level. We saw very few. Eventually we turned the cabin lights back on and briefly tested the forward lights that throw a beam in front of the observation window. Formless plankton streamed past, giving us a sensation of great speed.
>
> There were minor incidents such as the small leak that always developed in one of the hull connectors—a place where wires from lights and instruments on the outside of the sphere pass through the hull. The leak started at about 10,000 feet. It was an old friend, a tiny drip, drip, drip. I timed the drips and found no change from before, which meant that it had not become more serious. We expected it to disappear at about 15,000 feet, when the water pressure packed the plastic sealer in more tightly—and it did.

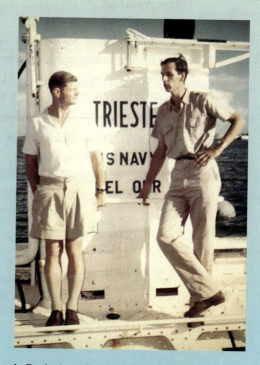

b Dr. Andreas Rechnitzer and Jacques Piccard aboard *Trieste* during trials in the South Pacific in early 1960. *Trieste* would set an unbeatable depth record a few days later.

---

[1] The position of the Challenger Deep is marked in Figure 4.24, and its depth in relation to Mount Everest is shown in Figure 4.25a.

At 27,000 feet we checked our rate of descent to 2 feet per second by dumping some shot ballast. We were not too sure of the underwater currents here, and we did not want to go crashing into a wall of the trench by mistake. As we neared 30,000 feet I started thinking about the changes we had planned to make when we got to within 1,000 feet of the bottom—which we now were expecting to find only another 3,500 feet below us. I was running through a mental checklist when we heard and felt a powerful crack. The sphere rocked as though we were on land and going through a mild earthquake.

We waited anxiously for what might happen next. Nothing did. We flipped off the instruments and the underwater telephone so that we could hear better. Still nothing happened. We switched the instruments back on and studied the dials that would tell us if something critical had occurred. No, we had our equilibrium and were descending exactly as before.

We checked our speed to half a foot a second and continued. At that rate, time and distance pass very slowly, and I think for the first time in the dive both of us had the feeling of awe that comes from exploring the totally unknown.

I did not take my eyes off the fathometer, and Jacques never stopped watching out of the tiny port-hole with its weak probe of light. No bottom was in sight. . . . Soon Jacques could see a difference in the effect of our light in the water, as the rays reflected off the bottom. As we approached the floor I called the fathometer readings to Jacques in fathoms: "Thirty . . . twenty . . . ten . . ." At eight he called that he could see the bottom.

At 1:10 P.M. we sank gently onto the soft floor. A great cloud of silt rose around us. The fifteen-man, Navy civilian/military team had set a record that could not be broken.

The voyagers took temperature readings and spotted a pair of bottom fish and some crustaceans before ascending. With the possible exception of the moon, no place visited by humans has been more hazardous to explore.　　　　　　　　　　　　　　　4-1

Source: Walsh, D. 1979. "Twenty Years After the *Trieste* Dive." *Oceans*, June, 37–41.

KEY CONCEPTS

1. Seafloor features result from a combination of tectonic activity and the processes of erosion and deposition. New bathymetric devices being deployed to study these features include side-scan sonar and satellites using sensitive radar for altimetry.

2. Near shore the features of the ocean floor are similar to those of the adjacent continents, because they share the same granitic basement. The transition to basalt marks the true edge of the continent and divides ocean floors into two major provinces. The submerged outer edge of a continent is called the continental margin. The deep-sea floor beyond the continental margin is properly called the ocean basin.

3. Features of the continental margins include continental shelves, continental slopes, submarine canyons, and continental rises. Features of the deep-ocean basins include oceanic ridges, hydrothermal vents, abyssal plains and hills, seamounts, guyots, trenches, and island arcs.

CHAPTER AT A GLANCE

Bathymetry
Echo Sounding
Multibeam Systems
Satellite Altimetry

The Topography of Ocean Floors

Continental Margins
Continental Shelves
Continental Slopes
Submarine Canyons
Continental Rises

Deep-Ocean Basins
Oceanic Ridges
Hydrothermal Vents
Box 4.1: Going Deep
Abyssal Plains and Abyssal Hills
Seamounts and Guyots
Trenches and Island Arcs

The Grand Tour

In February 2000, the *Mars Global Surveyor* spacecraft completed a map of the surface of our planetary neighbor. The photos from orbit were clear and detailed, and objects as small as about 2 meters (7 feet) across could be seen almost anywhere on the planet. There was no ocean and few storms to spoil the view.

Mapping Earth is much more difficult, because water and clouds hide more than three-quarters of the surface. Until surprisingly recently, we knew more about the global contours of the moon and the inner planets than we knew about our own home. Thanks to modern bathymetry, our view is clearing.

BATHYMETRY 4-2

The discovery and study of ocean floor contours is called **bathymetry** (*bathy* = deep + *meter* = measure). The earliest known bathymetric studies were carried out in the Mediterranean by a Greek named Posidonius in 85 B.C. He and his crew let out nearly 2 kilometers (1.25 miles) of rope until a stone tied to the end of the line touched bottom. Bathymetric technology had not improved by the time Sir James Clark Ross obtained soundings of 4,893 meters (16,054 feet) in the South Atlantic in 1818. In the 1870s the researchers aboard HMS *Challenger* added the innovation of a steam-powered winch to raise the line and weight, but the method was the same (**Figure 4.1**). The *Challenger* crew made 492 bottom soundings and confirmed Matthew Maury's earlier discovery of the Mid-Atlantic Ridge.

Echo Sounding 4-3

The sinking of RMS *Titanic* in 1912 stimulated research that finally ended slow, laborious weight-on-a-line efforts. By April 1914, Reginald A. Fessenden, a former employee of Thomas Edison, had developed an "Iceberg Detector and Echo Depth Sounder." The detector worked by directing a powerful sound pulse ahead of a ship and then listening for an echo from the submerged portion of an iceberg. It was easy to direct the beam downward to sense the distance to the bottom.[2] It might take most of a day to lower and raise a weighted line, but echo sounders could take many bottom recordings in a minute.

When the First World War intervened, Fessenden's talents were turned to enemy submarine detection devices. But when peace returned, he continued his echo sounder research. In June 1922, an echo sounder based on his designs made the first continuous profile across an ocean basin aboard USS *Stewart,* a U.S. Navy vessel. Using an improved echo sounder based on his design, the German research vessel *Meteor* made 14 profiles across the Atlantic from 1925 to 1927. The wandering path of the Mid-Atlantic Ridge was revealed, and its obvious coincidence with coastlines on both sides of the Atlantic stimulated the discussions that culminated in our present understanding of plate tectonics.

Echo sounding wasn't perfect, as **Figure 4.2a** shows. The ship's exact position was sometimes uncertain. The speed of

---

[2] Figure 1.21 shows this method of depth detection.

Figure 4.1 An illustration from the *Challenger Report* (1880). Seamen are handling the steam winch used to lower a weight on the end of a line to the seabed to find ocean depth.

sound through seawater varies with temperature, pressure, and salinity, and those variations made depth readings slightly inaccurate. Simple depth sounder images (such as that shown in **Figure 4.2b**) were also unable to resolve the fine detail that oceanographers needed to explore seabed features. Even so, researchers using depth sounder tracks had painstakingly compiled the first comprehensive charts of the ocean floor by 1959. (A portion of one of those astonishing charts is shown in Figure 4.18a.)

Two new techniques—made possible by improved sensors and high-speed computers—have been perfected to minimize inaccuracies and speed the process of bathymetry. The features discussed in this chapter have been studied using these (and other) systems, any of which is surely an improvement over lowering rocks into the ocean!

Multibeam Systems 4-4

Like an echo sounder, a multibeam system bounces sound off the seafloor to measure ocean depth. Unlike a simple echo sounder, a multibeam system may have as many as 121 beams radiating from a ship's hull. Fanning out at right angles to the direction of travel, these beams can cover a 120° arc (**Figure 4.3a**). Typically, a pulse of sound energy is sent toward the seabed every 10 seconds. Listening devices record sounds reflected from the bottom, but only from the narrow corridors

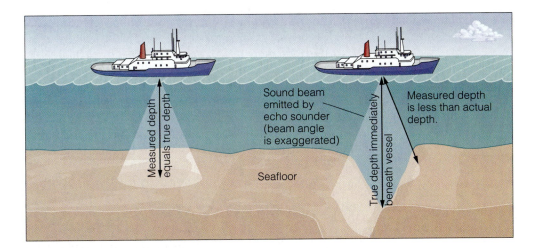

a

b

Figure 4.2 (**a**) The accuracy of an echo sounder can be affected by water conditions and bottom contours. The pulses of sound energy, or "pings," from the sounder spread out in a narrow cone as they travel from the ship. When depth is great, the sounds reflect from a large area of seabed. Because the first sound of the returning echo is used to sense depth, measurements over deep depressions are often inaccurate. (**b**) An echo sounder trace. A sound pulse from a ship is reflected off the seabed and returns to the ship (as in Figure 1.21). Transit time provides a measure of depth. For example, it takes about 2 seconds for a sound pulse to strike the bottom and return to the ship when water depth is 1,500 meters (4,900 feet). Bottom contours are revealed as the ship sails a steady course. In this trace, the horizontal axis represents the course of the ship, the vertical axis represents water depth. The ship has sailed over a small submarine canyon.

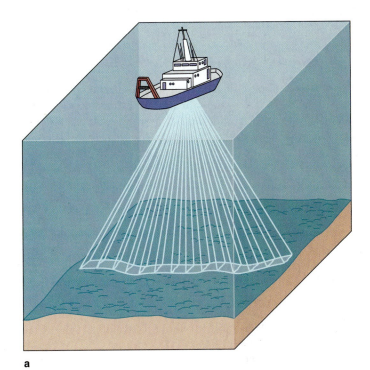

a

b

Figure 4.3 (**a**) A multibeam echo sounder uses as many as 121 beams radiating from a ship's hull. Fanning out at right angles to the direction of travel, these beams can cover a 120° arc and measure a swath of bottom about 3.4 times as wide as the water is deep. Typically, a "ping" is sent toward the seabed every 10 seconds. Listening devices record sounds reflected from the bottom, but only from the narrow corridors corresponding to the outgoing pulse. Thus, a multibeam system is much less susceptible to contour error than the single-beam device shown in Figure 4.2b. (**b**) A multibeam reading of a fragment of sea floor near the East Pacific Rise south of the tip of Baja California, Mexico. The uneven coverage reflects the path of the ship across the surface. Detailed analysis requires sailing a careful pattern, rather like one would use to mow a lawn evenly. Computer processing provides extraordinarily detailed images, such as those in Figures 4.11, 4.12, and 4.18b.

corresponding to the outgoing pulse. Successive observations build a continuous swath of coverage beneath the ship. By "mowing the lawn"—moving the ship in a coverage pattern similar to one you would follow in cutting grass—researchers can build a complete map of an area (**Figure 4.3b**). Fewer than 200 research vessels are equipped with multibeam systems. At the present rate, charting the entire seafloor in this way would require more than 125 years.

Satellite Altimetry
4-5

Satellites cannot measure ocean depths directly, but they can measure small variations in the elevation of surface water. Using about a thousand radar pulses each second, the U.S. Navy's *Geosat* satellite (**Figure 4.4a**) measured its distance from the ocean surface to within 0.03 meters (1 inch)! Because the precise position of a satellite can be calculated, the average height of the ocean surface can be known with great accuracy.

Disregarding waves or tides or currents, the ocean surface can vary from the ideal smooth (ellipsoid) shape by as much as 200 meters (660 feet). This is because the pull of gravity varies across the surface of Earth, depending on the nearness (or distance) of massive parts of Earth. An underwater mountain or ridge "pulls" water toward it from the sides, forming a mound of water over itself (**Figure 4.4b**). For example, a typical undersea volcano with a height of 2,000 meters (6,600 feet) above the seabed and a radius of 20 kilometers (32 miles) would produce a 2-meter (6.6-foot) rise in the ocean surface. (This mound cannot be seen with the unaided eye because the slope of the surface is very gradual.) The large features of the seabed are amazingly and accurately reproduced in the subtle standing irregularities of the sea surface (**Figure 4.4c**)!

Geosat and its successor, *TOPEX/Poseidon,* have allowed the rapid mapping of the world ocean floor. Hundreds of previously unknown features have been discovered through the data they have provided.

THE TOPOGRAPHY OF OCEAN FLOORS
4-6

Most people think an ocean basin is shaped like a giant bathtub. They imagine that the continents drop off steeply just beyond the surf zone and that the ocean is deepest somewhere out in the middle. As is clear in **Figure 4.5,** bathymetric studies have shown this is not the case.

Why? As you read in the last chapter, plate tectonics theory suggests that Earth's surface is not a static arrangement of continents and ocean, but a dynamic mosaic of jostling lithospheric plates. The lighter continental lithosphere floats in isostatic equilibrium above the level of the heavier lithosphere of the ocean basins. The great density of the seabed partly explains why more than half of Earth's solid surface is at least 3,000 meters (10,000 feet) below sea level (**Figure 4.6**).

Notice in **Figure 4.7** the transition between the thick (and less dense) granitic rock of the continents and the relatively thin (and more dense) basalt of the deep-sea floor. Near shore the features of the ocean floor are similar to those of the adjacent continents, because they share the same granitic basement. The transition to basalt marks the *true* edge of the continent and divides ocean floors into two major provinces. The submerged outer edge of a continent is called the **continental margin.** The deep-sea floor beyond the continental margin is properly called the **ocean basin.**

CONTINENTAL MARGINS
4-7

You learned in Chapter 3 that lithospheric plates converge, diverge, or slip past each other. As you might expect, the submerged edges of continents—continental margins—are greatly influenced by this tectonic activity. Continental margins facing the edges of *diverging* plates are called **passive margins** because relatively little earthquake or volcanic activity is now associated with them. Because they surround the Atlantic, passive margins are sometimes referred to as *Atlantic-type* margins. Continental margins near the edges of *converging* plates (or near places where plates are slipping past each other) are called **active margins** because of their earthquake and volcanic activity. Because of their prevalence in the Pacific, active margins are sometimes referred to as *Pacific-type* margins.

Active and passive margins west and east of South America are shown in **Figure 4.8.** Note that active margins coincide with plate boundaries but passive margins do not. Passive margins are also found outside the Atlantic, but active margins are confined mostly to the Pacific.

Continental margins have two main divisions: a shallow, nearly flat continental *shelf* close to shore and a more steeply sloped continental *slope* to seaward.

Continental Shelves
4-8

The shallow, submerged extension of a continent is called the **continental shelf.** Continental shelves, extensions of the adjacent continents, are underlain by granitic continental crust. They are much more like the continent than like the deep-ocean floor, and they may have hills, depressions, sedimentary rocks, and mineral and oil deposits similar to those on the dry land nearby. Earth's continental shelves are shown in **Figure 4.9.** Taken together, the area of the continental shelves is 7.4% of Earth's ocean area.

Figure 4.10 shows a passive-margin continental shelf characteristic of Atlantic Ocean edges. The broad shelf extends far from shore in a gentle incline, typically 1.7 meters per kilometer (about 9 feet per mile), much less than the slope of a well-drained parking lot. Shelves along the margin of the Atlantic Ocean often reach 350 kilometers (220 miles) in width, and end at a depth of about 140 meters (460 feet), where a steeper drop-off begins.

a

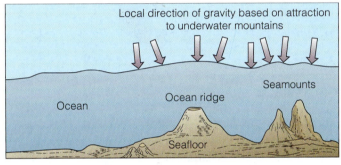

b

Figure 4.4 (**a**) *Geosat,* a U.S. Navy satellite that operated from 1985 through 1990, provided measurements of sea surface height from orbit. Moving above the ocean surface at 7 kilometers (4 miles) a second, *Geosat* bounced 1,000 pulses of radar energy off the ocean every second. Height accuracy was within 0.03 meters (1 inch)! (**b**) Distortion of the sea surface above a seabed feature occurs when the extra gravitational attraction of the feature "pulls" water toward it from the sides, forming a mound of water over itself. (**c**) This view of the complex system of ridges, trenches, and fracture zones on the South Atlantic seafloor east of the tips of South America (SA) and Antarctica's Palmer Peninsula (ANT) was derived in large part from measurements of the sea surface obtained by *Geosat* and declassified by the U.S. Navy in 1995.

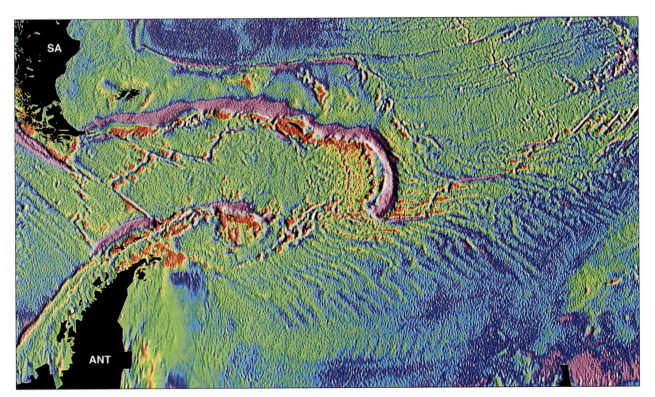

c

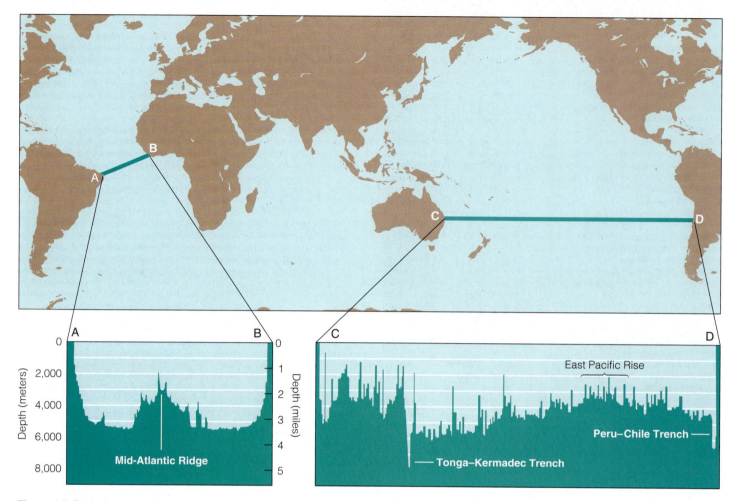

Figure 4.5 Typical cross sections of the Atlantic and Pacific ocean basins. The vertical scale has been greatly exaggerated.

The passive-margin shelves of the Atlantic Ocean formed as the fragments of Pangaea were carried away from each other by seafloor spreading. The continental lithosphere, thinned during initial rifting, cooled and contracted as it moved away from the spreading center, submerging the trailing edges of the continents and forming the shelves.

Most of the material comprising a shelf comes from erosion of the adjacent continental mass. Rivers assist in passive shelf building by transporting huge amounts of sediments to the shore from far inland. In some places the sediments accumulate behind natural dams formed by ancient reefs or ridges of granitic crust (see again Figure 4.7). The weight of the sediment isostatically depresses the continental edges and allows the sediment load to grow even thicker. Sediment at the outer edge of a shelf can be up to 15 kilometers (9 miles) thick and 150 million years old.

The width of a shelf is usually determined by proximity to a plate boundary. You can see in Figure 4.8 that the shelf at the *passive* margin (east of South America) is broad, but the shelf at the *active* margin (west of South America) is very narrow. The widest shelf, 1,280 kilometers (800 miles) across, lies north of Siberia in the tectonically quiet Arctic Sea. Shelf width

depends not only on tectonics but also on marine processes: Fast-moving ocean currents can sometimes prevent sediments from accumulating. For example, the east coast of Florida has a very narrow shelf (**Figure 4.11**) because there is no natural offshore dam formed by ridges of granitic crust, and because the swift current of the nearby Gulf Stream scours surface sediment away. Shelf width also depends on the position of sea level—shelves are narrow when global sea level is low, broad when it is high.

The shelves of the active Pacific margins are generally not as broad and flat as Atlantic shelves. An example is the abbreviated shelf off the west coast of South America, where the steep western slope of the Andes Mountains continues nearly uninterrupted beneath the sea into the depths of the Peru–Chile Trench (see again Figure 4.8). Active-margin shelves have more varied topography than passive-margin shelves; the character of continental shelves at an active margin may be determined more by faulting, volcanism, and tectonic deformation than by sedimentation.

Because of their gentle slope, continental shelves are greatly influenced by changes in sea level. Around 18,000 years ago—at the height of the last **ice age** (period of extensive

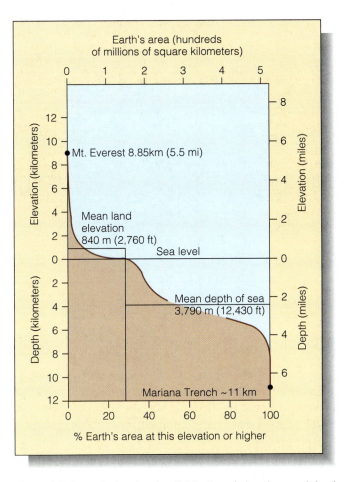

Figure 4.6 A graph showing the distribution of elevations and depths on Earth. This graph is not a land-to-sea profile of Earth, but rather a plot of the area of Earth's surface above any given elevation or depth below sea level. Note that more than half of Earth's solid surface is at least 3,000 meters (10,000 feet) below sea level. The average depth of the ocean (3,790 meters or 12,430 feet) is much greater than the average elevation of the continents (840 meters or 2,760 feet).

glaciation)—massive ice caps covered huge regions of the continent. The water that formed the thick ice sheets was derived from the ocean, and sea level fell about 125 meters (410 feet) below its present position. The continental shelves were almost completely exposed, and the surface area of the continents was about 18% greater than it is today. Rivers and waves cut into the sediments that had accumulated during periods of higher sea level, and they transported some coarse sediments to their present locations at the shelves' outer edges. Sea level began to rise again when the ice caps melted, and sediments again began to accumulate on the shelves. (More on the history and effects of sea level change will be found in the discussion of coasts in Chapter 11.)

The continental shelves have been the focus of intense exploration for natural resources. Because shelves are the submerged margins of continents, any deposits of oil or minerals along a coast are likely to continue offshore. Water depth over shelves averages only about 75 meters (250 feet), so large areas of the shelves are accessible to mining and drilling activities. Many of the techniques used to find and exploit natural resources on land can also be used on the continental shelves. Resource development requires intense scientific investigations, and our understanding of the geology of the shelves has benefited greatly from the search for offshore oil and natural gas.

Continental Slopes 4-9

The **continental slope** is the transition between the gently descending continental shelf and the deep-ocean floor. Continental slopes are formed of sediments that reach the built-out edge of the shelf and are transported over the side. At active margins, a slope may also include marine sediments scraped off a descending plate during subduction (**Figure 4.12**). The

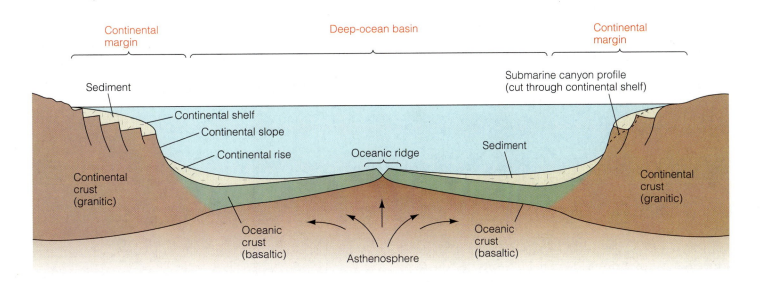

Figure 4.7 Cross section of a typical ocean basin flanked by passive continental margins. (The vertical scale has been greatly exaggerated to emphasize the basin contours.)

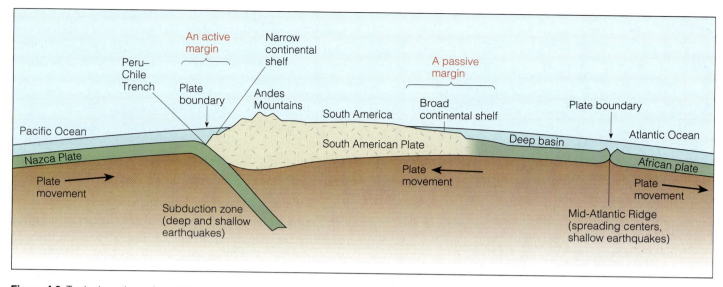

Figure 4.8 Typical continental margins bordering the leading (tectonically active) and trailing (passive) edges of a moving continent. (The vertical scale has been exaggerated.)

inclination of a typical continental slope is about 4° (or 70 meters per kilometer, 370 feet per mile), equal to the steepest road slope allowed on the interstate highway system. As Figure 4.10b implies, even the steepest of these slopes is not precipitous: A 25° slope is the greatest incline yet discovered. In general, continental slopes at active margins are steeper than those at passive margins. Continental slopes average about 20 kilometers (12 miles) wide and end at the continental rise, usually at a depth of about 3,700 meters (12,000 feet). The bottom of the continental slope is the true edge of a continent.

The **shelf break** marks the abrupt transition from continental shelf to continental slope. The depth of water at the shelf break is surprisingly constant—about 140 meters (460 feet) worldwide—but there are exceptions. The great weight of ice on Antarctica, for example, has isostatically depressed that continent, and the depth at the shelf break is 300–400 meters (1,000–1,300 feet). The shelf break in Greenland is similarly depressed.

Submarine Canyons **4-10**

Submarine canyons cut into the continental shelf and slope, often terminating on the deep-sea floor in a fan-shaped wedge of sediment (**Figure 4.13**). More than a hundred submarine canyons worldwide nick the edge of nearly all of Earth's continental shelves. The canyons generally trend at right angles

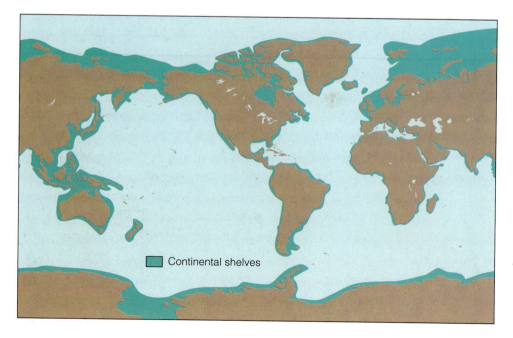

Figure 4.9 The worldwide distribution of continental shelves. The continental islands of Ireland and Great Britain are part of the continent of Europe, and Alaska connects to Siberia.

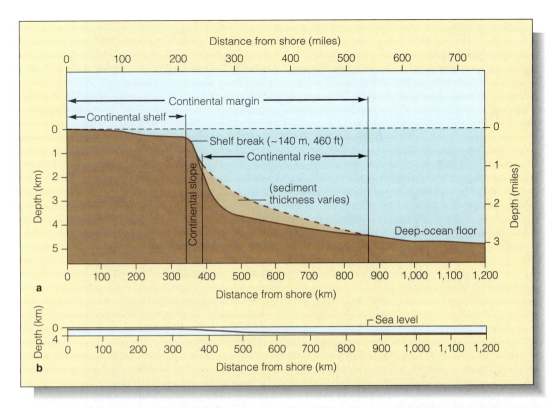

Figure 4.10 The features of a passive continental margin. (**a**) Vertical exaggeration 50:1. (**b**) No vertical exaggeration.

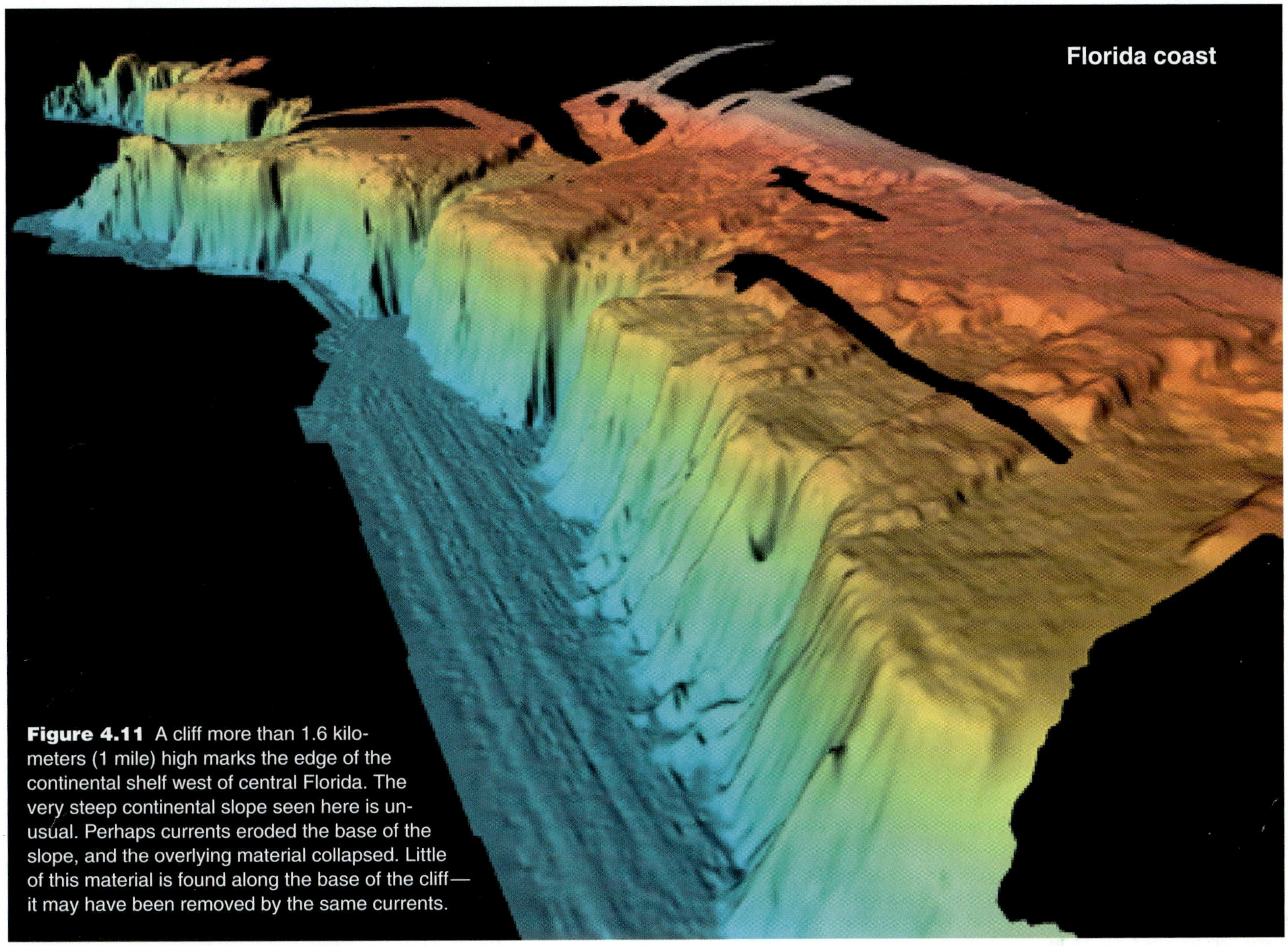

Figure 4.11 A cliff more than 1.6 kilometers (1 mile) high marks the edge of the continental shelf west of central Florida. The very steep continental slope seen here is unusual. Perhaps currents eroded the base of the slope, and the overlying material collapsed. Little of this material is found along the base of the cliff — it may have been removed by the same currents.

Florida coast

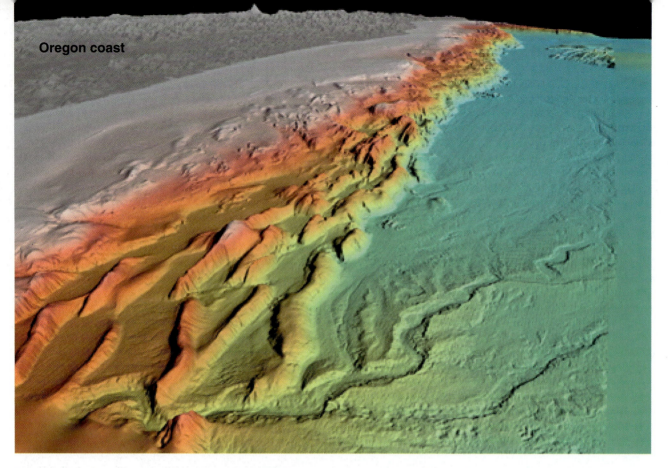

Oregon coast

Figure 4.12 Folded ridges of sediment cover the ocean floor west of Oregon. The crinkles result from the collision between the North American Plate and the Juan de Fuca Plate. Like a bulldozer, the North American Plate scrapes sediments from the subducting Juan de Fuca Plate and piles it into folds. To the south (the upper right), part of the Juan de Fuca Plate breaks through the sediment.

to the shoreline (and shelf edge), sometimes beginning very close to shore. Congo Canyon actually extends into the African continent as a deep estuary at the mouth of the Congo River. These enigmatic features are quite large. In fact, submarine canyons similar in size and profile to Arizona's Grand Canyon have been discovered!

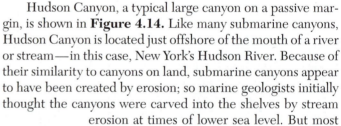
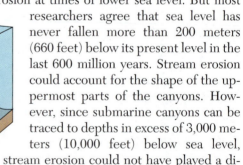

Figure 4.13 A submarine canyon.

Labels in figure: Canyon heads, Continental shelf, Shelf break, Continental slope, Distribution channel, Deep-sea fan

Hudson Canyon, a typical large canyon on a passive margin, is shown in **Figure 4.14.** Like many submarine canyons, Hudson Canyon is located just offshore of the mouth of a river or stream—in this case, New York's Hudson River. Because of their similarity to canyons on land, submarine canyons appear to have been created by erosion; so marine geologists initially thought the canyons were carved into the shelves by stream erosion at times of lower sea level. But most researchers agree that sea level has never fallen more than 200 meters (660 feet) below its present level in the last 600 million years. Stream erosion could account for the shape of the uppermost parts of the canyons. However, since submarine canyons can be traced to depths in excess of 3,000 meters (10,000 feet) below sea level, stream erosion could not have played a direct role in cutting their lower depths. What, then, caused the submarine canyons to form? As is so often the case in the history of marine science, the answer to this riddle was associated with a peculiar event. A sharp earthquake struck the Grand Banks south of Newfoundland at 1531 (3:31 p.m.) Greenwich time on 18 No-

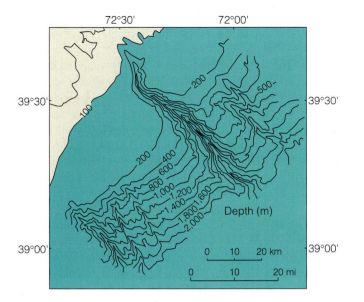

Figure 4.14 A topographic map of Hudson Canyon.

Figure 4.15 A turbidity current flowing down a submerged slope off the island of Jamaica. The propeller of a submarine caused the turbidity current by disturbing sediment along the slope.

vember 1929. The quake's **epicenter,** the point on the surface of Earth directly above the focus of an earthquake, was located on the continental slope at a water depth of about 2,200 meters (7,200 feet). Ordinarily such an earthquake would not attract much attention, but a number of important transatlantic communication cables lay on the ocean floor south of the earthquake area. One by one these cables failed. Officials had expected an occasional random failure, but a *sequence* of

Figure 4.16 A continuous cascade of sediment at the head of San Lucas submarine canyon (off the coast of Baja California, Mexico), which may be eroding the narrow gorge in conjunction with occasional turbidity currents.

cable failures, from north to south along 480 kilometers (300 miles) of seafloor in the mid-Atlantic, was baffling.

Local landslides or sediment liquefaction triggered by the earthquake probably caused the first cable breaks, and although no one thought of it right away, an abrasive underwater "avalanche" almost certainly caused the more distant breaks. These avalanche-like sediment movements, called **turbidity currents,** are caused when turbulence mixes sediments into water above a sloping bottom. The sediment-filled water is denser than the surrounding water; so the thick, muddy fluid runs down the slope at speeds of up to 27 kilometers (17 miles) per hour. **Figure 4.15** is a rare photograph of a turbidity current.

What is the connection between turbidity currents and submarine canyons? Most geologists believe that the canyons have been formed by abrasive turbidity currents plunging down the canyons. Small amounts of debris may cascade continuously down the canyons (**Figure 4.16**), but earthquakes can shake loose huge masses of sediment that rush down the edge of the shelf, scouring the canyon deeper as they go. In this way the canyons can be cut to depths far below the reach of streams even during the low sea levels of the ice ages.

Continental Rises   4-11

Along passive margins, the oceanic crust at the base of the continental slope is covered by an apron of accumulated sediment called the **continental rise** (see again Figure 4.10). Sediments from the shelf slowly descend to the ocean floor along the whole continental slope, but most of the sediments that form the continental rise are transported to the area by turbidity currents. The width of the rise varies from about 100 to

1,000 kilometers (63 to 630 miles), and its slope is gradual—about one-eighth that of the continental slope. One of the widest and thickest continental rises has formed in the Bay of Bengal at the mouths of the Ganges–Brahmaputra River, the most sediment-laden of the world's great rivers.

DEEP-OCEAN BASINS

4-12

Away from the margins of continents, the structure of the ocean floor is quite different. Here the seafloor is a blanket of sediments up to 5 kilometers (3 miles) thick overlying basaltic rocks. Deep-ocean basins constitute more than half of Earth's surface.

The deep-ocean floor consists mainly of oceanic ridge systems and the adjacent sediment-covered plains. Deep basins may be rimmed by trenches or by masses of sediment. Flat expanses are interrupted by islands, hills, active and extinct volcanoes, and active zones of seafloor spreading. The sediments on the deep floor reflect the history of the surrounding continents, the biological productivity of the overlying water, and the ages of the basins themselves.

Oceanic Ridges

4-13

If the ocean evaporated, the oceanic ridges would be Earth's most remarkable and obvious feature. An **oceanic ridge** is a mountainous chain of young basaltic rock at the active spread-ing center of an ocean. Stretching 65,000 kilometers (40,000 miles), more than 1½ times Earth's circumference, oceanic ridges girdle the globe like seams surrounding a softball (**Figure 4.17**). The rugged ridges, which often are devoid of sediment, rise about 2 kilometers (1.25 miles) above the seafloor. In places they project above the surface to form islands such as Iceland, the Azores, and Easter Island. Oceanic ridges and their associated structures account for 22% of the world's solid surface area (in comparison, all the land above sea level accounts for 29%). Although these features are often called mid-ocean ridges, less than 60% of their length actually exists along the centers of ocean basins.

As we saw in our discussion of plate tectonics, the rift zones associated with oceanic ridges are sources of new ocean floor where lithospheric plates diverge. The oceanic ridges are widest where they are most active. The youngest rock is located at the active ridge center, and rock becomes older with distance from the center. As the lithosphere cools, it shrinks and subsides. Slowly spreading ridges have a steeper profile than rapidly spreading ones because slowly diverging seafloor cools and shrinks closer to the spreading center. **Figure 4.18a**, a bathymetric map of the North Atlantic, clearly shows the great extent of the Mid-Atlantic Ridge, a typical oceanic ridge. **Figure 4.18b**, a multibeam image, provides a detailed look at the young central rift.

As can be seen in Figure 4.18a, the Mid-Atlantic Ridge is offset at more or less regular intervals by transform faults. A

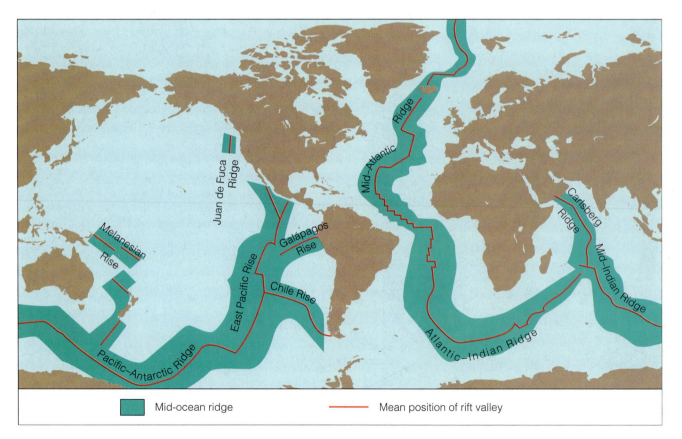

Figure 4.17 The oceanic ridge system.

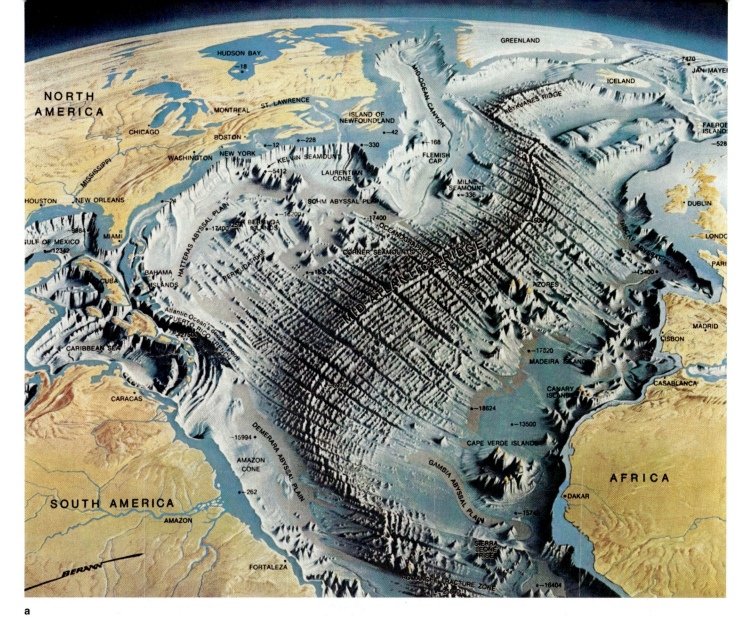

a

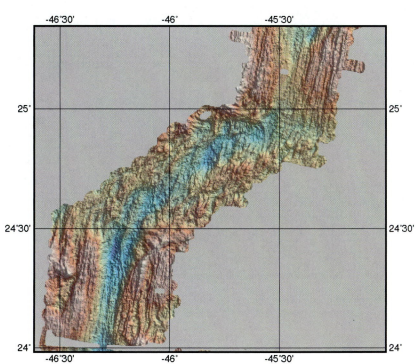

b

Figure 4.18 (**a**) Map of a portion of the Atlantic Ocean floor showing some major oceanic features: mid-ocean ridge, transform faults, fracture zones, submarine canyons, seamounts, continental rises, trenches, and abyssal plains. Depths are in feet. The map is vertically exaggerated. (**b**) The fine structure of the central portion of the Mid-Atlantic Ridge between Florida and western Africa. The depressed central valley (the spreading center, shown in blue) is clearly visible in this multibeam image.

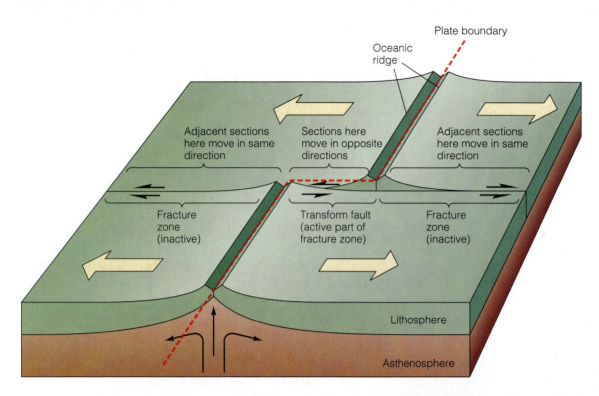

Adjacent sections here move in same direction

Sections here move in opposite directions

Adjacent sections here move in same direction

Oceanic ridge

Plate boundary

Fracture zone (inactive)

Transform fault (active part of fracture zone)

Fracture zone (inactive)

Lithosphere

Asthenosphere

Figure 4.19 Transform faults and fracture zones along an oceanic ridge. Transform faults are fractures along which lithospheric plates slide horizontally past one another. Transform faults are the active part of fracture zones.

fault is a fracture in the lithosphere along which movement has occurred, and **transform faults** are fractures along which lithospheric plates slide horizontally (**Figure 4.19**). When segments of a ridge system are offset, the fault connecting the axis of the ridge is a transform fault. Shallow earthquakes are common along transform faults. Since the ocean floor cannot expand evenly on the surface of a sphere, plate divergence on the spherical Earth can only be irregular and asymmetrical, and transform faults and fracture zones result.

Transform faults are the active part of **fracture zones.** Extending outward from the ridge axis, fracture zones are seismically inactive areas that show evidence of past transform fault activity. While segments of a lithospheric plate on either side of a transform fault move in *opposite* directions from each other, the plate segments adjacent to the outward segments of a fracture zone move in the *same* direction, as Figure 4.19 shows.

Hydrothermal Vents

 4-14

Some of the most exciting features of the ocean basins are the **hydrothermal vents.** Hot springs were discovered on oceanic ridges in 1977 by Robert Ballard and J. F. Grassle of the Woods Hole Oceanographic Institution. Diving in *Alvin* at 3 kilometers (1.9 miles) near the Galápagos Islands along the East Pacific Rise (an oceanic ridge), they came across rocky chimneys up to 20 meters (66 feet) high, from which dark, mineral-laden water was blasting at 350°C (660°F) (**Figure 4.20**). Only the great pressure at this depth prevented the escaping water from flashing to steam. These *black smokers,* as they were quickly nicknamed, fascinate marine geologists. It is believed that water descends through fissures and cracks in the ridge

Figure 4.20 A black smoker discovered at a depth of about 2,800 meters (9,200 feet) along the East Pacific Rise.

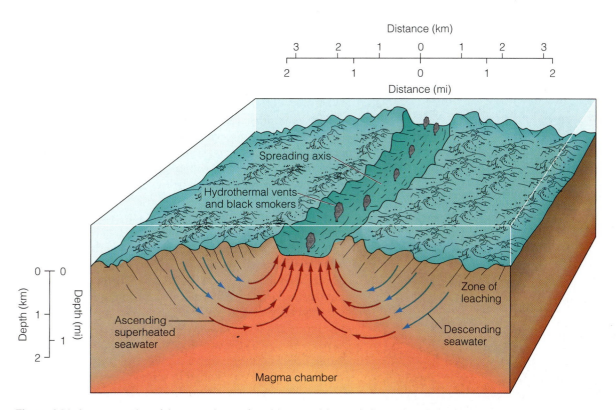

Figure 4.21 A cross section of the central part of a mid-ocean ridge—similar to that shown in Figure 4.18b—showing the origin of hydrothermal vents. Cool water (blue arrows) is heated as it descends toward the hot magma chamber, leaching sulfur, iron, copper, zinc, and other materials from the surrounding rocks. The heated water (red arrows) returning to the surface carries these elements upward, discharging them at the hydrothermal springs on the seafloor. The areas around the vents support unique communities of organisms (see Chapter 14).

floor until it comes into contact with very hot rocks associated with active seafloor spreading. There the superheated, chemically active water dissolves minerals and gases and escapes upward through the vents by convection (**Figure 4.21**).

Since that first discovery, vents have been found on the Mid-Atlantic Ridge east of Florida, in the Sea of Cortez east and south of Baja California, and on the Juan de Fuca Ridge off the coast of Washington and Oregon. Scientists now believe that hydrothermal vents may be very common on oceanic ridges, especially in zones of rapid seafloor spreading. In July 1990 vents were discovered in fresh water, at the bottom of Lake Baikal in southern Siberia. This discovery suggests that the world's oldest and deepest lake may someday become part of the ocean as Asia slowly breaks apart.

Not all vents form chimneys of mineral deposits—some are simply cracks in the seabed, or porous mounds, or broad segments of ocean ridge floor through which warm, mineral-laden water percolates upward. Cooler vents result when hot, rising water mixes with cold bottom water before reaching the surface. Water temperature in the vicinity of most hydrothermal vents averages 8–16°C (46–61°F), much warmer than usual for ocean bottom water, which has an average temperature of 3–4°C (37–39°F). We will study the unusual communities of animals that populate the vents in Chapter 14.

A volume of water equal to the volume of the world ocean is thought to circulate through the hot oceanic crust at spreading centers every 1 to 10 million years! The heat and chemicals issuing from these structures may play important roles in the chemical composition of seawater and the atmosphere, and in the formation of mineral deposits.

Abyssal Plains and Abyssal Hills 4-15

A quarter of Earth's surface consists of abyssal plains and abyssal hills. *Abyssal* is an adjective derived from a Greek word meaning "without bottom." While this is obviously not literally true, you can appreciate how the term came into use following the *Challenger* expedition's laborious soundings of these extremely deep areas!

Abyssal plains are flat, featureless expanses of sediment-covered ocean floor found on the periphery of all oceans. They are most common in the Atlantic, less so in the Indian Ocean, and relatively rare in the active Pacific, where peripheral trenches trap most of the sediments flowing from the continents. They lie between the continental margins and the oceanic ridges about 3,700 to 5,500 meters (12,000 to 18,000 feet) below the surface (see again Figure 4.18a). The Canary Abyssal Plain, a huge plain west of the Canary Islands in the North

Technological advances have permitted researchers to explore ocean basins in person. The sophistication and cost of deep-diving devices increases with their depth capability: Scuba outfits cost only a few hundred dollars, but a human-carrying vehicle suitable for a trip to the abyssal plains costs millions.

With the proper mix of gases and adequate training, a scuba diver can descend to about 60 meters (200 feet). With special mixtures of breathing gases and dry suits inflated with argon to avoid heat loss, professional divers have reached a depth of 300 meters (984 feet). Below that depth a small research submarine is better equipped to withstand the cold and pressure. One of the newest and most promising is the maneuverable, compact, one-person *DeepWorker 2000* (**Figure a**) developed by Nuytco Research, Ltd. The sub is "flown" by a pilot seated in a comfortable chair and surrounded by a strong plastic bubble allowing 360° visibility. Using foot pedals and a control stick, *DeepWorker 2000* can descend to 610 meters (2,000 feet) and remain on station for up to 12 hours. A small fleet of these quick and maneuverable subs will explore the continental shelf and upper continental slope for five years in association with the National Geographic Society's Sustainable Seas Expeditions.

For descents deeper than mid-slope, however, stronger and more costly submersibles like *Alvin* must be used. *Alvin* (**Figure b**) was commissioned in 1964 and operated jointly

a *DeepWorker 2000,* among the newest and most capable of human-carrying research vehicles. Described by its inventor as ". . . an underwater sports car," *DeepWorker 2000* and its sister vessels embarked in 1999 on a five-year exploration of the continental shelf and upper continental slope. Equipped with high-resolution sonar, video cameras, and two robotic arms that can extend 3.7 meters (12 feet) from the vehicle, the maneuverable little sub can take a researcher to a depth of 610 meters (2,000 feet) for up to 12 hours.

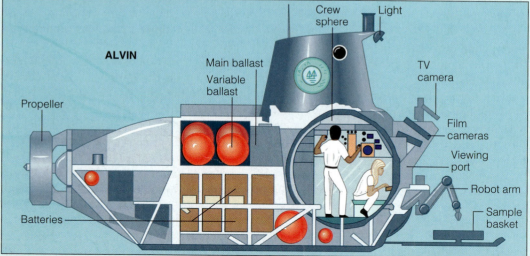

b *Alvin,* the best-known and oldest of the six deep-diving manned research submarines now in operation, has made more than 3,500 dives. *Alvin* carries three people and is capable of diving to 4,000 meters (13,120 feet).

c *Shinkai 6500,* built in 1981 by Mitsubishi Heavy Industries, is the world's newest deep-diving person-carrying submersible. Three people can be accommodated in the vehicle's titanium pressure hull. *Shinkai 6500* reached and exceeded its design depth of 6,500 meters (21,320 feet) on 11 August 1989.

by the Woods Hole Oceanographic Institution, the National Science Foundation, the Office of Naval Research, and the National Oceanic and Atmospheric Administration. The most famous of the research submersibles, *Alvin* has explored the Mid-Atlantic Ridge near the Azores at 2,700 meters (9,000 feet), measuring rock temperatures, collecting water samples for chemical analysis, and taking photographs.

The deepest-diving vehicle capable of carrying a human crew is Mitsubishi Heavy Industries' *Shinkai 6500* (**Figure c**). *Shinkai 6500* safely descended to a depth of 6,527 meters (21,409 feet) on 11 August 1989, and thus can reach all but the deepest trench floors, or about 98% of the world ocean bottom.

4-16

Atlantic, has an area of about 900,000 square kilometers (350,000 square miles).

Abyssal plains are extraordinarily flat. A 1947 survey by the Woods Hole Oceanographic Institution ship *Atlantis* found that a large Atlantic abyssal plain varies no more than a few meters in depth over its entire area. Such flatness is caused by the smoothing effect of the layers of sediment, which often exceed 1,000 meters (3,300 feet) in thickness. Most of the sediment that forms the abyssal plains appears to be of terrestrial or shallow-water origin, not derived from biological activity in the ocean above. Some of it may have been transported to the plains by winds or turbidity currents. These deep sediment layers mask irregularities in the underlying ocean crust, but a powerful type of echo sounder can "see" through this sediment to reveal the complex topography of the basaltic basin floor below (**Figure 4.22**). The broad basaltic shoulders of the Mid-Atlantic Ridge extend beneath this cloak of sediment almost as far as the bordering continental slopes.

Abyssal plain sediments may not be thick enough to cover the underlying basaltic floor near the edges bordering the oceanic ridges. Here the plains are punctuated by **abyssal hills**—small, sediment-covered extinct volcanoes or intrusions of once-molten rock, usually less than 200 meters (650 feet) high (one of which is seen extending above the flat surface in Figure 4.22). These abundant features are associated with seafloor spreading; they form when newly formed crust moves away from the center of a ridge, stretches, and cracks. Some blocks of the crust drop to form valleys, and others remain higher as hills. Lava erupting from the ridge flows along the fractures, coating the hills. This helps explain why abyssal hills occur in lines parallel to the flanks of the nearby oceanic ridge, and why they occur most abundantly in places where the rate of seafloor spreading is fastest. Abyssal plains and hills account for nearly all of the area of deep-ocean floor that is not part of the oceanic ridge system.

Seamounts and Guyots

4-17

The ocean floor is dotted with thousands of volcanic projections that do not rise above the surface of the sea. These projections are called **seamounts.** Seamounts are circular or elliptical, more than a kilometer (0.6 mile) in height, with relatively steep slopes of 20° to 25°. (Abyssal hills, in contrast, are much more abundant, less than a kilometer high, and not as steep.) Seamounts may be found alone or in groups of from 10 to 100. Though many form at hot spots (see Figure 3.23), most are thought to be submerged inactive volcanoes that formed at spreading centers (see **Figure 4.23**). Movement of the lithosphere away from spreading centers has carried them outward and downward to their present positions. As many as 10,000 seamounts are thought to exist in the Pacific, about half the world total.

Guyots are flat-topped seamounts that once were tall enough to approach or penetrate the sea surface. Generally they are confined to the west-central Pacific. The flat tops suggest that they were eroded by wave action when they were

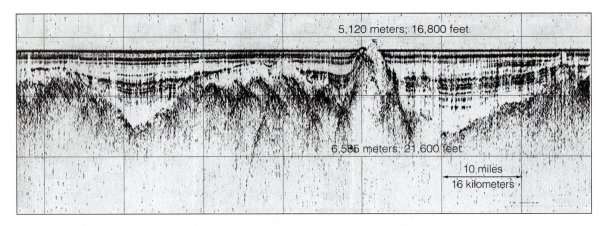

Figure 4.22 The deep, smooth sediments of the Atlantic's Northern Madeira Abyssal Plain bury 100-million-year-old mountains.

near sea level. Their plateaulike tops eventually sank too deep for wave erosion to continue wearing them down. Like the more abundant seamounts, most guyots were formed near spreading centers and transported outward and downward as the seafloor moved away from a spreading center and cooled (see again Figure 3.23). As noted in Chapter 3, chains of seamounts and guyots are compelling evidence for the theory of plate tectonics.

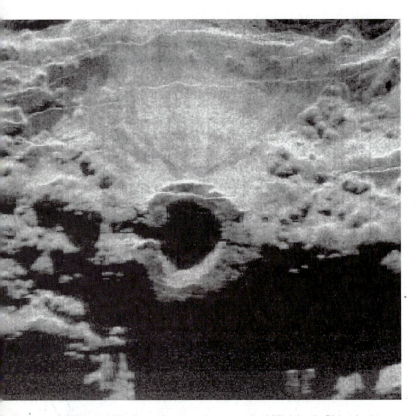

Figure 4.23 An undersea volcano on the Mid-Atlantic Ridge is "photographed" by a side-scan sonar device towed 3,000 meters (10,000 feet) below the surface

Trenches and Island Arcs 4-18

A **trench** is an arc-shaped depression in the deep-ocean floor. These flat-bottomed creases in the seafloor occur where a converging oceanic plate is subducted. The water temperature at the bottom of a trench is slightly cooler than the near-freezing temperatures of the adjacent flat ocean floor, reflecting the fact that trenches are underlain by old, relatively cold ocean crust sinking into the upper mantle. Trenches (and their associated island arcs topped by erupting volcanoes) are among the most active geological features on Earth. Great earthquakes and tsunami (huge waves we will discuss in Chapter 9) often originate in them. **Figure 4.24** shows the distribution of the ocean's major trenches. Not surprisingly, most are around the edges of the active Pacific.

Trenches are the deepest places in Earth's crust, 3 to 6 kilometers (1.9 to 3.7 miles) deeper than the adjacent basin floor. The ocean's greatest depth is the Mariana Trench of the western Pacific, where *Trieste* set the world's deep-diving record (described in the chapter opener). Near the place where *Trieste* came to rest, the ocean bottom is 11,022 meters (36,163 feet) below sea level, 20% deeper than Mount Everest is high (see **Figure 4.25**). The Mariana Trench is about 70 kilometers (44 miles) wide and 2,550 kilometers (1,600 miles) long, typical dimensions for these structures.

Trenches are curving chains of V-shaped indentations. The trenches are curved because of the geometry of plate interactions on a sphere. The convex sides of these curves generally face the open ocean (see again Figure 4.24). The trench walls on the island side of the depressions are steeper than those on the seaward side, indicating the direction of plate subduction. The sides of trenches become steeper with depth, normally reaching angles of about 10° to 16° before flattening to a floor underlain by thick sediment. (Parts of the concave wall of the Kermadec–Tonga Trench are the world's steepest at 45°.) No continental rise occurs along coasts with trenches, because the sediment that would form the rise ends up at the bottom of the trench.

Island arcs, curving chains of volcanic islands and seamounts, are almost always found paralleling the concave edges

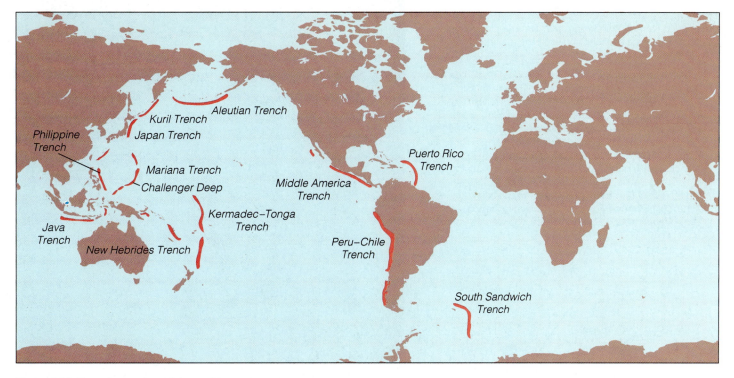

Figure 4.24 Oceanic trenches of the world.

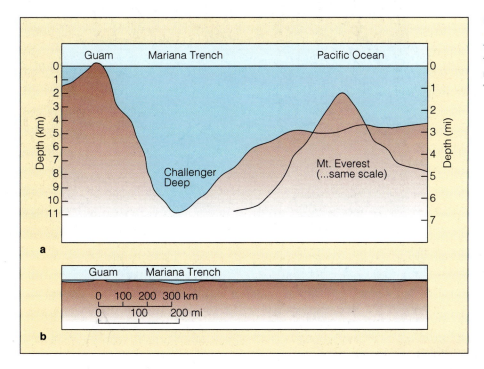

Figure 4.25 The Mariana Trench. (**a**) Comparing the Challenger Deep and Mount Everest at the same scale shows that the deepest part of the Mariana Trench is about 20% deeper than the mountain is high. (**b**) The Mariana Trench shown without vertical exaggeration.

of trenches. As you may remember from Chapter 3, trenches and island arcs are formed by tectonic and volcanic activity associated with subduction. The descending lithospheric plate contains some materials that melt as the plate sinks into the mantle. These materials rise to the surface as magmas and lavas that form the chain of islands behind the trench. The Aleutian Islands, most Caribbean islands, and the Marianas Islands are island arcs.

THE GRAND TOUR   4-19

Researchers at the National Oceanic and Atmospheric Administration have generated a map of the world ocean floor based on satellite observations of the shape of the sea surface (**Figure 4.26**). The graphic shows all the features discussed in this chapter. These features—and a basic understanding of the geological reasons for their existence—will help you recall

Figure 4.26 A technological *tour de force,* derived from data provided to the National Geophysical Data Center from satellites and shipborne sensors, shows all the features discussed in this chapter. Key to features:

(1) Aleutian Trench, (2) Hawaiian Islands, (3) Juan de Fuca Ridge, (4) Clipperton Fracture Zone, (5) Peru–Chile Trench, (6) East Pacific Rise, (7) Mariana Trench, Challenger Deep, (8) Mid-Atlantic Ridge, (9) South Sandwich Trench, (10) Red Sea, (11) Easter Island, (12) Galapagos Rift, (13) Iceland, (14) Gulf of Aden, (15) Pacific–Antarctic Ridge, (16) Azores, (17) Tristan de Cunha, (18) Kermadec Trench, (19) Tonga Trench, (20) Hatteras Abyssal Plain, (21) Grand Banks, (22) Mid-Indian Ridge, (23) Atlantic–Indian Ridge, (24) Maldive Islands, (25) Eltanin Fracture Zone, (26) Emperor Seamounts, (27) Java Trench, (28) Great Barrier Reef, (29) Kuril Trench, (30) Japan Trench.

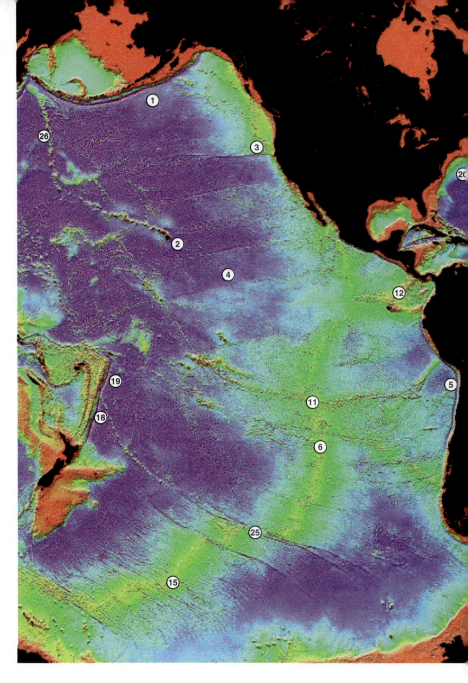

the dramatic nature and history of the seafloor that we have discussed in the past two chapters.

QUESTIONS FROM STUDENTS

1. You mentioned that Walsh and Piccard saw animals at a depth of 11,022 meters (6.85 miles). The pressure there is more than 8 tons per square inch. How could any living thing withstand such crushing pressure?

It seems impossible until you think about your own situation. Humans live at the bottom of a surprisingly heavy "ocean" of air: A column of air 1 inch square extending from sea level to the top of the atmosphere weighs about 6.7 kilograms (14.7 pounds). This pressure presents no problem to us because we

have the same pressure inside of us pushing out. The animals at great depths also have the same pressure inside as outside; that is, they live in perfect hydrostatic balance. They notice the weight of water above them no more than you notice the weight of air above you.

2. Turbidity currents seem important in forming canyons and distributing deep sediments over abyssal plains. Has anybody ever seen a turbidity current in action?

Yes, surprisingly. In the late 1940s the Dutch geologist Philip Kuenen produced turbidity currents in his laboratory by pouring muddy water into a trough with a sloping bottom. His observations confirmed nineteenth-century reports that the muddy Rhône River continued to flow in a dense stream

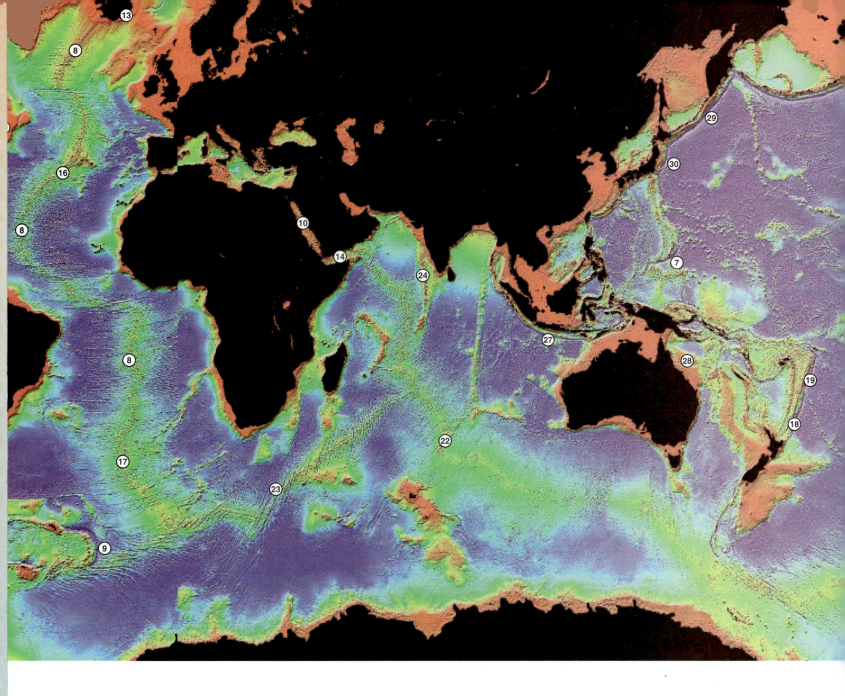

along the bottom of Lake Geneva. In the 1960s Robert Dill and Francis Shepard viewed sandfalls in southern California's Scripps Canyon, and French researchers have recently photographed these currents in the Mediterranean.

3. Why is Iceland one of only a few places in the world where oceanic crust is found above sea level?

Oceanic crust is thin and dense. Because of isostasy, the ocean floor lies at a lower elevation than the thicker, less dense, higher continents. Water filled the lower elevations first, submerging nearly all the basaltic basement we now call ocean floor. In areas of rapid seafloor spreading, or at hot spots, peaks are occasionally pushed toward the ocean surface by the large volume of mantle material rising from below. Large quantities of erupted magma (lava) then build the crests above sea level

to form islands like Iceland. The Azores is another place on the Mid-Atlantic Ridge where this is happening.

CHAPTER SUMMARY

The ocean floor can be divided into two regions: continental margins and deep ocean basins. The continental margin, the relatively shallow ocean floor nearest the shore, consists of the continental shelf and the continental slope. The continental margin shares the structure of the adjacent continents, but the deep ocean floor away from land has a much different origin and history. Prominent features of the deep ocean basins include rugged oceanic ridges, flat abyssal plains, occasional deep trenches, and curving chains of volcanic islands. Most of

Gibraltar—the narrow passage between the Mediter-ranean and the Atlantic—was blocked by an uplift of the land—a dam, in effect. Water could no longer flow into the Mediterranean; so the sea east of Gibraltar gradually evaporated, leaving behind a desert of gypsum-rich salts—the future M-reflector. Half a million years later, however, either the strait eroded or subsided, or the level of the Atlantic rose—and the ocean flooded back into the basin. The types of fossil microorganisms in the drill cores suggest the rate of refilling was very rapid. In little more than a century, waters equal to a thousand Nia-garas falling across the Gibraltar precipice refilled the Mediterranean Sea!

At least that's how Ryan and Hsü reconstructed the tale of the drill cores. But some scientists who have stud-ied the same evidence doubt that the Mediterranean was ever completely dry. They suggest that most of the floor was studded with alkaline lakes or that perhaps the entire bottom was covered by very shallow, very salty water. The M-reflector could still have formed in this scenario.

Sediments have many stories to tell, some of them still open to interpretation. In a sense, sediments are the memory of the ocean because they provide a record—if we can read it—of the recent history of the ocean, and therefore of the planet itself.   🔲 **5-1**

SEDIMENT

Sediment is particles of organic or inorganic matter that accumulate in a loose, unconsolidated form. The particles originate from the weathering and erosion of rocks, from the activity of living organisms, from volcanic eruptions, from chemical processes within the water itself, and even from space. Most of the ocean floor is being slowly dusted by a continuing rain of sediments. Accumulation rates on the deep-sea floor vary from a few centimeters per year to the thickness of a dime every thousand years.

Marine sediments occur in a broad range of sizes and types. Beach sand is sediment; so is the mud of a quiet bay and the mix of silt and tiny shells found on the continental margins. Less familiar sediments are the fine clays of the deep-ocean floor, the biologically derived oozes of abyssal plains, and the nodules and coatings that form around hard objects on the seafloor. The origin of these materials—and the distribution and sizes of the particles—depends on a combination of physical and biological processes.

What do sediments look like? That depends on where you look. **Figure 5.1** shows the Mid-Atlantic Ridge, 48°N, at a depth of 2,629 meters (8,626 feet). Sponges and some relatives of coral grow from young rocky outcrops only lightly powdered with sediment. Contrast that rough ridge with the smooth seafloor shown in **Figure 5.2**. The sediment there is about 35 meters (115 feet) thick and marked by the tracks of brittle stars. These widely distributed organisms feed on surface bacteria and fallen particles of organic sediment.

The surface of the sediment is not always smooth. Where bottom currents are swift and persistent, they can cause ripples like those on a streambed (**Figure 5.3**).

The extraordinary thickness of some layers of marine sediment can be seen in **Figure 5.4**, a seismic profile of the eastern edge of a seamount in the North Atlantic's Sohm Abyssal Plain south of Nova Scotia. The sediment at the eastern boundary of this profile covers the oceanic crust to a depth of more than 1.8 kilometers (1.1 miles).

The colors of marine sediments are often quite striking. Sediments of biological origin are white or cream-colored,

Figure 5.1 Sediment near the crest of the Mid-Atlantic Ridge. The rock outcrops are dusted only lightly with sediment; crinoids and gorgonians (relatives of coral) attach to the rocks.

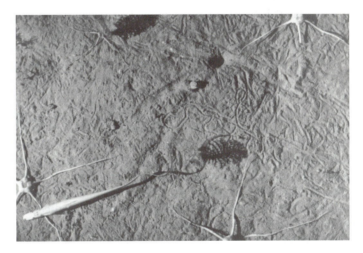

Figure 5.2 Brittle stars and their tracks on the continental slope off New England. The depth is 1,476 meters (4,842 feet).

Figure 5.3 Ripples on the sediment beneath the swift Antarctic Circumpolar Current in the northern Drake Passage. The depth here is 4,010 meters (13,153 feet).

particles by size from relatively large grains near the coast to relatively small grains near the shelf break.

There are exceptions, however. Shelf deposits are subject to further modification and erosion as sea level fluctuates: Larger particles may be moved toward the shelf edge when sea level is low, as it was in periods of extensive glaciations—ice ages. Poorly sorted sediments are also found as glacial deposits. In polar regions, glaciers and ice shelves give rise to icebergs. These carry particles of all sizes, and when they melt they distribute their mixtures of rocks, gravel, sand, and silt onto high-latitude continental margins and deep-ocean floor. Turbidity currents also disrupt the orderly sorting of sediments on the continental margin by transporting coarse-grained particles away from coastal areas and onto the deep-ocean floor.

The rate of sediment deposition on continental shelves is variable, but it is almost always greater than the rate of sediment deposition in the deep ocean. Near the mouths of large rivers, 1 meter (about 3 feet) of sediment may accumulate every thousand years. Along the east coast of the United States, however, many large rivers terminate in estuaries, which trap most of the sediment brought to them. The continental shelf of eastern North America is therefore covered mainly by sediments laid down during the last period of glaciation, when sea level was lower.

In addition to terrigenous material, the continental margins almost always contain biogenous sediments. Biological productivity in coastal waters is often quite high, and the skeletal remains of creatures living on the bottom or in the water above mix with the terrigenous sediments and dilute them.

Sediments can build to impressive thickness on continental shelves. In some cases, shelf sediments undergo **lithification**: They are converted into sedimentary rock by pressure or by cementation. If these lithified sediments are thrust above sea level by tectonic forces, they can form mountains or plateaus. The top of Mount Everest, the world's tallest peak, is a shallow-water biogenic marine limestone (a calcareous rock). Much of the Colorado Plateau, with its many stacked layers, was formed by sedimentary deposition and lithification beneath a shallow continental sea beginning about 570 million years ago. The Colorado River has cut and exposed the uplifted beds to form the Grand Canyon. Hikers walking from the canyon rim down to the river pass through spectacular examples of continental shelf sedimentary deposits. Their journey takes them deep into an old ocean floor!

The Sediments of Deep-Ocean Basins

5-11

Deep-ocean sediment thickness is highly variable. When averaged, the Atlantic Ocean bottom is covered by sediments to a thickness of about 1 kilometer (3,300 feet), while the Pacific floor has an average sediment thickness of less than 0.5 kilometer (1,650 feet). There are three reasons for this difference. First, the Atlantic Ocean is smaller in area. Second, the Atlantic is fed by a greater number of rivers laden with sediment. Third, in the Pacific Ocean many oceanic trenches trap sediments moving toward basin centers. Beyond this, the composition and thickness of pelagic sediments also vary with location, being thickest on the abyssal plains and thinnest (or absent) on the oceanic ridges.

TURBIDITES Dilute mixtures of sediment and water periodically rush down the continental slope in turbidity currents, the erosive force of which is thought to help cut submarine canyons (see again Figure 4.15). These underwater avalanches of thick, muddy fluid can reach the continental rise and often continue moving onto an adjacent abyssal plain before eventually coming to rest. The resulting deposits are called **turbidites**; they are poorly sorted layers of terrigenous sand interbedded with the finer sediments typical of the deep-sea floor.

CLAYS About 38% of the deep seabed is covered by clays and other fine terrigenous particles. As we have seen, the finest terrigenous sediments are easily transported by wind and water currents. Microscopic waterborne particles and tiny bits of wind-borne dust and volcanic ash settle slowly to the deep-ocean floor, forming fine brown, olive-colored, or reddish clays. As Table 5.1 shows, the velocity of particle settling is related to particle size, and clay particles usually fall very slowly indeed. Terrigenous sediment accumulation on the deep-ocean floor is typically about 2 millimeters (⅛ inch) every thousand years.

OOZES Seafloor samples taken farther from land usually show a greater proportion of biogenous sediments than those obtained near the continental margins. This is not because biological productivity is higher farther from land (the opposite is usually true), but because there is less terrigenous material far from shore, and thus the deposits contain a greater proportion of biogenous material.

Deep-ocean sediment containing at least 30% biogenous material is called an **ooze** (surely one of the most descriptive terms in the marine sciences). Oozes are named after the dominant remnant organism constituting them. The organisms contributing their remains to deep-sea oozes are small, single-celled, drifting, plantlike organisms and the single-celled animals that feed on them. The hard shells and skeletal remains of these creatures are of relatively dense glasslike silica or calcium carbonate (limey) substances. When these organisms die, their shells settle slowly toward the bottom, mingle with fine-grained terrigenous silts and clays, and accumulate as ooze. The silica-rich residues give rise to **siliceous ooze,** the calcium-containing material to **calcareous ooze.**

Oozes accumulate slowly, at a rate of about 1 to 6 centimeters (½ to 2½ inches) per thousand years. But they collect almost ten times more quickly than deep-ocean terrigenous clays. The accumulation of any ooze therefore depends on a delicate balance between the abundance of organisms at the surface, the rate at which they dissolve once they reach the bottom, and the rate of accumulation of terrigenous sediment.

Calcareous ooze forms mainly from shells of the amoeba-like **foraminifera** (**Figure 5.7a** and **b**), small drifting mollusks called **pteropods,** and tiny algae known as **coccolithophores** (**Figure 5.7c**). Although these creatures live in nearly all surface ocean water, calcareous ooze does not accumulate

a

b

Figure 5.7 Organisms that contribute to calcareous ooze. (**a**) A living foraminiferan, an amoeba-like organism. The shell of this beautiful foram, genus *Hastigerina,* is surrounded by a bubble-like capsule. It is one of the largest of the planktonic species with spines, reaching nearly 5 centimeters (2 inches) in length. (**b**) a much smaller foraminiferan, the snail-like planktonic *Rosalina.* (**c**) Coccoliths, individual plates of coccolithophores, a form of planktonic algae. Because of their tendency to dissolve, calcareous oozes very rarely occur at bottom depths below 4,500 meters (14,800 feet).

c

everywhere on the ocean floor because the shells are dissolved by seawater. At great depths, seawater contains more CO_2 and becomes slightly acid.[1] This acidity, combined with the increased solubility of calcium carbonate in cold water under pressure, dissolves the shells. Below a certain depth, then, the tiny skeletons of calcium carbonate dissolve on the seafloor, so no calcareous oozes form. Calcareous sediment dominates the deep-sea floor at depths of less than about 4,500 meters (14,800 feet). Approximately 48% of the surface of deep-ocean basins is covered by calcareous oozes.

Siliceous (silicon-containing) ooze predominates at greater depths and in colder polar regions. Siliceous ooze is formed from the hard parts of another amoeba-like animal, the beautiful glassy **radiolarian** (**Figure 5.8a**), and from single-celled algae called **diatoms** (**Figure 5.8b**). After a radiolarian or diatom dies, its shell will also dissolve back into the seawater, but this dissolution occurs *much* more slowly than the dissolution of calcium carbonate. Slow dissolution, combined with very high diatom productivity in some surface

waters, leads to the buildup of siliceous ooze. Diatom ooze is most common in the Antarctic, because strong ocean currents and seasonal upwelling in this area support large populations of diatoms. Radiolarian oozes occur in equatorial regions, most notably in the zone of equatorial upwelling west of South America (as will be seen in Figure 5.11). About 14% of the surface of the deep-ocean floor is covered by siliceous oozes.

The very small particles that make up most of these deep-ocean sediments would need between 20 and 50 years to sink to the bottom. By that time they would have drifted a great lateral distance from their original surface position. But researchers have noted that the composition of deep-bottom sediments is usually similar to particle composition in the water directly above. How could such tiny particles fall quickly enough to avoid great horizontal displacement? The answer

---

[1] A discussion of acid–base balance can be found in Chapter 6.

http://www.garrisonessentials2.com

SEDIMENTS • • • • **91**

a

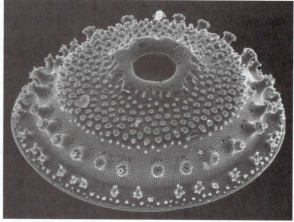

b

Figure 5.8 Scanning electron micrographs of siliceous oozes, which are most common at great depths. (**a**) Shells of radiolarians, amoeba-like organisms. Radiolarian oozes are found primarily in the equatorial regions. (**b**) A test (shell) of a diatom, a single-celled alga. Diatom oozes are most common at high latitudes.

appears to involve their compression into fecal pellets (**Figure 5.9**). While still quite small, the fecal pellets of small animals are much larger than the tiny individual skeletons of diatoms, foraminifera, and other plantlike organisms that they consume, so they fall much faster, reaching the deep-ocean floor in about two weeks.

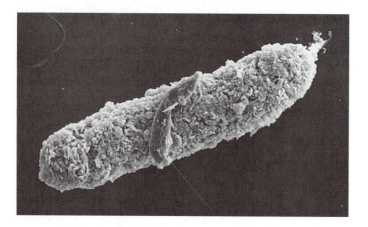

Figure 5.9 A fecal pellet of a small planktonic animal. The compressed pellet is about 80 micrometers in length and consists of the indigestible remains of microscopic plantlike organisms, mostly coccolithophores. Unaided by pellet packing, these remains might take months to reach the seabed, but compressed in this way, they can be added to the ooze in perhaps two weeks.

Some deep-sea oozes have been uplifted by geologic processes and are now visible on land. The calcareous chalk White Cliffs of Dover in eastern England are partially lithified deposits composed largely of foraminifera and coccolithophores. Fine-grained siliceous deposits called *diatomaceous earth* are mined from other deposits. This fossil material is a valued component in flat paints, pool and spa filters, and mildly abrasive car and tooth polishes.

HYDROGENOUS MATERIALS   Hydrogenous sediments also accumulate on deep-sea floors. They are associated with terrigenous or biogenous sediments and very rarely form sediments by themselves. Most hydrogenous sediments originate from chemical reactions that occur on particles of the dominant sediment.

The most famous hydrogenous sediments are manganese **nodules,** which were discovered by the hard-working crew of HMS *Challenger.* The nodules consist primarily of manganese and iron oxides but also contain small amounts of cobalt, nickel, chromium, copper, molybdenum, and zinc. They form in ways not fully understood by marine chemists, "growing" at an average rate of 1 to 10 millimeters (0.04 to 0.4 inch) per *million* years, one of the slowest chemical reactions in nature. Though most are irregular lumps the size of a potato, some nodules exceed 1 meter (3.3 feet) in diameter. Manganese nodules often form around nuclei such as sharks' teeth, bits of bone, microscopic algae, animal skeletons, and tiny crystals—

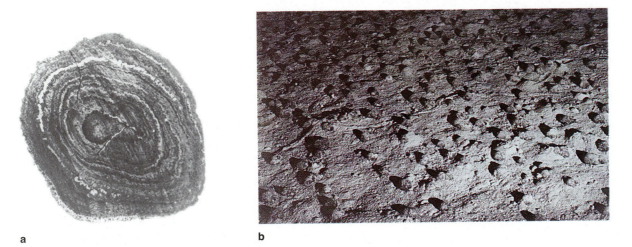

a   b

Figure 5.10 Manganese nodules. (**a**) A cross section cut through a manganese nodule, showing the concentric layers of manganese and iron oxides. This nodule is about 11 centimeters (4½ inches) long, a typical size. (**b**) Lemon-sized manganese nodules littering the abyssal Pacific.

as the cross section of a manganese nodule in **Figure 5.10a** shows. Bacterial activity may play a role in the development of a nodule. Between 20% and 50% of the Pacific Ocean floor may be strewn with nodules (**Figure 5.10b**).

Why don't these heavy lumps disappear beneath the constant rain of accumulating sediment? Possibly the continuous churning of the underlying sediment by creatures living there keeps the dense lumps on the surface, or perhaps slow currents in areas of nodule accumulation waft particulate sediments away. For now, the market value of the minerals in manganese nodules makes them too expensive to recover. As techniques for deep-sea mining become more advanced and raw material prices grow higher, however, the nodules' concentration of valuable materials will almost certainly be exploited.

EVAPORITES **Evaporites** are an important group of hydrogenous deposits that include many salts important to humanity. These salts precipitate as water evaporates from isolated arms of the ocean or from landlocked seas or lakes. For thousands of years people have collected sea salts from evaporating pools or deposited beds. Evaporites are forming today in the Gulf of California, the Red Sea, and the Persian Gulf. The first evaporites to precipitate as water's salinity increases are the carbonates, such as calcium carbonate (from which limestone is formed). Calcium sulfate, which gives rise to gypsum, is next. (The M-reflector at the bottom of the Mediterranean consists of a layer of gypsum-rich evaporites.) Crystals of sodium chloride (table salt) will form if evaporation continues.

The Distribution of Sediments   5-12

Figure 5.11 is a simplified look at the distribution of marine sediments. Notice especially the lack of radiolarian deposits in much of the deep North Pacific; the strand of siliceous oozes

extending west from equatorial South America; and the broad expanses of the Atlantic, South Pacific, and Indian Ocean floors covered by calcareous oozes. The broad, deep, relatively old Pacific contains extensive clay deposits, most delivered in the form of airborne dust. As you might expect, the poorly sorted glacial deposits are found only at high latitudes.

This map summarizes more than a century of effort by marine scientists. Studies of sediments will continue because of their importance to natural resource development and because of the details of Earth's history that remain locked beneath their muddy surfaces.

STUDYING SEDIMENTS   5-13

Deep-water cameras have enabled researchers to photograph bottom sediments. The first of these cameras was simply lowered on a cable and triggered by a trip wire. Other more elaborate cameras have been taken to the seafloor on towed sleds or deep submersibles.

Actual samples usually provide more information than photographs do. HMS *Challenger* scientists used weighted, wax-tipped poles and other tools attached to long lines to obtain samples, but today's oceanographers have more sophisticated equipment. Shallow samples may be taken using a **clamshell sampler** (named because of its method of operation, not its target; see **Figure 5.12**). Deeper samples are taken by a **piston corer** (**Figure 5.13**), a device capable of punching through as much as 25 meters of sediment and returning an intact plug of material. Using a rotary drilling technique similar to that used to drill for oil, the drilling ship *Glomar Challenger* has returned much longer core segments, some more than 1,100 meters (3,600 feet) in length! *Glomar Challenger* drilled more than a thousand holes for the Deep Sea Drilling Project, a program begun in 1968 with funds

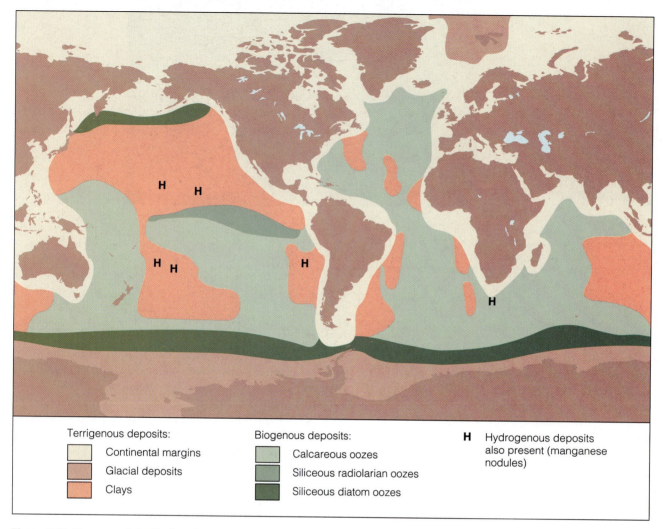

Terrigenous deposits:

Continental margins

Glacial deposits

Clays

Biogenous deposits:

Calcareous oozes

Siliceous radiolarian oozes

Siliceous diatom oozes

H Hydrogenous deposits also present (manganese nodules)

Figure 5.11 The general distribution of sediments on the ocean floor.

a

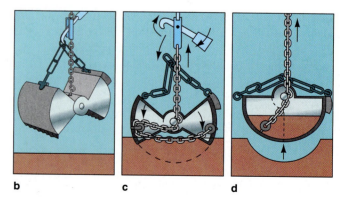

b　　c　　d

Figure 5.12 (**a**) A clamshell sampler (**b**) before sampling, (**c**) during sampling, and (**d**) after the sample has been taken. Note that the sample is relatively undisturbed.

The Ocean Drilling Program (ODP), currently the largest and most successful multinational earth science research project, is the direct successor to the Deep Sea Drilling Project (DSDP), which began in 1968. The two drilling ships commissioned for the projects pioneered drilling technology to retrieve cores, long cylinders of sediment and rock extracted from beneath the seafloor. The cores arrive on the deck of *JOIDES Resolution*—the ship currently in active service—in 9.5-meter (30-foot) sections encased in plastic sheaths. Immediately after retrieval, a core is marked to indicate its original location on the seafloor, coded to distinguish top from bottom, measured, and cut into sections for study and storage. Each section is slit lengthwise; one half is stored for the ODP archives, and the other is taken to the first of many shipboard laboratories for study.

Paleontologists then examine the sediments at the bottom of the core to determine the age of the oldest material. Other researchers measure its density, strength, molecular composition, radioactivity, and ability to conduct heat. Magnetic specialists read paleomagnetic data from the core to determine the ages of the rock fragments and the latitude at which they probably formed. Sensors are also lowered into the hole from which the core was removed, to gather additional information on the physical and chemical properties of the site.

ODP scientists have recently recovered fragments of the oldest remaining seafloor. The sample is about 175 million years old, a relic of the Middle Jurassic period, when the continents were massed in one huge cluster. From this and other cores they have learned about cycles of global climate change, information that will be useful in evaluating the present potential for global warming. They have also discovered how fluids move through the lithosphere, found evidence of an ice-free Antarctica, and noted the influence of plate tectonics on worldwide weather and current patterns. From trapped pollen grains, they have even been able to tell what land plants were thriving on Earth at the time the core sediments were laid down. As analysis technology evolves, stored cores are restudied and new information obtained.

The size and shape of a typical deep-core section can be seen in **Figure a.** This core was taken 250 meters (820 feet) beneath the seafloor off the northwestern coast of Australia by *JOIDES Resolution* on 22 July 1988. The ocean in the area was around 2,000 meters (6,500 feet) deep.

The material in the core progresses from sandstone (near the bottom of the core, at the right side of the photo) to fine claystone (near the top). Foraminiferans and microfossils are abundant in the sandstone, and small squidlike fossils, easily visible to the unaided eye, are present in the claystone. The sediments date from the Late Jurassic period, about 140 million years ago. The progression of slowly deposited sediments and fossils suggests that the seabed in this area has slowly subsided, possibly because of tectonic effects.

Details in a core from DSDP Hole 480, leg 64, are shown in **Figure b.** This core was obtained by the now-retired drilling ship *Glomar Challenger* on 1 January 1979

a Researchers examine a deep core taken off the northwestern coast of Australia.

b This core from the Gulf of California shows thin alternating bands of clay and ooze. Voids in the core show where samples were removed for study.

near Guaymas in the Gulf of California. The gulf is an area of active seafloor spreading, and researchers were interested in sampling the bottom in this place where basaltic magma is intruding into soft, wet, young sediments. The sediments here are very deep and have been accumulating at the extremely rapid rate of about 1,200 meters (4,000 feet) per million years.

Glomar Challenger arrived in the area on 29 December 1978 and used a conventional core barrel to penetrate 444 meters (1,456 feet) into diatomaceous ooze and mudstones. Nearly 273 meters (900 feet) of core were recovered from the hole drilled before Hole 480. Because of the disturbance caused by the conventional coring process, researchers were unable to determine the details of fine structure tantalizingly seen in a few lengths of that core.

On 31 December, *Glomar Challenger* moved to a new location 7 kilometers (4.4 miles) northwest. Hole 480, source of the core segment pictured here, was begun later that day. Unlike the previous hole, Hole 480 was drilled using a newly designed hydraulic corer developed by three DSDP engineers. The new device allowed 80% recovery of a core containing essentially undisturbed laminated diatomaceous ooze and muds.

As can be seen in Figure b, the core consists of thin alternating brown and gray bands of sediments. These are believed to be annual couplets, formed about 2 million years ago in response to two seasonal events that still occur in the gulf: the winter rains that introduce terrigenous clays into the region, and diatom blooms produced by seasonal upwelling and northwest winds. Preservation of these fine details depends upon a low level of free oxygen at the seafloor. Burrowing animals would normally churn these fine layers into mush, but because in a low-oxygen environment these animals are absent, the thin alternating sheets of clay and diatom tests (shells) are preserved.

Note that parts of the core have already been removed for study. Trenches across the width of the core mark places where sets of layers have been removed for isotope analysis. A larger sample for paleomagnetic study was taken from the square depression in the sample. Considering the vast expense and skill necessary to retrieve and analyze them, these small, gray, gritty bits of sediment are probably more costly than their weight in diamonds! 5-14

from the National Science Foundation. These cores are stored in core libraries, a very valuable scientific resource (**Figure 5.14**). Analysis of sediments and fossils from the Deep Sea Drilling Project cores helped verify the theory of plate tectonics. It has also shed light on the evolution of life forms and helped researchers to decipher the history of changes in Earth's climate over the last 100,000 years. A newer and larger ship, *JOIDES Resolution* (**Figure 5.15**), is carrying this work forward as part of the Ocean Drilling Program, an international research consortium.

Powerful new continuous seismic profilers have also been used to determine the thickness and structure of layers of sediment on the continental shelf and slope (see, for example, Figure 5.4), and to assist in the search for oil and natural gas. Recent improvements in computerized image processing of the echoes returning from the seabed now permit detailed analysis of these deeper layers.

SEDIMENTS AS HISTORICAL RECORDS　　5-15

Because the deep-sea sediment record is ultimately destroyed in the subduction process, the ocean's sedimentary "memory" is not as long as early marine scientists thought it would be. Still, modern studies of deep-sea sediments using seafloor samples, cores obtained by deep drilling, and continuous seismic profiling have demonstrated that these deposits contain a remarkable record of relatively recent ocean history. The analysis of layered sedimentary deposits in the ocean (or on land) represents the discipline of **stratigraphy** (*stratum* = layer, *graph* = a drawing). Deep-sea stratigraphy utilizes variations in the composition of rocks, microfossils, depositional patterns, geochemical character, and physical character (density and such) to trace or correlate distinctive sedimentary layers from place to place, establish the age of the deposits, and interpret changes in ocean and atmospheric circulation, productivity, and other aspects of past ocean behavior. In turn, these sorts of studies and the advent of deep-sea drilling have given rise to the emerging science of **paleoceanography** (*palaios* = ancient), the study of the ocean's past.

Early attempts to interpret ocean and climate history from evidence in deep-sea sediments occurred in the 1930s through 1950s as cores became available. These initial studies relied primarily on down-core variations in the abundance and distribution of glacial marine sediments, calcareous and siliceous oozes, and temperature-sensitive microfossils. Modern paleoceanographic studies continue to focus on these same features, but with much greater understanding of their significance and aided by seismic imaging of the deposits over large areas. In addition, newer and more precise methods of dating deep-sea sediments have allowed events to be placed in a proper time context. Finally, the appearance of instruments capable of analyzing very small variations in the relative abundances of the stable isotopes of oxygen preserved within the carbonate shells of microfossils found in deep-sea sediments has allowed

a

Figure 5.14 Sediment cores in storage. Cores are sectioned longitudinally, placed in trays, and stored in hermetically sealed cold rooms. The Gulf Coast Repository of the Ocean Drilling Program, located at Texas A&M University (pictured here), stores about 75,000 sections taken from more than 80 kilometers (50 miles) of cores recovered from the Pacific and Indian Oceans. Smaller core libraries are maintained at the Scripps Institution in California (Pacific and Indian Oceans) and at the Lamont–Doherty Earth Observatory in New York State (Atlantic Ocean).

Figure 5.15 *JOIDES Resolution,* the deep-sea drilling ship operated by the Joint Oceanographic Institutions for Deep Earth Sampling. The vessel is 124 meters (407 feet) long, with a displacement of more than 16,000 tons. The rig can take samples to a depth of 9,150 meters (30,000 feet) below sea level.

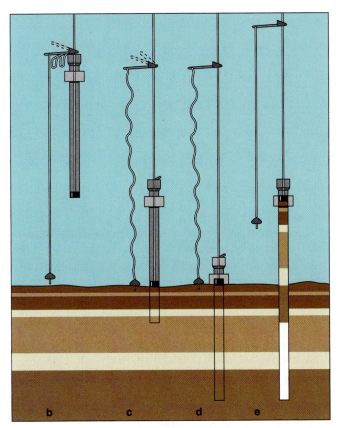

b c d e

Figure 5.13 (**a**) A piston corer. (**b**) The corer is allowed to fall to the bottom. (**c**) The corer reaches the bottom and continues, forcing a sample partway into the cylinder. (**d**) Tension on the cable draws a small piston within the corer toward the top of the cylinder, and the pressure of the surrounding water forces the corer deeper into the sediment. (**e**) The corer and sample being hauled in.

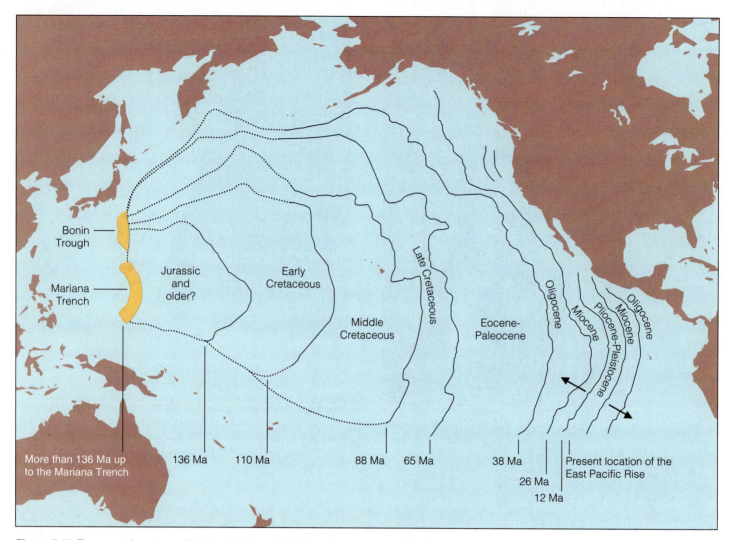

Figure 5.16 The ages of portions of the Pacific Ocean floor, based on core samples of sediments just above the basalt seabed, in millions of years ago (Ma, mega-annum). The youngest sediments are found near the East Pacific Rise and the oldest close to the eastern side of the troughs and trenches. Contrast this figure with Figure 3.27.

scientists to interpret changes in the temperature of surface and deep water over time. These same data are also used to estimate variations in the volume of ice stored in continental ice sheets, and thus track the ice ages. Other geochemical evidence contained in the shells of marine microfossils, including variations in carbon isotopes and trace metals such as cadmium, provide insights into ancient patterns of ocean circulation, productivity of the marine biosphere, and ancient upwelling. These sorts of data have already provided quantitative records of the glacial-interglacial climatic cycles of the past 2 million years. Future drilling and analysis of deep-sea sediments are poised to extend our paleoceanographic perspective much farther back in time.

Figure 5.16 shows the ages of different portions of the Pacific Ocean floor using data obtained largely from analyses of the overlying sediment. Note that sediments get older with increasing distance from the East Pacific Rise spreading center.

Global analysis of marine sediments in the modern basins can shed light on unexpected details of the last 180 million years of Earth's history. One of the most significant events is the unexplained extinction of up to 52% of known marine animal species (and the dinosaurs) at the end of the Cretaceous period 65 million years ago. Many researchers now believe that a sudden and violent impact of one or more asteroids or comets caused this catastrophe, which you'll read about in Chapter 12. The clouds of dust and ash thrown into the atmosphere by any of these events would have drastically reduced incident sunlight and greatly affected the lives of organisms and the photosynthetic base of ecosystems. Oceanographers are currently searching for evidence of the cause of the Cretaceous extinctions in layers of deep sediments.

Earth might not be the only planet where marine sediments have left historical records. As you read in Chapter 2, Mars probably had an ocean between 3.2 and 1.2 billion years

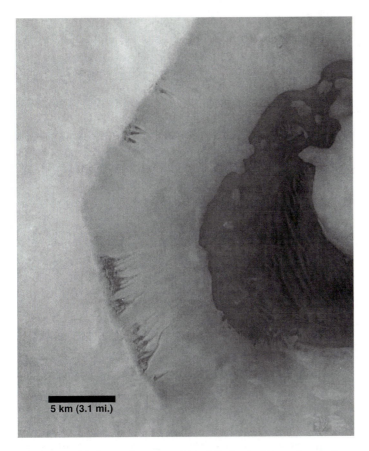

Figure 5.17 Marine sediments on Mars? Dark windblown dunes of sandy sediments rest on the floor of an ancient bay in the southern Martian lowlands. At some time in the past, water seeped out of layers within the cliffs (to the left) and flowed downhill, flooding part of the basin. The zones of contrast between the dark floor material and the lighter material of the slopes suggests the formation of bays and peninsulas. The appearance of dunes within the crater may be coincidental, or the sand may have been generated by wind and wave action. The lack of superimposed fresh impact craters suggests this process may have been active relatively recently. Photograph by *Mars Global Surveyor,* May 1998.

ago. In May 1998 *Mars Global Surveyor* photographed sediments that look suspiciously marine near the edge of an ancient bay (**Figure 5.17**). One can only wonder what stories they will tell.

QUESTIONS FROM STUDENTS

1. What economic benefit, if any, comes from studying sediments?

Study of sediments has brought practical benefits. In 1998 an estimated 32% of the world's crude oil and 28% of its natural gas were extracted from the sedimentary deposits of continental shelves and continental rises. Offshore hydrocarbons currently generate annual revenues in excess of $125 billion. Deposits within the sediments of continental margins account for about one-third of the world's estimated oil and gas reserves.

In addition to oil and gas, in 1998 sand and gravel valued at more than $450 million were taken from the ocean. This is about 1% of world needs. Commercial mining of manganese nodules has also been considered. In addition to manganese, these nodules contain substantial amounts of iron and other industrially important chemical elements. The high iron content of these nodules has prompted a proposal to rename them *ferromanganese* nodules. We will investigate these resources in more detail in Chapter 15.

2. Where are sediments thickest?

Sediments are thickest close to eroding land and beneath biologically productive neritic waters, and thinnest over the fast-spreading oceanic ridges of the eastern South Pacific. The thickest accumulations of sediment can be found along and beneath the continental margins (especially on continental rises). Some are typically more than 1,500 meters (5,000 feet) thick. Remember, much of the rocky material of the Grand Canyon was once marine sediment atop an ancient seabed. The Grand Canyon is nearly 2 kilometers (1¼ miles) deep, and the uppermost layer of sedimentary rock has already been eroded completely away!

3. What's the relationship between deep-sea animals and the sediments on which they live?

Though microscopic bacteria and benthic foraminifera may be very abundant on the seabed, visible life is not abundant on the bottom of the deep ocean. There are no plants at great depths because there is no light, but animals do live there. Some, like the brittle stars in Figure 5.2, move slowly along the surface searching for bits of organic matter to eat. Others burrow through the muck in search of food particles. Worms eat quantities of sediment to extract any nutrients that may be present and then deposit strings of fecal material as they move forward. The deeps may be uninviting places, but life is tenacious and survives even in this hostile environment.

CHAPTER SUMMARY

The ocean floor is covered in most places by layers of sediment. The sediment is comprised of particles from land, from biological activity in the ocean, from chemical processes within water, and even from space. The blanket of marine sediment is thickest at the continental margins and thinnest over the active oceanic ridges.

Sediments may be classified by particle size, source, location, or color. Terrigenous sediments, the most abundant, originate on continents or islands near them. Biogenous sediments are composed of the remains of once-living organisms. Hydrogenous sediments are precipitated directly from seawater. Cosmogenous sediments, the ocean's rarest, come to the seabed from space.

The position and nature of sediments provide important clues to Earth's recent history, and valuable resources can sometimes be recovered from them.

authigenic sediment
biogenous sediment
calcareous ooze
clamshell sampler
clay
coccolithophore
cosmogenous sediment
diatom
evaporite
foraminiferan
hydrogenous sediment
lithification
microtektite
neritic sediment
nodule

ooze
paleoceanography
pelagic sediment
piston corer
poorly sorted sediment
pteropod
radiolarian
sand
sediment
siliceous ooze
silt
stratigraphy
terrigenous sediment
turbidite
well-sorted sediment

STUDY QUESTIONS

1. In what ways are sediments classified?

2. List the four types of marine sediments. Explain the origin of each.

3. How are neritic sediments generally different from pelagic ones?

4. Is the thickness of ooze always an accurate indication of the biological productivity of surface water in a given area? (Hint: See next question.)

5. What happens to the calcium carbonate skeletons of small organisms as they descend to great depths? How do the siliceous components of once-living things compare?

6. What sediments accumulate most rapidly? Least rapidly?

7. Can marine sediments tell us about the history of the ocean from the time of its origin? Why?

8. How do paleoceanographers infer water temperatures, and therefore terrestrial climate, from sediment samples?

9. Where are sediments thickest? Are there any areas of the ocean floor free of sediments?

10. Are sediments commercially important? In what ways?

FOR FURTHER STUDY

Dietz, R. S. 1978. "IFO's (Identified Flying Objects)." *Sea Frontiers* 24 (no. 6): 341–6. Discusses the origin of microtektites.

Gehrels, T. T. 1996. "Collisions with Asteroids and Comets." *Scientific American,* March, 54–61. The odds are dis-cussed along with the consequences. For more on this topic, see Box 12.1 on mass extinctions.

Heezen, B., and C. Hollister. 1971. *The Face of the Deep.* New York: Oxford University Press. A superb treatise on the seabed with many wonderful photographs.

Hsü, K. J. 1983. *The Mediterranean Was a Desert: A Voyage of the* Glomar Challenger. Princeton: Princeton University Press. An accessible presentation of evidence for the author's theory that the Mediterranean dried up during the Pliocene epoch.

Kennett, J. P. 1982. *Marine Geology.* Englewood Cliffs, NJ: Prentice-Hall. Excellent and complete general refer-ence, especially strong in sediments.

Linklater, E. 1972. *The Voyage of the* Challenger. New York: Doubleday. An interesting description of the first sys-tematic sediment sampling.

Mack, W. N., and E. A. Leistikow. 1996. "Sands of the World." *Scientific American,* August, 62–7. A pictorial survey of surprisingly beautiful sediments.

Mylroie, J. E., and J. L. Carew. 1995. "Depositional Model and Stratigraphy for the Quaternary Geology of the Ba-hama Islands." Geological Society of America Special Paper 300.

Oceanus 36, no. 4 (Winter 1993–94) is dedicated to the topic of 25 years of ocean drilling. Articles include reports by drilling teams of different nations, results from paleo-ceanographers, findings about plate tectonics and sedi-mentary processes, and a summary of the latest in deep-sea drilling technology and its spin-offs.

Prospero, J. 1985. "Records of Past Continental Climates in Deep-Sea Sediments." *Nature* 315: 279–80. Subtitle could be "History in the Mucking."

Siebold, E., and W. H. Berger. 1993. *The Sea Floor: An Introduction to Marine Geology.* 2d ed. New York: Springer-Verlag. Methods and data are described in this update. Includes new material on physical marine resources.

Siever, R. 1988. *Sand.* New York: Freeman, Scientific Ameri-can Library.

Spencer, D. W., and S. Honjo. 1978. "Particles and Particle Fluxes in the Ocean." *Oceanus* 21 (no. 1): 20–6. The role of fecal pellets in sediment deposition.

Stanley, D. J. 1990. "Med Desert Theory Is Drying Up." *Oceanus* 33 (no. 1): 14–23. Recent information suggests that Hsü's vision of a deep desert basin in the Mediter-ranean 5.5 million years ago may not be correct.

For additional readings, go to InfoTrac College Edi-tion, your online research library at:a

http://infotrac.thomsonlearning.com/

ICEBERGS

Water occurs in three states: liquid, solid, and gas. Oceanographers are most familiar with water's liquid form, but about 6% of the world ocean is covered by ice. A small fraction of this ice is contained in the fantastic shapes of icebergs. Southern ocean icebergs originate as huge ice sheets attached to the Antarctic continent; lengths of up to 8 kilometers (5 miles) are not unusual, with flat tops rising 45 meters (150 feet) above sea level. Arctic icebergs are typically smaller in area but can be higher, pinnacled (or "castellated," after their castle turret shapes), and extraordinarily beautiful:

On the afternoon of the day we knew the storm had passed, I stood on the starboard side of the bridge at a window, with the heavy protective glass lowered. I rested my forearms on the sill, feeling the warmth of the bridge heaters around my legs and a slipstream of cool air past my face. The first icebergs we had seen, just north of the Strait of Belle Isle, listing and guttered by the ocean, seemed immensely sad, exhausted by some unknown calamity. We sailed past them.

I occasionally drew back from the starboard window to make a sketch, or to bring the binoculars up to my eyes. I marveled as much at the behavior of light around the

Penguins hitch a ride on an iceberg off the Antarctic coast.

icebergs as I did at their austere, implacable progess through the water. They took their color from the sun, and from the clouds and the water. But they also took their dimensions from the light: the stronger and more direct it was, the greater the contrast upon the surface of the ice, of the ice itself with the sea. And the more finely etched were the dull surfaces of their walls. The bluer the sky, the brighter their outline against it. . . .

Where the walls entered the water, the surf pounded them, creating caverns, grottoes, and ice bridges, strengthening an impression of sea cliffs. At the waterline the ice gleamed aquamarine against its own gray-white walls above. Where meltwater had filled cracks or made ponds, the pools and veins were milk-blue, or shaded to brighter marine blues, depending on the thickness of the ice. If the iceberg had recently fractured, its new face glistened greenish blue— the greens in the older, weathered faces were grayer. In twilight the ice took on the colors of the sun: rose, reddish yellows, watered purples, soft pinks. The ice both reflected the light and trapped it within its crystalline corners and edges, where it intensified. . . . How utterly still, unorthodox, and wondrous they seem. 6-1

Source: Barry Lopez, *Arctic Dreams* (New York: Scribner's, 1986). Reprinted with permission.

KEY CONCEPTS

1. The polar nature of the water molecule makes water an unusually good solvent. Every element present in the crust and atmosphere is also present in the ocean.

2. The ocean is in chemical equilibrium. Neither the proportion nor the amount of most dissolved substances changes significantly through time.

3. The thermal properties of water are responsible for the mild physical conditions at Earth's surface.

4. Changes in salinity and temperature greatly influence water density. Ocean water is usually layered by density, with the densest water on or near the seabed.

5. Sound and light in the ocean are affected by the physical properties of water, with refraction and absorption effects playing important roles.

CHAPTER AT A GLANCE

The Water Molecule

The Dissolving Power of Water

Seawater
Salinity: Dissolved Solids and Water Together
The Source of the Ocean's Salts
The Principle of Constant Proportions
Determining Salinity
Chemical Equilibrium and Residence Times
Mixing Time

Dissolved Gases
Nitrogen
Oxygen
Carbon Dioxide

Acid–Base Balance

Water and Heat
Temperature and Density
Freezing Water
Evaporating Water
Seawater and Pure Water

Global Thermostatic Effects

The Density Structure of the Ocean

Refraction, Light, and Sound
Light in the Ocean
Sound in the Ocean
The Sofar Layer
Sonar

Water is familiar and abundant, and we don't always appreciate its unusual characteristics. If liquid methane or ammonia or any other flowing substance dominated the surface of Earth, physical conditions here would be strikingly different. Indeed, life as we know it would be impossible.

As you may recall from Chapter 2, most of Earth's surface waters are thought to have escaped from the crust and mantle through the process of outgassing. Outgassing of substances other than water, and the chemical dissolution by water of crustal material, have added salts and other solids and gases to the ocean. In this chapter we will investigate the structure of pure water, look at the chemistry of seawater with its many dissolved substances, and discuss some of water's physical properties and the implications of these properties for the ocean and Earth as a whole.

THE WATER MOLECULE   6-2

A **molecule** is a group of atoms held together by chemical bonds. **Chemical bonds,** the energy relationships between atoms that hold them together, are formed when **electrons**—tiny negatively charged particles found toward the outside of an atom—are shared between atoms or moved from one atom to another. Water's familiar chemical formula, H_2O, shows that two atoms of hydrogen (H) are present for each atom of oxygen (O). Because of the way a water molecule's oxygen electrons are distributed, the overall geometry of the molecule is a bent or angular shape. The angle of the chemical bonds formed by the two hydrogen atoms and the central oxygen atom is about 105°. The formation of a water molecule is depicted in **Figure 6.1.**

The angular shape of the water molecule makes it electrically asymmetrical, or **polar.** Each water molecule can be thought of as having a positive (+) end and a negative (−) end because positively charged particles at the center of the hydrogen atoms—called **protons**—are left partially exposed when the negatively charged electrons bond more closely to oxygen. The polar water molecule acts something like a magnet; its positive end attracts particles having a negative charge, and its negative end attracts particles having a positive charge. When water comes in contact with compounds whose elements are held together by the attraction of opposite electrical charges (most salts, for example), the polar water molecule will separate that compound's component elements from each other. This explains why water can dissolve so many other compounds so easily.

The polar nature of water also permits it to attract other water molecules. When a hydrogen atom in one water molecule is attracted to the oxygen atom of an adjacent water molecule, a **hydrogen bond** forms. The resulting loosely held webwork of water molecules is shown in **Figure 6.2.** Hydrogen bonds greatly influence the properties of water by enabling individual water molecules to stick to each other, a property called **cohesion.** Cohesion gives water an unusually high surface tension that results in a surface "skin" capable of support-

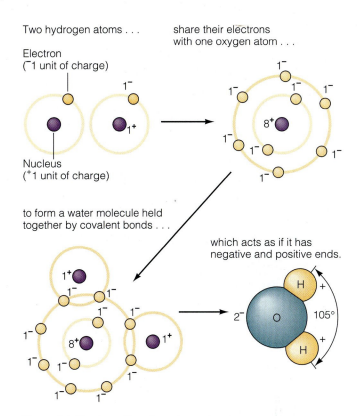

Figure 6.1 The formation of a water molecule.

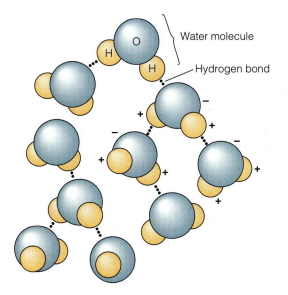

Figure 6.2 Hydrogen bonds in liquid water. The attractions between adjacent polar water molecules form a webwork of hydrogen bonds. These bonds are responsible for cohesion and adhesion, the properties of water that cause surface tension and wetting. Hydrogen bonds between water molecules also make it difficult for individual molecules to escape from the surface.

ing needles, razor blades, and even walking insects. **Adhesion,** the tendency of water to stick to other materials, allows water to adhere to solids—that is, to make them wet. Cohesion and adhesion are the causes of capillary action, the tendency of water to spread through a towel when one corner is dipped in water.

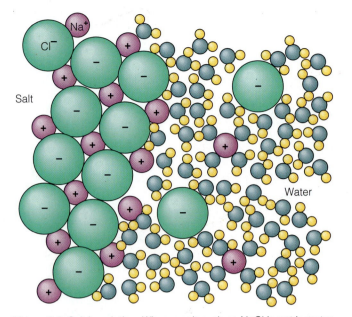

Figure 6.3 Salt in solution. When a salt such as NaCl is put in water, the positively charged hydrogen end of the polar water molecule is attracted to the negatively charged Cl⁻ ion, and the negatively charged oxygen end is attracted to the positively charged Na⁺ ion. The ions are surrounded by water molecules that are attracted to them, and become solute ions in the solvent (right side of figure).

THE DISSOLVING POWER OF WATER 6-3

Water is a powerful solvent—it will eventually dissolve nearly any substance. No wonder, then, that seawater and most other liquids in nature are water solutions. Water's dissolving power is due to the polar nature of the water molecule. Consider how water dissolves sodium chloride (or NaCl), the most common salt.[1] In solid crystals of this salt, the sodium atoms have lost electrons, and the chloride atoms have gained them. The resulting charged atoms are called **ions.** When liquid water is present, the polarity of water causes the sodium ion (Na^+) to separate from the chloride ion (Cl^-) (**Figure 6.3**). The ions move away from the salt crystal, permitting water to attack the next layer of NaCl. *Note that NaCl does not exist as "salt" in seawater*—its components are separated when salt crystals dissolve and joined when crystals re-form.

By contrast, oil doesn't dissolve in water even if the two are very thoroughly shaken together. When oil is dispersed in water it forms a mixture, because molecules of oil are nonpolar in character. This means that oil has no positive or negative charges to attract the polar water molecule. In a way this is fortunate—living tissues would readily dissolve in water if the oils within their membranes didn't blunt water's powerful attack.

SEAWATER 6-4

The total quantity (or concentration) of dissolved inorganic solids in water is its **salinity.** The ocean's salinity varies from about 3.3% to 3.7% by weight, depending on such factors as evaporation, precipitation, and fresh water runoff from the continents, but the average salinity is usually given as 3.5%. The world ocean contains some 5,000 trillion kilograms (5.5 trillion tons) of salt. If the ocean's water evaporated completely, leaving its salts behind, the dried residue could cover the entire planet with an even layer 45 meters (150 feet) thick! Most of the dissolved solids in seawater are salts that have been separated into ions. Sodium and chloride are the most abundant.

Salinity: Dissolved Solids and Water Together 6-5

About 3.5% of seawater consists of dissolved substances. Boiling away 100 kilograms of seawater theoretically produces a residue weighing 3.5 kilograms. Variations of 0.1% are significant, and oceanographers prefer to use parts-per-thousand notation (‰) rather than percent (%, parts per hundred) in discussing these materials.[2] The seven ions named in the rightmost pie in **Figure 6.4** make up more than 99% of this residual material. When seawater evaporates, its ionic components combine in many different ways to form table salt (NaCl), epsom salts ($MgSO_4$), and other mineral salts.

Seawater also contains minor constituents. The ocean is sort of an "Earth tea"—every element present in the crust and the atmosphere is also present in the ocean, though sometimes in extremely small amounts. Only 14 elements have concentrations in seawater larger than 1 part per million (ppm). Elements present in amounts less than 0.001‰ (1 part per million) are known as **trace elements.**

The Source of the Ocean's Salts 6-6

Remembering the effectiveness of water as a solvent, you might think that the ocean's saltiness has resulted from the ability of rain, groundwater, or crashing surf to dissolve crustal rock. Much of the sea's dissolved material originated in that way, but is crustal rock the source of all the ocean's solutes? An easy way to find out would be to investigate the composition of salts in river water and compare these figures to those of the ocean as a whole. If crustal rock is the only source, the salts in the ocean should be like those of concentrated river water. But they are not. River water is usually a dilute solution of calcium and bicarbonate ions, while the principal ions in seawater are chloride and sodium. The magnesium content of seawater would also be higher if seawater were simply concentrated river water. The proportions of salts in isolated salty inland

---

[1] Na, the chemical symbol for sodium, comes from *natrium*, the Latin name for sodium.

[2] Note that 3.5% = 35‰. If you began with 1,000 kilograms of seawater, you would expect 35 kilograms of residue.

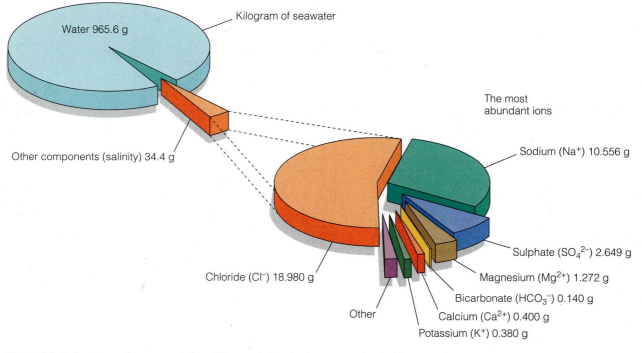

Figure 6.4 A diagrammatic representation of the most abundant components of a kilogram of seawater at 35‰ salinity. Note that the specific ions are represented in grams per kilogram, equivalent to parts per thousand (‰).

lakes, such as Utah's Great Salt Lake or the Dead Sea, are indeed much different from the proportions of salts in the ocean. Thus weathering and erosion of crustal rocks cannot be the sole source of sea salts.

The components of ocean water whose proportions are *not* accounted for by the weathering of surface rocks are called **excess volatiles.** To find the source of these excess volatiles we must look to Earth's deeper layers. The upper mantle appears to contain more of the substances found in seawater (including the water itself) than are found in surface rocks, and their proportions are about the same as found in the ocean. As you read in Chapter 3, convection currents slowly churn Earth's mantle, causing the movement of tectonic plates. This activity allows some deeply trapped volatile substances to escape to the exterior, outgassing through volcanoes and rift vents. These excess volatiles include carbon dioxide, chlorine, sulfur, hydrogen, fluorine, nitrogen, and, of course, water vapor. This material, along with residue from surface weathering, accounts for the chemical constituents of today's ocean.

Some of the salts that form from the ocean's dissolved materials are hybrids of the two processes of weathering and outgassing. Table salt, sodium chloride, is an example. The sodium ions come from the weathering of crustal rocks, while the chlorine ions come from the mantle by way of volcanic vents and outgassing from mid-ocean rifts. As for the lower-than-expected quantity of magnesium ions in the ocean, recent research at a spreading center east of the Galápagos Islands suggests that mid-ocean rifts may play a role in reducing the magnesium content and increasing the calcium content of sea-

water. The water that circulates through new ocean floor at these sites is apparently stripped of magnesium and a few other elements. The magnesium seems to be incorporated into mineral deposits, but calcium is added as hot water dissolves adjacent rocks.

The Principle of Constant Proportions 6-7

In 1865 the chemist Georg Forchhammer noted that although the total *amount* of dissolved solids (salinity) might vary between samples, the *ratio* of major salts in samples of seawater from many locations was constant. In other words, the percentage of various salts in seawater is the same in samples from many places regardless of how salty the water is. This constant ratio is known as **Forchhammer's principle,** or the **principle of constant proportions.** Forchhammer was also the first to observe that seawater contains fewer silica and calcium ions than concentrated river water, and the first to realize that removal of these compounds by marine animals and plants to form shells and other hard parts might account for some of the difference.

Determining Salinity 6-8

Water's salinity by weight would seem an easy property to measure. Why not simply evaporate a known weight of seawater and weigh the residue? This simple method yields imprecise results because some salts will not release all the molecules of water associated with them. If these salts are heated

a

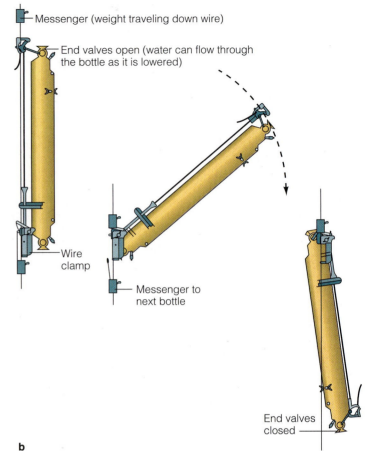

Messenger (weight traveling down wire)

End valves open (water can flow through the bottle as it is lowered)

Wire clamp

Messenger to next bottle

End valves closed

b

c

Figure 6.5 Sampling bottles. (**a**) A Nansen bottle being prepared for deployment. (**b**) When the messenger (a small weight) slides down the line, it triggers the bottle to flip upside down, closing the end valves at the same time. Nansen bottles are being replaced by sampling bottles that are able to hold greater volumes of water and are more resistant to additional triggering, such as the 10-liter Niskin bottles shown in (**c**). Each Niskin bottle is remotely triggered by a signal from the research ship when the array reaches a predetermined depth.

to drive off the water, other salts (carbonates, for example) will decompose to form gases and solid compounds not originally present in the water sample.

Modern analysis depends instead on determining the sample's chlorinity. **Chlorinity** is a measure of the total weight of chloride, bromide, and iodide ions in seawater. Because chlorinity is comparatively easy to measure, and because the proportion of chlorinity to salinity is constant, marine chemists have devised the following formula to determine salinity:

$$\text{Salinity in ‰} = 1.80655 \times \text{chlorinity in ‰}$$

Chlorinity is about 19.4‰, so salinity is around 35‰.

Seawater samples can be obtained by methods ranging from tossing a clean bucket over the side of a ship to sophisticated tube-and-pump systems. Typically water samples are collected using a string of sampling bottles similar to those perfected early in this century by the Norwegian scientist and explorer Fridtjof Nansen. A **Nansen bottle** (and its modern equivalent, a **Niskin bottle**) is open at both ends (**Figure 6.5**). A series of bottles is lowered on a wire to known depths. Each bottle is then triggered to close by a brass weight (appropriately called a "messenger") or an electrical signal sent down the line. The bottles are hauled to the surface and their contents analyzed.

Figure 6.6 This portable salinometer reads temperature, pH, and dissolved oxygen as well as conductivity. Designed to be lowered over the side of a research vessel at the end of a line, the self-powered device contains a small pump that passes water over sensors. This model takes two readings every second and operates down to about 200 meters (660 feet). Data may be retrieved by a cable connected to the research vessel or stored in the salinometer's memory to be retrieved when it is brought back aboard ship. A microprocessor contained within the white tube converts conductivity to salinity, and, when connected to a portable computer, displays the results of all readings as graphs.

Until recently, marine chemists used a delicate chemical procedure involving a silver nitrate solution to measure the chlorinity of seawater. Conversion to salinity was made by a set of mathematical tables. The procedure was calibrated against a standard sample of seawater of precisely known chlorinity. Today's marine scientists use an electronic device called a **salinometer** that measures the electrical conductivity of seawater (**Figure 6.6**). Conductivity varies with the concentration and mobility of ions present and with water temperature. Circuits in the salinometer adjust for water temperature, convert conductivity to salinity, and then display salinity. Salinometers are also calibrated against a sample of known conductivity and salinity. The best salinometers can determine salinity to an accuracy of 0.001%. Some salinometers are designed for remote sensing—the electronics stay aboard ship while the sensor coil is lowered over the side.

Chemical Equilibrium and Residence Times

6-9

If outgassing and the chemical weathering of rock are continuing processes, shouldn't the ocean become progressively saltier with age? Landlocked seas and some lakes usually become saltier as they grow older, but the ocean does not. The ocean

appears to be in **chemical equilibrium**; that is, the proportion *and amounts* of dissolved salts per unit volume of ocean are nearly constant. Evidently, whatever goes in must come out somewhere else.

Geologists in the 1950s developed the concept of a "steady state ocean." The idea suggests that ions are added to the ocean at the same rate as they are being removed. This theory helps explain why the ocean is not growing saltier. The idea led to the concept of **residence time,** the average length of time an element spends in the ocean. Residence time for a particular element can be calculated by this equation:

$$\text{Residence time} = \frac{\text{Amount of element in the ocean}}{\text{Rate at which the element is added to (or removed from) the ocean}}$$

Additions of salts from the mantle or from the weathering of rock are balanced by subtractions of minerals being bound into sediments. Dissolved salts precipitate out of the water, and the hard parts of living organisms containing silicon and calcium carbonate drift slowly down to the seabed. Some of these sediments are removed from the ocean and drawn into the mantle at subduction zones by the cycling of crustal plates. Input from runoff and outgassing equals outfall (binding into sediments) for each dissolved component.

The residence time of an element depends on its chemical activity. Atoms (or ions) of some elements, such as aluminum and iron, remain in seawater for a relatively short time before becoming incorporated in sediments; others, such as chloride, sodium, and magnesium, remain in water for millions of years. The approximate residence times for the major constituents of seawater are shown in **Table 6.1**. Because of

Table 6.1	Approximate Residence Times for Constituents of Seawater
Constituent	**Residence Time (years)**
Chloride (Cl^-)	100,000,000
Sodium (Na^+)	68,000,000
Magnesium (Mg^{2+})	13,000,000
Potassium (K^+)	12,000,000
Sulfate (SO_4^{2-})	11,000,000
Calcium (Ca^{2+})	1,000,000
Carbonate (CO_3^{2-})	110,000
Silicon (Si)	20,000
Water (H_2O)	4,100
Manganese (Mn)	1,300
Aluminum (Al)	600
Iron (Fe)	200

Sources: Data from Broecker and Peng, *Tracers in the Sea*, 1982; Bruland, "Trace Elements in Seawater," in *Chemical Oceanography*, vol. 8, 1983; Riley and Skirrow, *Chemical Oceanography*, vol. 1, 1975.

evaporation and precipitation, water in the ocean has a residence time of about 4,100 years.

Mixing Time 6-10

If constituent minerals are added to ocean water at rates that are less than the ocean's mixing time, they will become evenly distributed throughout the ocean. Because of the vigorous activity of currents, the **mixing time** of the ocean is thought to be approximately 1,600 years—thus, the ocean has been mixed hundreds of thousands of times during its long history. The relatively long residence times of seawater's major constituents assure thorough mixing, which is the basis of Forchhammer's principle of constant proportions.

DISSOLVED GASES 6-11

Gases in the air readily dissolve in seawater at the ocean's surface. Plants and animals living in the ocean require these dissolved gases to survive—no marine animal has the ability to break down water molecules to obtain oxygen directly, and no marine plant can manufacture enough carbon dioxide to support its own metabolism. In order of their relative abundance, the major gases found in seawater are nitrogen, oxygen, and carbon dioxide (see **Table 6.2**). Because of differences in their solubility in water and air, the proportions of dissolved gases in the ocean are very different from the proportions of the same gases in the atmosphere.

Unlike solids, gases dissolve most readily in *cold* water. A cubic meter of chilly polar water usually contains a greater volume of dissolved gases than a cubic meter of warm tropical water.

Nitrogen 6-12

About 48% of the dissolved gas in seawater is nitrogen. (In contrast, the atmosphere is slightly more than 78% nitrogen by volume.) The upper layers of ocean water are usually saturated with nitrogen; that is, additional nitrogen will not dissolve. Living organisms require nitrogen to build proteins and other important biochemicals, but they cannot use the free nitrogen in the atmosphere and ocean directly. It must first be "fixed" into usable chemical forms by specialized organisms. Though some species of bottom-dwelling bacteria can manufacture usable nitrates from the nitrogen dissolved in sea-water, most of the nitrogen compounds needed by living organisms must be recycled among the organisms themselves.

Oxygen 6-13

About 36% of the gas dissolved in the ocean is oxygen, but there is about a hundred times more gaseous oxygen in Earth's atmosphere than is dissolved in the whole ocean. An average of 6 milligrams of oxygen is dissolved in each liter of seawater (that is, 6 parts of oxygen per million parts of seawater, by weight). Yet this small amount of oxygen is a vital resource for animals that extract oxygen with gills. The primary source of the ocean's dissolved oxygen is its photosynthetic plants. Since photosynthesis requires sunlight, most of the available oxygen lies near the ocean's surface. Because marine plants are so abundant and photosynthetically active, much more free oxygen diffuses from ocean to atmosphere than moves from atmosphere to ocean.

Carbon Dioxide 6-14

The amount of carbon dioxide (CO_2) in the atmosphere is very small (0.03%) because CO_2 is in great demand by photosynthetic plants as a source of carbon for growth. Carbon dioxide is very soluble in water, though; the proportion of dissolved CO_2 in water is about 15% of all dissolved gases. Because CO_2 combines chemically with water to form a weak acid (H_2CO_3, carbonic acid), water can hold perhaps a thousand times more carbon dioxide than either nitrogen or oxygen at saturation. Carbon dioxide is quickly used by marine plants, so dissolved quantities of CO_2 are almost always much less than this theoretical maximum. Even so, at the present time there is about 60 times as much CO_2 dissolved in the ocean as in the atmosphere. Much more CO_2 moves from atmosphere to ocean than from ocean to atmosphere, in part because some dissolved CO_2 forms carbonate ions, which are locked into sediments, minerals, and the shells and skeletons of living organisms.

Table 6.2 Major Gases in the Atmosphere and Ocean			
Gas	Percent of Gas in Atmosphere, by Volume	Percent of Dissolved Gas in Seawater, by Volume	Concentration in Seawater in Parts per Million, by Weight
Nitrogen (N_2)	78.08%	48%	10–18 ppm
Oxygen (O_2)	20.95%	36%	0–13 ppm
Carbon dioxide (CO_2)[a]	0.035%	15%	64–107 ppm

[a] Also present in seawater as carbonic acid, carbonate ions, and bicarbonate ions.
Sources: Data from Weihaupt, *Exploration of the Oceans*, 1979; Hill, *The Sea: Composition of Seawater*, 1963.

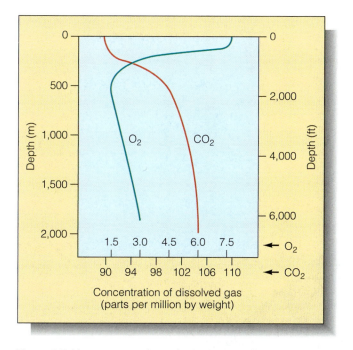

Figure 6.7 How concentrations of oxygen and carbon dioxide vary with depth. Oxygen is abundant near the surface because of the photosynthetic activity of marine plants. Oxygen concentration decreases below the sunlit layer because of the respiration of marine animals and bacteria, and because of the oxygen consumed by the decay of tiny dead organisms slowly sinking through the area. In contrast, because plants use carbon dioxide during photosynthesis, surface levels of CO_2 are low. Because photosynthesis cannot take place in the dark, CO_2 given off by animals and bacteria tends to build up at depths below the sunlit layer. CO_2 also increases with depth because its solubility increases as pressure increases and temperature decreases.

Figure 6.7 illustrates how carbon dioxide and oxygen concentrations vary with depth. Carbon dioxide concentrations increase with increasing depth, but oxygen concentrations usually decrease through the middle depths and then rise again toward the bottom. High concentrations of oxygen at the surface are usually associated with the photosynthesis of plants in the ocean's brightly lit upper layer. Since plants require carbon dioxide for metabolism, surface CO_2 concentrations tend to be low. A decrease in oxygen below the sunlit upper layer is usually due to the respiration of bacteria and marine animals, which leads to higher concentrations of carbon dioxide. Oxygen levels are slightly higher in deeper water because fewer animals are present to take up oxygen reaching these depths and because oxygen-rich polar water that sinks from the surface is the greatest source of deep water.

ACID–BASE BALANCE
6-15

Water can separate to form hydrogen ions (H^+) and hydroxide ions (OH^-). These two ions are present in equal concentrations in pure water. An imbalance in the proportion of ions produces an acidic or basic solution. An **acid** is a substance that *releases* a hydrogen ion in solution; a **base** is a substance that *combines with* a hydrogen ion in solution. Basic solutions are also called **alkaline** solutions.

Acidity or alkalinity, determined by the concentration of hydrogen ions in a solution, is measured in terms of the **pH scale.** An excess of hydrogen ions (H^+) in a solution makes that solution acidic. An excess of hydroxide ions (OH^-) makes a solution alkaline. **Figure 6.8** shows a pH scale and the pH of a few familiar solutions. The scale is logarithmic, which means that a change of one pH unit represents a tenfold change in hydrogen ion concentration. Thus, a modern nonphosphate detergent is a thousand times more alkaline than seawater, and black coffee is a hundred times more acidic than pure water. Pure water, which is neutral (neither acidic nor alkaline) has a pH of 7; lower numbers indicate greater acidity (more H^+ ions) and higher numbers indicate greater alkalinity (fewer H^+ ions).

Seawater is slightly alkaline; its average pH is about 7.8. This seems odd because of the large amount of CO_2 dissolved in the ocean. If dissolved CO_2 combines with water to form carbonic acid, why is the ocean mildly alkaline and not slightly acidic? When dissolved in water, CO_2 is actually present in several different forms. Carbonic acid (H_2CO_3) is only one of these. In water solutions, some carbonic acid breaks down to produce the hydrogen ion (H^+), the bicarbonate ion (HCO_3^-), and the carbonate ion (CO_3^{2-}). This behavior acts to **buffer** seawater, preventing broad swings of pH when acids or bases are introduced.

Though seawater remains slightly alkaline, it is subject to some variation. In areas of rapid plant growth, for example, pH will rise because CO_2 is used by the plants for photosynthesis. Because temperatures are generally warmer at the surface, less CO_2 can dissolve in the first place. Thus, surface pH in warm productive water is usually around 8.5.

At middle depths and in deep water, more CO_2 may be present. Its source is the respiration of animals and bacteria. With cold temperatures, high pressure, and no photosynthetic plants to remove it, this CO_2 will lower the pH of seawater, making it more acidic with depth. Thus, deep, cold seawater below 4,500 meters (15,000 feet) has a pH of around 7.5. This lower pH can dissolve calcium-containing marine sediments — you may recall from Chapter 5 that sediments containing calcium carbonate are rarely found in deep water. A drop to pH 7 can occur at the deep-ocean floor when bottom bacteria consume oxygen and produce hydrogen sulfide.

WATER AND HEAT
6-16

Heat is energy produced by the random vibration of atoms or molecules. Water molecules in hot water vibrate more rapidly than water molecules in cold water. Heat and temperature are not the same thing. Heat tells us *how many* and *how rapidly* molecules are vibrating. Temperature records only *how rapidly* the molecules of a substance are vibrating. **Temperature** is an object's response to an input (or removal) of heat. The

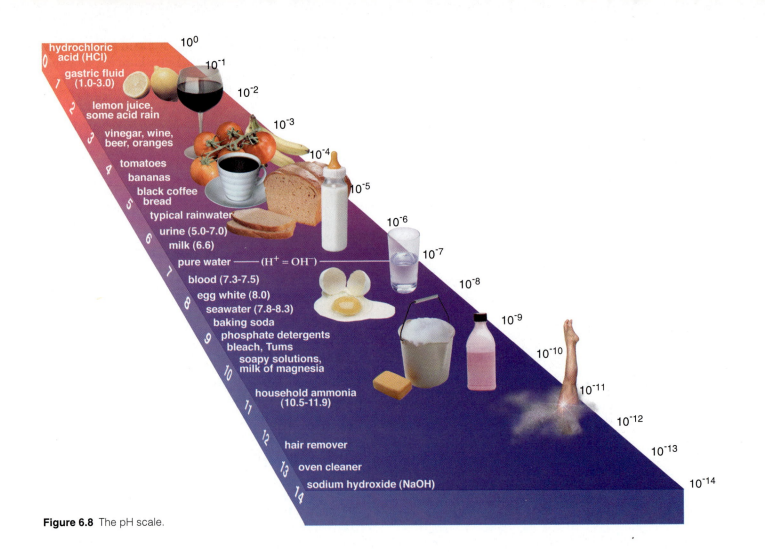

0	hydrochloric acid (HCl)	10^0
1	gastric fluid (1.0-3.0)	10^{-1}
2	lemon juice, some acid rain	10^{-2}
3	vinegar, wine, beer, oranges	10^{-3}
4	tomatoes bananas	10^{-4}
5	black coffee bread	10^{-5}
	typical rainwater	
6	urine (5.0-7.0) milk (6.6)	10^{-6}
7	pure water —— ($H^+ = OH^-$)	10^{-7}
	blood (7.3-7.5)	10^{-8}
8	egg white (8.0) seawater (7.8-8.3) baking soda	
9	phosphate detergents bleach, Tums	10^{-9}
10	soapy solutions, milk of magnesia	10^{-10}
	household ammonia (10.5-11.9)	10^{-11}
11		10^{-12}
12	hair remover	10^{-13}
13	oven cleaner	
14	sodium hydroxide (NaOH)	10^{-14}

Figure 6.8 The pH scale.

amount of heat required to bring a substance to a certain temperature varies with the nature of that substance.[3]

Temperature is measured in **degrees.** One degree Celsius (°C) = 1.8 degrees Fahrenheit (°F). Though we are more familiar with the older Fahrenheit scale (named after the eighteenth-century German inventor of the mercury thermometer), Celsius degrees are more useful in science because they are based on two of pure water's most significant properties: its freezing point (0°C) and its boiling point (100°C). The Celsius scale—until recently called the centigrade scale because of the 100° interval between base points—was invented by Anders Celsius, a Swedish astronomer, in 1742.

Specific heat is a measure of the heat required to raise the temperature of 1 gram of a substance by 1°C (1.8°F). Different substances have different specific heats—*not all substances respond to identical inputs of heat by rising in temperature the same number of degrees.* Specific heat is measured in calories. A **calorie** is the amount of heat required to raise the temperature of 1 gram (0.035 ounces) of pure water by 1°C.[4]

The specific heat of water is among the highest of all known substances, so water can absorb (or release) large amounts of heat while changing relatively little in temperature. Anyone who waits by a stove for water to boil knows much about water's specific heat! By contrast with water, ethyl alcohol has a much lower specific heat. If both liquids absorb heat from identical stove burners at the same rate, pure ethyl alcohol, the active ingredient in alcoholic beverages, will rise in temperature about twice as fast as an equal mass of water. Beach sand has an even lower specific heat—sand requires as little as 0.2 calories to rise 1°C (1.8°F), so beaches can get too hot to stand on with bare feet while the water remains pleasantly cool.

Temperature and Density 6-17

The uniqueness of pure water becomes even more apparent when we consider the effect of a temperature change on water's **density** (its mass per unit of volume). The density of pure

---

[3] This example will help: Which has a higher temperature—a candle flame or a bathtub of hot water? The flame. Which contains more heat? The tub. The molecules in the flame vibrate very rapidly, but there are relatively few of them. The molecules of water in the tub vibrate more slowly, but there are a great many of them, so the total amount of heat energy in the tub is greater.

[4] A nutritional calorie, the unit we see on cereal boxes, equals 1,000 of these calories. A gram is about ten drops of seawater.

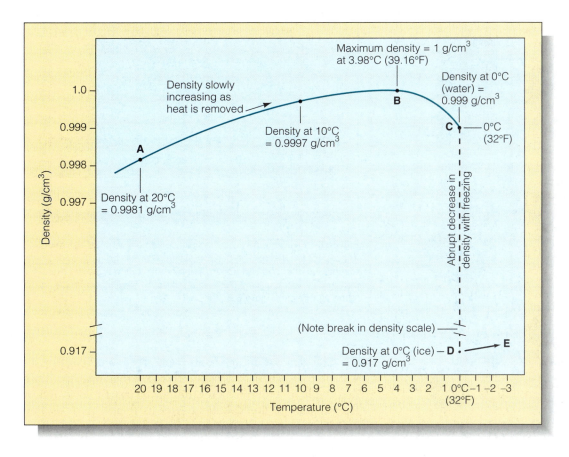

Figure 6.9 The relationship of density to temperature for pure water. Note that points C and D both represent 0°C (32°F) but different densities and thus different states of water. Because the density of ice is less than the density of liquid water, ice floats.

water is 1 gram per cubic centimeter (1 g/cm^3); granite rock is heavier with a density of about 2.7 g/cm^3, and air lighter with a density of about 0.0012 g/cm^3. Most substances become denser (weigh more per unit of volume) as they get colder. Cold air sinks and warm air rises because the rapidly vibrating molecules of warm air are forced farther apart and therefore occupy more space than the same number of cool air molecules. The cold air falls because it is denser. Like air, pure water generally becomes denser as heat is removed and temperature falls, but water's density curve shows a curious quirk as the temperature approaches the freezing point.

A **density curve** shows the relationship between the temperature of a substance and its density. Most substances become progressively denser as they cool—their temperature-density relationships are linear (that is, appear as a straight line on graphs). But **Figure 6.9** shows the unusual temperature–density relationship of pure water. Imagine heat being removed from some water in a freezer. Initially, the water is at room temperature, point A on the graph. As expected, the density of water increases as its temperature drops along the line from point A toward point B. Approaching point B the density increase slows, reaching a maximum at point B of 1 g/cm^3 at 3.98°C (39.2°F). Oddly, water then becomes slightly *less* dense as cooling continues, until point C (0°C, 32°F) is reached. At point C the water begins to freeze—to change state by crystallizing to ice.

State is an expression of the internal form of a substance. Water exists in three states: liquid, gas (water vapor), and solid (ice). If the freezer continues to remove heat from the water at point C, the water will change from liquid to solid state. Through this transition from water to ice—from point C to point D—the density of the water *decreases* abruptly. Ice is therefore lighter than an equal volume of water. Ice increases in density as it gets colder than 0°C, but no matter how cold it gets, ice never reaches the density of liquid water. Being less dense than water, ice "freezes over" as a floating layer instead of "freezing under" like the solid forms of virtually all other liquids.

As we'll see in a moment, the implications of water's high specific heat, and the ability of ice to float, are vital in maintaining Earth's moderate surface temperature. But first let's take a closer look at the transition from point C to point D in Figure 6.9.

Freezing Water

6-18

During the transition from liquid to solid state at the freezing point, the bond angle between the oxygen and hydrogen atoms in water expands from about 105° to slightly more than 109°. This change allows ice to form a crystal lattice (**Figure 6.10**). The space taken by 24 water molecules in the solid lattice could be occupied by 27 water molecules in liquid state,

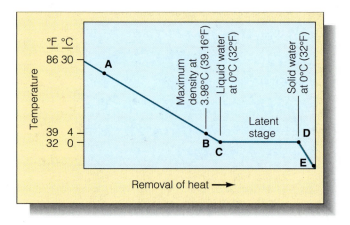

Figure 6.11 A graph of temperature versus heat removal as water freezes (or ice melts). The horizontal line between points C and D represents the latent heat of fusion, when heat is being removed but temperature is not changing. (Note that points A-E on this graph are the same as those in Figure 6.9.)

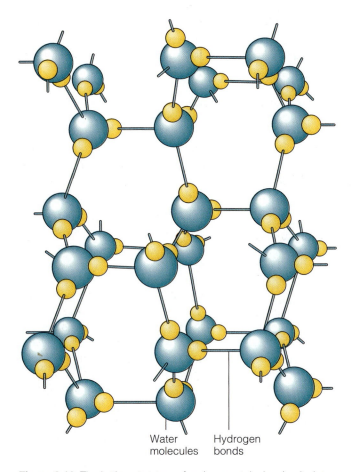

Water molecules Hydrogen bonds

Figure 6.10 The lattice structure of an ice crystal, showing its hexagonal arrangement at the molecular level. The space taken by 24 water molecules in the solid lattice could be occupied by 27 water molecules in liquid state, so water expands about 9% as the crystal forms. Because molecules of liquid water are packed less efficiently, ice is less dense than liquid water and floats.

so water expands about 9% as the crystal forms. Because molecules of liquid water are packed less efficiently, ice is less dense than liquid water and floats. Ice at 0°C (32°F) weighs only 0.917 g/cm³ where liquid water at 0°C weighs 0.999 g/cm³.

The transition from liquid water to ice crystal (point C to point D in Figure 6.9) requires continued removal of heat energy—the change in state does not occur instantly throughout the mass when the cooling water reaches 0°C (32°F). Again, consider water in a freezer. **Figure 6.11,** a plot of heat removal versus temperature, details the water's progress to ice. As in Figure 6.9, point A represents warm water just placed in the freezer. The removal of heat does not stop when the water reaches point C, but the decline in temperature does. Even though heat continues to be removed, the water will not get colder until all of it has changed state from liquid (water) to solid (ice). Heat may therefore be removed from water without the water dropping in temperature when water is changing state—that is, when it is freezing. Indeed, the continued removal of heat is what makes the state change possible. Heat is released as bonds form to make ice, and that heat must be removed to allow more ice to form.

The removal of heat from point A to point C in Figures 6.9 and 6.11 produces a *measurable* lowering of temperature detectable by a thermometer. Removing just 1 calorie of heat from a gram of liquid water causes its temperature to drop 1°C. This detectable decrease in heat is called **sensible heat** loss (because it can be sensed by a thermometer). But the loss of heat as water freezes between points C and D is not measurable by a thermometer. Removing a calorie of heat from freezing water at 0°C (32°F) won't change its temperature at all; 80 calories of heat energy must be removed per gram of pure water at 0°C (32°F) to form ice. This heat is called the **latent heat of fusion** (from *latere* = to be hidden). The straight line between points C and D in Figure 6.11 represents water's latent heat of fusion.

No more ice crystals can form when all the water in the freezer has turned to ice. If the removal of heat continues, the ice will get colder and will soon reach the temperature inside the freezer, point E in Figures 6.9 and 6.11.

Latent heat of fusion is also a factor during thawing. When ice melts, it *absorbs* large quantities of heat (the same 80 calories per gram) and does not change in temperature until the entire mass has turned to liquid. This is why ice is so effective in cooling drinks.

Evaporating Water

6-19

When water evaporates, individual water molecules diffuse into the air. Since each water molecule is hydrogen-bonded to adjacent molecules, heat energy is required to break those bonds and allow the molecule to break away from the surface. Evaporation cools a moist surface because departing molecules of water vapor carry this energy away with them. (This is how perspiring cools us when we're hot. The heat energy required to evaporate water from our skin is taken away from our bodies, cooling us.)

Hydrogen bonds are numerous, and the amount of energy required to break them—known as the **latent heat of evaporation**—is high. At 585 calories per gram at 20°C (68°F), water has the highest latent heat of evaporation of any known substance. As before, the term *latent* applies to heat input that does not cause a temperature change but does produce a change of state—in this case from liquid to gas.

About 1 meter (3.3 feet) of water evaporates each year from the surface of the ocean, a volume of water equivalent to 334,000 cubic kilometers (80,000 cubic miles). The great quantities of solar energy that cause this evaporation are carried from the ocean by the escaping water vapor. When a gram of water vapor condenses back into liquid water, the same 585 calories is again available to do work. As we shall see in the next chapter, winds, storms, ocean currents, and wind waves are all powered by that heat.

Why the big difference between water's latent heat of *fusion* (80 calories per gram) and its latent heat of *evaporation* (585 calories per gram)? Only a small percentage of hydrogen bonds are broken when ice melts, but *all* must be broken during evaporation. Breaking these bonds requires additional energy in proportion to their number.

Seawater and Pure Water 6-20

The solids dissolved in seawater change its thermal characteristics, lowering its specific heat by about 4%. Only 0.96 calorie of heat energy is needed to raise the temperature of seawater by 1°C.

The dissolved solids also interfere with the formation of the ice lattice, acting as "antifreeze" to lower the freezing point. The saltier the water, the lower the freezing point. Seawater at 35‰, typical ocean salinity, freezes at −1.91°C (28.6°F). Seawater's density simply increases smoothly with decreasing temperature until it freezes. The crystals that form are pure water ice, with the seawater salts excluded. The leftover cold, salty water is very dense and sinks toward the ocean floor.

Seawater evaporates more slowly than fresh water under identical circumstances because the dissolved salts tend to attract and hold water molecules. The latent heat of evaporation, however, is essentially the same for both fresh water and seawater. Salts are left behind as seawater evaporates. The remaining cool, salty water is also very dense.

GLOBAL THERMOSTATIC EFFECTS 6-21

The **thermostatic properties** (*therme* = heat + *stasis* = standing still) of water are those properties that act to moderate changes in temperature. Water temperature rises as sunlight is absorbed and changed to heat, but as we've seen, water has a very high specific heat, so its temperature will not rise very much even if a large quantity of heat is added. This tendency of a substance to resist change in temperature with the

gain or loss of heat energy is called **thermal inertia.** To investigate the impact of water's thermostatic properties on conditions at Earth's surface we need to look at the planet's overall heat balance.

Only about 1 part in 2.2 billion of the sun's radiant energy is intercepted by Earth, but that amount averages 7 million calories per square meter per day at the top of the atmosphere, or, for Earth as a whole, an impressive 17 trillion kilowatts (23 trillion horsepower)! About half of this light reaches the surface. As can be seen in **Figure 6.12,** on a global basis about 26% of the incoming light is reflected back into space from clouds or scattered by ice, water droplets, or other particles in the atmosphere. Another 19% is absorbed by water vapor, dust, and CO_2 in the atmosphere. About 4% bounces off the shining sea surface, ice and snow, or rocks and soil. Only 51% is absorbed by Earth's land and water surface. How much light penetrates the ocean depends greatly on the angle at which it approaches, sea state (surface turbulence), the presence of an ice covering or light-colored foam, and other factors.

The 51% of solar energy striking land and sea is converted to heat, then transferred into the atmosphere by conduction, radiation, and evaporation. The atmosphere, like the land and ocean, eventually radiates this heat back into space in the form of long-wave (infrared) radiation. The heat input and outflow "account" for Earth can be thought of as a **heat budget.** As in your personal financial budget, income must eventually equal outgo. Over long periods of time the total *incoming* heat (plus that from earthly sources) equals the total *outgoing* heat, so Earth is in **thermal equilibrium**; it is growing neither significantly warmer nor colder.[5] Heat input comes mainly from the sun; heat outflow can only occur as heat radiates into the cold of space.

Liquid water's thermal characteristics prevent broad swings of temperature during day and night, and, through a longer span, during winter and summer. Heat is stored in the ocean during the day and released at night. A much greater amount of heat is stored through the summer and given off during the winter. If our ocean were made of alcohol—or almost any other liquid—summer temperatures would be much hotter, and winters bitterly cold. Sea ice in the polar regions also contributes to thermal inertia. Because water expands and floats when it freezes, ice can absorb the morning warmth of the sun, melt, then refreeze at night, giving back to the atmosphere the heat it stored through the daylight hours. The *heat content* of the water changes throughout the day; its *temperature* does not. The same principle applies to the seasonal formation and melting of polar ice. More than 18,000 cubic kilometers (4,300 cubic miles) of polar ice thaws and refreezes each year. Seasonal extremes are moderated by the immense amounts of heat energy that are alternately absorbed and released without a change in temperature. Without these

---

[5] Changes in heat balance do occur over short periods of geological time. Increasing amounts of carbon dioxide and methane in Earth's atmosphere may be contributing to an increase in surface temperature called the *greenhouse effect.* More on this subject can be found in Chapter 15.

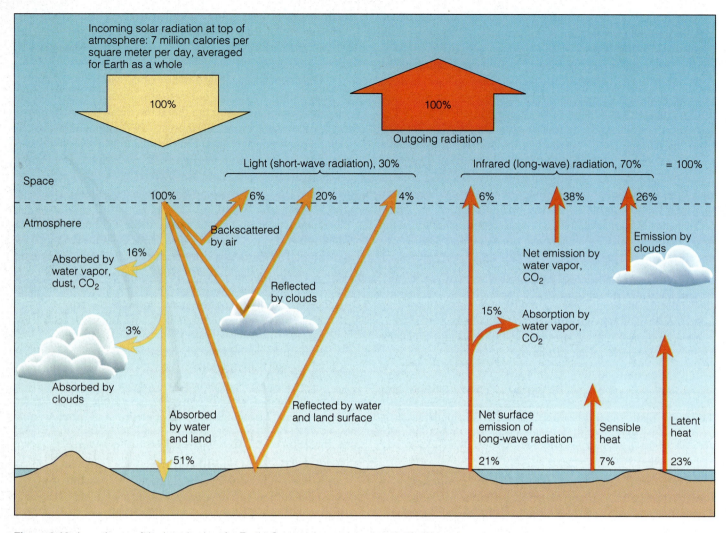

Figure 6.12 An estimate of the heat budget for Earth. On an average day, about half of the solar energy arriving at the upper atmosphere is absorbed at Earth's surface. Light (short-wave) energy absorbed at the surface is converted into heat. Heat leaves Earth as infrared (long-wave) radiation. Since input equals output over long periods of time, the heat budget is balanced.

properties of ice, temperatures on Earth's surface would change dramatically with minor changes in atmospheric transparency or solar output.

THE DENSITY STRUCTURE OF THE OCEAN

6-22

The density of water is mainly a function of its salinity and temperature. Cold, salty water is denser than warm, less salty water. The density of seawater varies between 1.020 and 1.030 g/cm³, indicating that a liter of seawater weighs between 2% and 3% more than a liter of pure water (1.000 g/cm³) at the same temperature. Seawater's density increases with increasing salinity, increasing pressure, and decreasing temperature. **Figure 6.13** shows the relationship between temperature, salinity, and density. Notice that two samples of water can have the *same* density at different combinations of temperature and salinity.

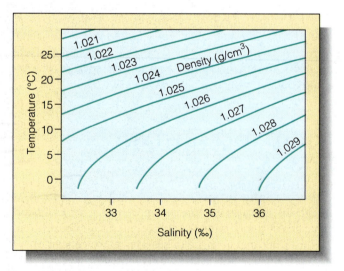

Figure 6.13 The complex relationship between the temperature, salinity, and density of seawater. Note that two samples of water can have the *same* density at *different* combinations of temperature and salinity.

Ocean water tends to form into stable layers with the heaviest water at the bottom, a form of density stratification. Curiously, even the deepest of these layers originates at the ocean's surface. Very cold and salty water produced during the formation of sea ice at the polar ocean surface is denser than the surrounding water and sinks until it reaches a layer of water of equal density, or the seabed. In a few marginal basins, the most notable being the Mediterranean Sea, evaporation can also produce salty, dense water that sinks toward the bottom. In contrast, warm fresh water entering the ocean at a river mouth is much less dense and floats for miles above the cooler salty layers below. Early explorers of South America were amazed to find a surface layer of fresh water far out at sea and discovered the Amazon River by following this fresh surface layer to its source.

Much of the ocean is divided into three density zones. The **surface zone,** or **mixed layer,** is the upper layer of ocean, in which temperature and salinity are relatively constant with depth because of the action of waves and currents. The surface zone consists of water in contact with the atmosphere and exposed to sunlight; it contains the ocean's least dense water and accounts for about 2% of total ocean volume. Depending on local conditions, the surface zone may reach a depth of 1,000 meters or be absent entirely. Beneath it is the **pycnocline** (*pyknos* = strong + *clinare* = slope, to lean), a zone in which density increases with increasing depth. This zone isolates surface water from the denser layer below. The pycnocline contains about 18% of all ocean water. The **deep zone** lies below the pycnocline at depths below about 1,000 meters (3,300 feet) in mid-latitudes (40°S to 40°N). There is little additional change in water density with increasing depth through this zone. This deep zone contains about 80% of all ocean water.

Temperature is the most important factor affecting density at mid-latitudes, because the change in temperature with depth is much more pronounced than the change in salinity with depth. **Figure 6.14** shows the general relationship of temperature to depth in the open sea. The surface zone is well mixed, with little decrease in temperature with depth; in the next layer, temperature drops rapidly with depth; beneath it lies the deep zone of cold, stable water. The middle layer, the zone in which temperature changes rapidly with depth, is called the **thermocline** (*therme* = heat + *clinare* = slope, to lean). This thermocline is the major cause of the pycnocline mentioned above.

Thermoclines are not identical in shape for all areas or latitudes. Temperate and tropical ocean areas cradle a warm surface layer whose bottom boundary is the top of the thermocline. Polar waters, which receive relatively little solar warmth, are not stratified by temperature and generally lack a thermocline. **Figure 6.15** contrasts polar, tropical, and temperate thermal profiles, showing that the thermocline is primarily a mid- and low-latitude phenomenon. Thermocline depth and intensity also vary with season, local conditions (storms and such), currents, and many other factors. Divers often notice minor thermoclines near the surface, but most temperate and tropical oceans contain a deep main thermocline also.

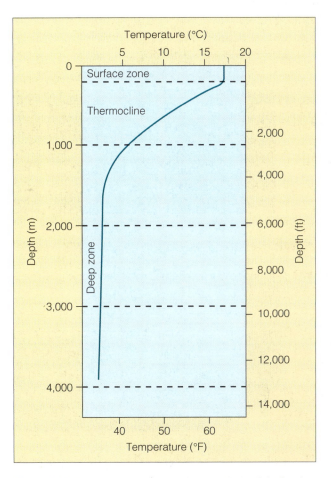

Figure 6.14 Typical temperature variation with depth in the deep ocean at mid-latitudes.

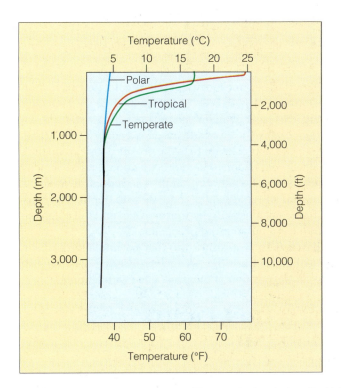

Figure 6.15 Typical temperature profiles at polar, tropical, and middle (temperate) latitudes. Note that polar waters lack a thermocline.

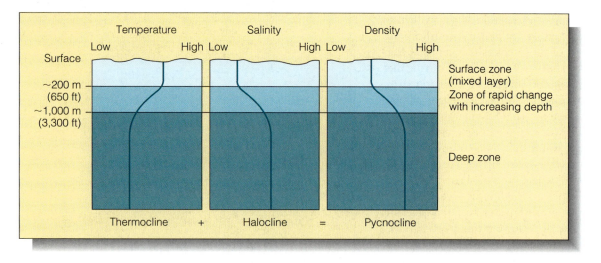

Figure 6.16 Formation of the pycnocline, which depends on the characteristics of the thermocline and the halocline.

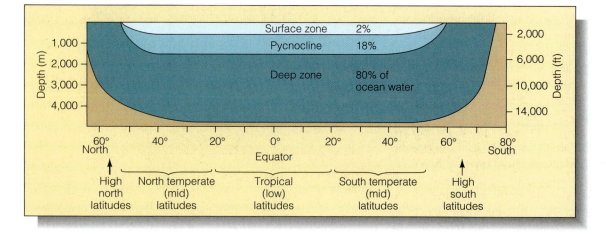

Figure 6.17 Variation in the average depth of vertical density zones with latitude.

Below the thermocline, water is very cold, ranging in temperature from −1°C to 3°C (30.5°F to 37.5°F). Because this deep and cold layer contains the bulk of ocean water, the average temperature of the world ocean is a chilly 3.9°C (39°F).

Dissolved salts contribute to the ocean's density structure, especially in cool regions where precipitation is rapid or along coasts exposed to fresh water runoff. The **halocline** (*halos* = salt) is a zone of rapid salinity increase with depth. The halocline often coincides with the thermocline, and the combination produces a pronounced pycnocline.

Figure 6.16 shows how the thermocline and the halocline combine to form the pycnocline. **Figure 6.17** records changes in the position of the pycnocline with latitude.

REFRACTION, LIGHT, AND SOUND

 6-23

The ring of light sometimes seen around the moon and the safe concealment of a submarine may not seem related, but both events depend on **refraction,** the bending of waves. Light

and sound are wave phenomena. When a light wave or a sound wave leaves a medium of one density—such as air—and enters a medium of a different density—such as water—at an angle other than 90°, it is bent from its original path. The reason for this bending is that light and sound waves travel at different speeds in different media.

The situation is analogous to a line of marchers walking along a desert highway with arms intertwined. The marchers can walk faster if they stay on the street than if they walk in the sand next to the street. Their speed on the pavement, then, is greater than their speed in the sand. As long as they stay on the street, they won't change direction. But if their marching angle gradually takes them off the edge (into a medium in which their speed is *lower*), the people who reach the sand first will suddenly slow down, and the line will pivot quickly off the street. They have been refracted. Their progress is depicted in **Figure 6.18a.** Note that the transition from one medium to another must occur at an angle other than 90° for refraction to occur—our marchers will not change direction if they march straight off the asphalt into the sand. (They will still slow down, however—see **Figure 6.18b.**)

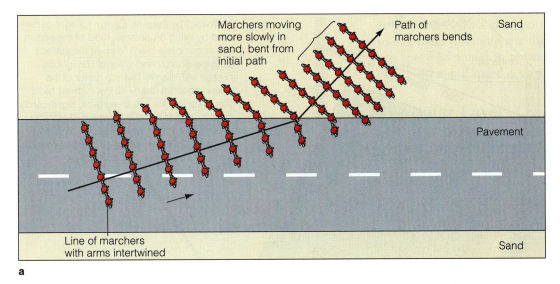

a

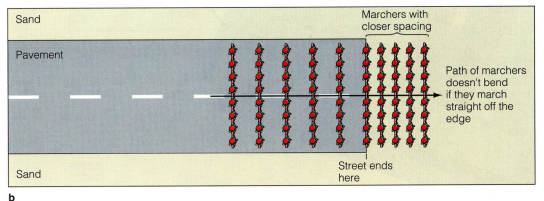

b

Figure 6.18 An analogy for refraction. The ranks of marchers represent light or sound waves; the pavement and sand represent different media. (**a**) If the marchers head off the pavement at an angle other than 90°, their path will bend (refract) as they hit the sand because some will be walking more slowly than others. (**b**) If they march straight off the pavement, the ranks will slow down but not bend as they hit the sand.

Examples of the refraction of light by water are all around you. A pencil sticking out of a glass of water looks bent because of refraction, the submerged steps of a swimming pool ladder look closer than they are because of refraction, and divers exaggerate the size of the fish that got away because refraction can magnify objects.

Light in the Ocean

6-24

As we've seen, sunlight has a difficult time reaching and penetrating the ocean—clouds and the sea surface reflect light, atmospheric gases and particles scatter and absorb it. Once past the sea surface, light is rapidly weakened by scattering and absorption. **Scattering** occurs as light is bounced between air or water molecules, dust particles, water droplets, or other objects before being absorbed. The greater density of water (along with the greater number of suspended and dissolved particles) makes scattering more prevalent in water than in air. The **absorption** of light is governed by the structure of the water molecules it happens to strike. When light is absorbed, molecules vibrate and the light's energy is converted to heat.

Even perfectly clear seawater is not perfectly transparent. If it were, the sun's rays would illuminate the greatest depths of the ocean and seaweed forests would fill its warmed basins.

The thin film of lighted water at the top of the surface zone is called the **photic zone** (*photos* = light). In clear tropical waters, the photic zone may extend to a depth of 200 meters (660 feet), but a more typical value for the open ocean is 100 meters (330 feet). All the production of food by photosynthetic marine plants occurs in this thin, warm surface zone. Here water is heated by the sun, heat is transferred from the ocean into the atmosphere and space, and gases are exchanged with the atmosphere. The thermostatic effects we've discussed function largely within this zone. Most of the ocean's life is found here. The photic zone may be extraordinarily thin, but it is also extraordinarily important.

The ocean below the photic zone lies in blackness. Except for light generated by living organisms, the region is perpetually dark. This dark water beneath the photic zone is called the **aphotic zone** (*a* = without + *photos* = light).

The light energy of some colors is converted into heat nearer to the surface than the light energy of other colors. **Figure 6.19** shows this differential absorption by color. Notice that after 1 meter (3.3 feet) of travel, only 45% of the light energy remains, most of it in the green and blue wavelengths. After 10 meters (33 feet), 85% of the light has been absorbed, and after 100 meters (330 feet), just 1% remains. The dimming light becomes bluer with depth because the red, yellow, and orange wavelengths have already been absorbed. Even in the

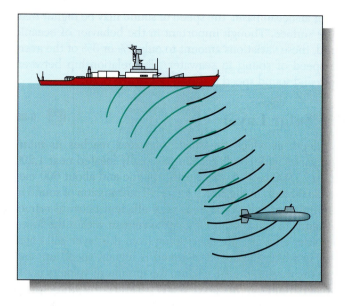

Figure 6.22 The principle of active sonar. Pulses of high-frequency sound are radiated from the sonar array of the sending vessel. Some of the energy of this ping reflects from the submerged submarine and returns to the sending vessel. The echo is analyzed to plot the position of the submarine.

As a result, loud noises made at this depth can be heard for thousands of kilometers. Navy depth charges detonated in the minimum-velocity layer in the Pacific have been heard 3,680 kilometers (2,280 miles) from the explosion. In a recent test, sound generated by a U.S. Navy ship in the Indian Ocean was heard at the Oregon coast!

In the early 1960s the U.S. Navy experimented with the use of sound transmission in the minimum-velocity layer as a lifesaving tool. Survivors in a life raft would drop a small charge into the water that was set to explode at the proper depth. A number of widely spaced listening stations ashore would compare the differences in the arrival times of the signal, then compute the position of the raft. The project—since abandoned in favor of radio beacons—was called **sofar** (for *sound fixing and ranging*). The minimum-velocity layer has come to be known as the **sofar layer.**

Sonar

6-27

Crews aboard surface ships and submarines employ active **sonar** (*sound navigation and ranging*), the projection and return through water of short pulses ("pings") of high-frequency sound to search for objects in the ocean (**Figure 6.22**). In a

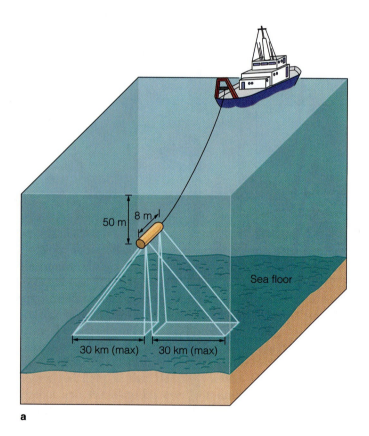

a

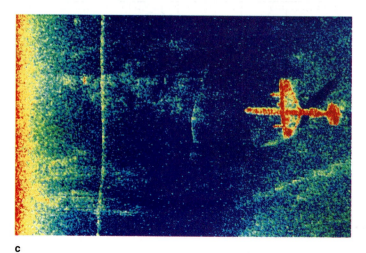

b

c

Figure 6.23 (**a**) Side-scan sonar in action. Sound pulses leave the submerged towed array (**b**), bounce off the bottom, and return to the device. One of the two slits through which sound pulses depart and return is visible on its side. Computers process the impulses into images: (**c**) A side-scan sonar record showing a downed World War II era PB4Y-2 Privateer at a depth of 50 meters (165 feet). The artificial colors enhance target detail and clarity.

modern system, electrical current is passed through crystals, which respond by producing powerful sound pulses pitched above the limit of human hearing. Some of the sound from the transmitter bounces off any object larger than the wavelength of sound employed and returns to a microphone-like sensor. Signal processors then amplify the echo and reduce the frequency of the sound to within the range of human hearing. An experienced sonar operator can tell the direction of the contact, its size and heading, and even something about its composition (whale or submarine or school of fish) by analyzing the characteristics of the returned ping.

Side-scan sonar is a type of active sonar. Operating with as many as 60 transmitter/receivers tuned to high sound frequencies, side-scan systems towed in the quiet water beneath a ship are sometimes capable of near-photographic resolution. Side-scan systems are used for geological investigations, archaeological studies, and locating downed ships and airplanes (**Figure 6.23**). The *multibeam system* you read about in Chapter 3 (see again Figures 4.3, 4.11, and 4.12), and the *echo sounder* (Figure 1.20), are also forms of active sonar.

The first human use of sea sound was passive: Mariners listened through their hulls to the whistles and clicks of whales and other animals. When submarine warfare emerged during World War I, the British invented a simple underwater listening device to detect noises made by enemy submarines. Operators would listen through a sensitive directional microphone for the telltale sounds of a propeller, a torpedo, or even a dropped wrench or slammed hatch cover. The first systems were primitive, but *passive sonar,* as these listening-only devices are now called, is currently undergoing a renaissance. Modern passive devices are much more sophisticated and sensitive than their World War I predecessors. Passive sonar confers the military benefit of surprise—unlike active sonar, in which a listener can hear the loud ping long before an operator aboard the sending vessel can hear his faint echo. Usually, it's safer just to listen. The combination of computerized signal processing, microphones towed at a distance from the listener's noisy ship, and better knowledge of ocean physics will improve the usefulness of both kinds of sonar.

Humans are not the only organisms to use sonar. Whales and other marine mammals use clicks and whistles to find food and avoid obstacles. We'll discuss this form of active sonar in Chapter 13.

QUESTIONS FROM STUDENTS

1. Why is water blue?

The absorption of red light by hydrogen bonds is what gives pure water—and thick ice—its pale bluish hue. Some of the blue light that remains after the red is absorbed is scattered back through the surface, and that's what we see.

2. You have noted that there is a hundred times more oxygen in the atmosphere than in the ocean. How can that be? Isn't water 86% oxygen by weight?

Yes, but the oxygen of water (H_2O) is bonded tightly to hydrogen atoms. Unlike atmospheric oxygen, the oxygen in water molecules cannot be used by organisms for respiration. Fish can't extract this oxygen with their gills. Marine animals must depend on dissolved oxygen, paired molecules of oxygen (O_2) present in the water and free to move through the gill membranes. Compared to the atmosphere, very little of this free oxygen is present in the ocean.

3. Some essential resources are relatively abundant in ocean water. Are there any prospects for chemical "mining" of seawater?

Yes. Commercial extraction of magnesium from seawater began in 1940. Seawater is treated with calcium hydroxide and hydrochloric acid to form magnesium chloride. This substance is dried and electrolytically separated into chlorine gas (which can be used to make more hydrochloric acid) and magnesium metal. Magnesium is an essential component of the aluminum alloys from which aircraft, beverage cans, and automobile parts are manufactured.

Many nonmetal resources are also obtained from seawater. For example, bromine—an element important in the production of motor fuels, photographic film, dyes, and insecticides—is extracted from concentrated brines or from crystallized sea salts. And, as we'll see in Chapter 15, the salts themselves are among the most important of the ocean's physical resources.

4. The longest day of the year in the Northern Hemisphere is around 22 June and the shortest around 22 December. That must mean the greatest amount of heat is reaching the north in June and the least in December. Why, then, do our warmest days occur in August or September and our coldest days in January or February?

Because of thermal inertia. There is a lag between maximum sunlight and maximum warmth because of water's great specific heat. The sun must shine on this watery planet for *many* weeks to raise the summer hemisphere's temperature. Of course, water also retains heat well; so the coldest days in the winter hemisphere come well after the darkest ones.

5. I can't tell where sound is coming from when I'm underwater. It seems to be coming from inside my head. Why is that?

Our normal sensation of stereo hearing depends in part on the difference in the arrival times of sound from one ear to the other. The speed of sound in water, however, is more than four times greater than its speed in air. Our brains are unable to sense arrival-time differences from sounds originating underwater. You'd need to be more than four times as sensitive—or have a head more than four times as large—to hear stereophonically underwater.

6. Why don't divers see shadows even when the sun is shining directly on the ocean at their location?

The greater density of water (along with the greater number of suspended and dissolved particles) makes scattering more prevalent in water than in air. A few meters below the surface, light in the ocean seems to come from all directions (rather than primarily from one direction), so shadows are rare.

7. If water has the highest latent heat of evaporation, why does a drop of alcohol make my hand feel colder when it evaporates than a drop of water does?

Heat *versus* temperature again. The alcohol makes your hand *colder,* but it evaporates much *more quickly.* The water stays around longer (takes longer to evaporate), and although it may not cause your skin to become as *cold,* it will remove more *heat.*

CHAPTER SUMMARY

Water, a chemical compound composed of two hydrogen atoms and one oxygen atom, is abundant on and within Earth. The polar nature of the water molecule produces some unexpected chemical properties, one of the most important of which is water's remarkable ability to dissolve more substances than any other natural solvent. Though most solids and gases are soluble in water, the ocean is in chemical equilibrium and neither the proportion nor amount of most dissolved substances changes significantly through time. Most of the properties of seawater are different from those of pure water because of the substances dissolved in the seawater.

The physical characteristics of the world ocean are largely determined by the physical properties of seawater. These properties include water's specific heat, density, salinity, and its ability to transmit light and sound.

The thermal properties of water are responsible for the mild physical conditions at Earth's surface. Liquid water is remarkably resistant to temperature change with the addition or removal of heat; and ice, with its large latent heat of fusion and low density, melts and re-freezes over large areas of the ocean to absorb or release heat with no change in temperature. These thermostatic effects, combined with the mass movement of water and water vapor, prevent large swings in Earth's surface temperature.

Changes in temperature and salinity greatly influence water density. Ocean water is usually layered by density, with the densest water on or near the bottom.

Sound and light in the sea are affected by the physical properties of water, with refraction and absorption effects playing important roles.

TERMS AND CONCEPTS TO REMEMBER

absorption	alkaline
acid	aphotic zone
adhesion	base
buffer	photic zone
calorie	polar molecule
chemical bond	principle of constant
chemical equilibrium	proportions
chlorinity	proton
cohesion	pycnocline
deep zone	refraction
degrees	residence time
density	salinity
density curve	salinometer
electron	scattering
excess volatiles	sensible heat
Forchhammer's principle	sofar
halocline	sofar layer
heat	sonar
heat budget	sound
ion	specific heat
hydrogen bond	state
latent heat of evaporation	surface zone
latent heat of fusion	temperature
mixed layer	thermal equilibrium
mixing time	thermal inertia
molecule	thermocline
Nansen bottle	thermostatic properties
Niskin bottle	trace elements
pH scale	

STUDY QUESTIONS

1. Why is water a polar molecule? What properties of water derive from its polar nature?

2. Other than hydrogen and oxygen, what are the most abundant elements in seawater?

3. How is salinity determined? How are modern methods dependent on the principle of constant proportions?

4. Which dissolved gas is represented in the ocean in much greater proportion than in the atmosphere? Why the disparity?

5. What factors affect seawater's pH? How does the pH of seawater change with depth? Why?

6. How is heat different from temperature?

7. How does water's high specific heat influence the ocean? Leaving aside its effect on beach parties, how do you think conditions on Earth would differ if our ocean consisted of ethyl alcohol?

8. Why does ice float? Why is this fact important to thermal conditions on Earth?

9. What factors affect the density of water? Why does cold air or water tend to sink? What is the role of salinity in water density?

10. How is the ocean stratified by density? What physical factors are involved? What names are given to the ocean's density zones?

11. What percent of incoming sunlight reaches the ocean? What happens to that light? Is the heat budget balanced? What would happen if it were not?

12. What factors influence the intensity and color of light in the sea? What factors affect the depth of the photic zone? Could there be a "photocline" in the ocean?

FOR FURTHER STUDY

Bowditch, N. 1966. *American Practical Navigator.* Washington, DC: U.S. Navy Hydrographic Office Publication 9. Summaries of the properties of light and sound in the ocean.

Charnock, H. 1971. "Physics of the Ocean." In *Deep Oceans,* ed. P. Herring and M. Clarke, 82–120. London: Praeger. Excellent reference in ocean physics for the interested layperson.

Jerlov, N. G., and E. S. Nielsen, eds. 1974. *Optical Aspects of Oceanography.* New York: Academic Press. A standard reference.

Kasting, J. F. 1998. "The Origins of Water on Earth." *Scientific American,* Fall special issue on "The Oceans," 16–22. Kasting suggests that the impact of comets early in Earth's history contributed most of Earth's surface water.

Kennish, M. J., ed. 1994. *Practical Handbook of Marine Science.* Boca Raton, FL: CRC Press. Invaluable compendium of oceanographic data. Source of trace element data, salinity proportions—almost everything, in fact.

Kerr, R. A. 1988. "Ocean Crust's Role in Making Seawater." *Science* 239 (no. 4837, 15 January): 260. The possible role of water circulation through active rift zones in reducing Mg^{2+} and increasing Ca^{2+}.

Lagerloef, G. S. E., C. T. Swift, and D. M. LeVine. 1995. "Sea Surface Salinity: The Next Remote Sensing Challenge." *Oceanography* 8 (no. 2): 44–50. Even salinity can be measured from space!

Libes, S. M. 1992. *An Introduction to Marine Biogeochemistry.* New York: Wiley. A balanced survey of the inorganic and organic aspects of marine chemistry.

Lynch, D. K. 1995. *Color and Light in Nature.* Cambridge, England: Cambridge University Press. A beautiful and intriguing book cataloguing interesting atmospheric and oceanic optical phenomena. Spectacular photographs.

Mellor, G. L. 1996. *Introduction to Physical Oceanography.* Woodbury, NY: American Institute of Physics. Technical, complete, recent.

Munk, W. 1991. "The Heard Island Experiment." *Oceanus* 34 (no. 1). The speed of sound in the ocean may be used to sense global ocean temperature, and thus the extent of global warming.

Oceanus 20, no. 2. (Spring 1977) is dedicated to sound in the sea.

Oceanus 35, no. 1. (Spring 1992) is dedicated to marine chemistry.

Schlee, S. 1973. *The Edge of an Unfamiliar World: A History of Oceanography.* New York: Dutton. Information on the invention of modern sonar.

Science 281, no. 5374 (10 July 1998) is dedicated to the chemistry and biology of the ocean.

Sverdrup, H. U. 1954. "Oceanography." In *The Earth as a Planet,* ed. C. P. Kuiper, 215–57. Chicago: University of Chicago Press. The thermostatic role of water on Earth; the importance of ocean and atmosphere in transporting heat.

Walton-Smith, F. G., ed. 1974. *Handbook of Marine Science,* vol. 1. Cleveland: CRC Press.

For additional readings, go to InfoTrac College Edition, your online research library at:

http://infotrac.thomsonlearning.com/

Atmospheric Circulation 7

RIDING THE WINDS

In March 1999 Brian Jones and Bertrand Piccard rode the winds around the world. Housed in a cylindrical capsule beneath a balloon 55 meters (180 feet) high, the adventurers were blown eastward in fast-moving ribbons of air—jet streams—that form between warm and cold air masses (**Figure a**). Their flight illustrates the general west-to-east flow of the winds at mid-latitude—a flow caused by Earth's uneven solar heating and daily rotation. This circulation of air is important in creating climate and weather, driving ocean currents, and steering storms.

Piccard, a Swiss psychiatrist, is the son of Jacques Piccard, inventor of the bathyscaphe and builder of *Trieste* (see Chapter 4 opener). Jones is a former British Royal Air Force pilot. Their journey in *Breitling Orbiter 3* began in a quiet valley between snow-capped peaks in Château-d'Oex, Switzerland, where compartments in the balloon were filled with helium and hot air (**Figure b**). It ended near the Egyptian oasis town of Mût 19 days and 21 hours later. They traveled about 45,800 kilometers (28,500 miles), setting records for distance and duration of flight.

The flight was not without its difficulties. Balloons necessarily go wherever the wind is blowing, but by adjusting altitude, the pilots could move between layers of air going in different directions. Sometimes they found themselves in winds headed the wrong way, or becalmed and going nowhere. Cruising at an average height of 6,400 meters (21,000 feet) meant they needed supplemental oxygen. Failure of a cabin heating system subjected them to 1.7C (33°F) temperatures for the last half of the trip.

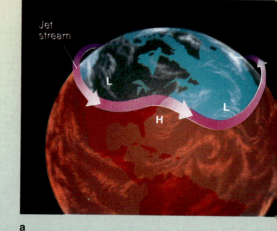

a

In March 1999, Bertrand Piccard and Brian Jones became the first people to fly around the world in a balloon. The balloonists took advantage of jet streams (**a**), swiftly flowing currents of air that move in a west-to-east direction. (**b**) *Breitling Orbiter 3* lifting off from a valley in Switzerland at the beginning of the flight, which lasted 19 days, 21 hours.

b

And they ventured outside the balloon once so Piccard could chip away at heavy ice that had formed on the cables and capsule. While waiting the six hours for Egyptian helicopter crews to find and retrieve them after landing, explorers Piccard and Jones must have amused themselves by reflecting on the world-spanning power of the winds, and, of course, contemplating the $1 million prize offered by the Anheuser-Busch corporation for the first nonstop circumnavigation of the globe by balloon. 🖳 7-1

. . . .

KEY CONCEPTS

1. The interaction of ocean and atmosphere moderates surface temperatures, shapes Earth's weather and climate, and creates most of the sea's waves and currents.

2. The atmosphere responds to uneven solar heating by flowing in three great circulating cells over each hemisphere. The flow of air within these cells is influenced by the rotation of Earth.

3. To observers on the surface, Earth's rotation causes moving air (or any moving mass) in the Northern Hemisphere to curve to the right of its initial path, and in the Southern Hemisphere to the left. The apparent curvature of path is known as the Coriolis effect.

4. Large storms are spinning areas of unstable air occurring between or within air masses. Extratropical cyclones originate at the boundary between air masses. Tropical cyclones, the most powerful of Earth's atmospheric storms, occur within a single humid air mass.

CHAPTER AT A GLANCE

Composition and Properties of the Atmosphere

Atmospheric Circulation
Uneven Solar Heating and Latitude
Uneven Solar Heating and the Seasons
Uneven Solar Heating and Atmospheric Circulation
The Coriolis Effect
The Coriolis Effect and Atmospheric Circulation Cells

Wind Patterns
Monsoons
Sea Breezes and Land Breezes
El Niño and La Niña

Storms
Air Masses
Extratropical Cyclones
Tropical Cyclones
Box 7.1: Monstrous Mitch

. . . .

Earth's atmosphere and its ocean are intimately intertwined, their gases and waters freely exchanged. Gases entering the atmosphere from the ocean have important effects on climate; and gases entering the ocean from the atmosphere can influence sediment deposition, the distribution of life, and some of the physical characteristics of the seawater itself. Water evaporated from the ocean surface and moved by the winds helps minimize worldwide extremes of surface temperature and, through rain, provides moisture for agriculture. The weather that so profoundly affects our daily lives is shaped at the junction of wind and water, and winds greatly influence the movement of seawater. In this chapter we will investigate the movement of air; in the next chapter, the movement of water.

COMPOSITION AND PROPERTIES OF THE ATMOSPHERE 🖳 7-2

The lower atmosphere is a nearly homogenous mixture of gases, most plentifully nitrogen (78.1%) and oxygen (20.9%). Air is never completely dry; **water vapor,** the gaseous form of water, can occupy as much as 4% of its volume. Sometimes liquid droplets of water are visible as clouds or fog; but more often the water is simply there—invisible in vapor form, having entered the atmosphere from the ground, plants, and the sea surface (**Figure 7.1**). The residence time of water vapor in the lower atmosphere is about ten days. Water leaves the atmosphere by condensing into dew, rain, or snow.

Air has mass. A 1-square-centimeter (0.16-square-inch) column of air, extending from sea level to the top of the atmosphere, weighs about 1.04 kilograms (2.3 pounds). A 1-square-foot column of air the same height weighs more than a ton.

Near Earth's surface, air is packed densely by its own weight. Air lifted from near sea level to a higher altitude is subjected to less pressure and will expand. As anyone knows who has felt the cool air rushing from an open tire valve, air becomes *cooler* when it *expands.* The opposite effect is also familiar: Air *compressed* in a tire pump becomes *warmer.* Air descending from high altitude toward sea level warms as it is compressed by the higher atmospheric pressure near Earth's surface.

The temperature and water content of air greatly influence its density. Because the molecular movement associated with heat causes a mass of warm air to occupy more space than an equal mass of cold air, warm air is less dense than cold air. But contrary to what we might guess, humid air is *less* dense than dry air at the same temperature—because molecules of water vapor weigh less than the nitrogen and oxygen molecules that the water vapor displaces.

Warm air can hold more water vapor than cold air. Water vapor in rising, expanding, cooling air will often condense into clouds (aggregates of tiny droplets) because the cooler air can no longer hold as much water vapor. If rising and cooling continues, the droplets may coalesce into raindrops or snowflakes. The atmosphere will then lose water as **precipitation,** liquid

Figure 7.1 Steam fog over the ocean indicates rapid evaporation. Water vapor is invisible, but as water vapor rises into cool air it can condense into visible droplets.

water or ice that falls from the air to Earth's surface. These rising-expanding-cooling and falling-compressing-heating relationships are important in understanding atmospheric circulation, weather, and climate.

ATMOSPHERIC CIRCULATION 7-3

About half of the energy radiated toward Earth from the sun is absorbed by Earth, but this energy is not distributed evenly across the planet's surface. The amount of solar energy reaching Earth's surface per minute varies with the angle of the sun above the horizon, the transparency of the atmosphere, and the local reflectivity of the surface. The most important factors that affect the angle of the sun above the horizon are latitude and season.

Uneven Solar Heating and Latitude 7-4

As can be seen in **Figure 7.2a**, sunlight striking the polar regions at a low angle spreads over a greater area, filters through more atmosphere, and is reflected to a large degree. The more nearly vertical angle at which the sunlight approaches in the tropics means the same amount of sunlight is distributed over a much smaller area, passes through less atmosphere, and is reflected only minimally. As you would expect, the tropics are warmer than the polar regions.

Uneven Solar Heating and the Seasons 7-5

At mid-latitudes the Northern Hemisphere receives about three times as much solar energy per day in June as it does in December. This difference is due to the 23½° tilt of Earth's rotational axis relative to the plane of its orbit around the sun (**Figure 7.2b**). As Earth revolves around the sun, the constant tilt of its rotational axis causes the Northern Hemisphere to lean *toward* the sun in June but *away* from it in December. The sun therefore appears higher in the sky in summer but lower in winter. The inclination of Earth's axis also causes days to become longer as summer approaches but shorter with the coming of winter. Longer days mean more time for the sun to warm Earth's surface.

Uneven Solar Heating and Atmospheric Circulation 7-6

As can be seen in **Figure 7.3**, near the equator the amount of solar energy received by Earth greatly exceeds the amount of heat radiated into space. In the polar regions the opposite is true. If nothing intervened, the tropical ocean would boil and the polar regions freeze solid. Global air and water circulation moves heat and thus prevents this unpleasant outcome.

Think of air circulation in a room with a hot radiator opposite a cold window (**Figure 7.4**). Air warms, expands, becomes less dense, and *rises* over the radiator. Air cools, contracts, becomes more dense, and *falls* near the cold glass

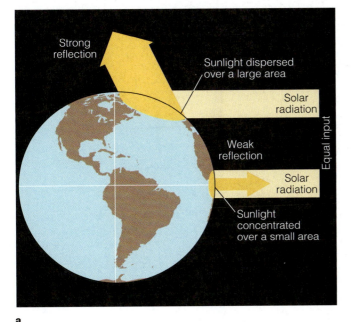

a

Figure 7.2 (a) How solar energy input varies with latitude. Equal amounts of sunlight are spread over a greater surface area near the poles than in the tropics. Ice near the poles reflects much of the energy that reaches the surface there. (b) The seasons (shown here for the Northern Hemisphere) are caused by variations in the amount of incoming solar energy as Earth makes its annual rotation around the sun on an axis tilted by 23½°. During the Northern Hemisphere winter, the Southern Hemisphere is tilted toward the sun and the Northern Hemisphere receives less light and heat. During the Northern Hemisphere summer, the situation is reversed.

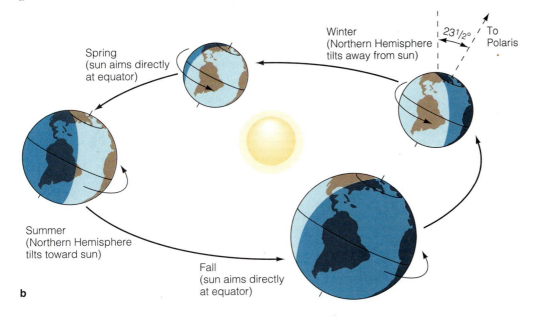

b

window. The circular current of air in the room, a **convection current,** is caused by the difference in temperature between the ends of the room. A similar process occurs over the surface of Earth. As we have seen, surface temperatures are higher at the equator than at the poles, and air can gain heat from warm surroundings. Since air is free to move over Earth's surface, it would be reasonable to assume that an air circulation pattern like the one shown in **Figure 7.5** would develop. In this ideal model, air heated in the tropics would expand and become less dense, rise to high altitude, turn poleward, and "pile up" as it converged near the poles. The air would then cool and contract by radiating heat into space, sink to the surface, and turn equatorward, flowing along the surface back to the tropics to complete the circuit.

But this is *not* what happens. Global circulation of air is governed by two factors: uneven solar heating *and* the rotation

of Earth. The eastward rotation of Earth on its axis deflects the moving air or water (or any moving object having mass) away from its initial course. This deflection is called the **Coriolis effect** in honor of **Gaspard Gustave de Coriolis,** the French scientist who worked out its mathematics in 1835. An understanding of the Coriolis effect is important to an understanding of atmospheric and oceanic circulation.

The Coriolis Effect 7-7

To an earthbound observer, any object moving freely across the globe appears to curve slightly from its initial path. In the Northern Hemisphere this curve is to the right from the expected path; in the Southern Hemisphere it is to the left. To earthbound observers the deflection is very real; it isn't caused

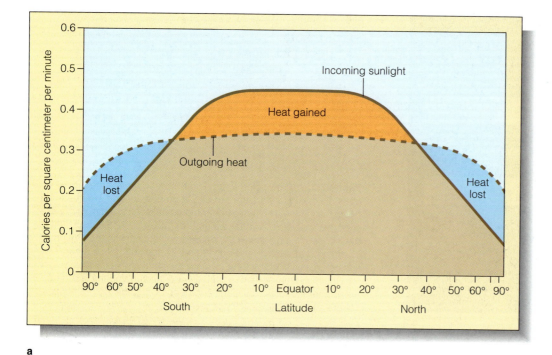

a

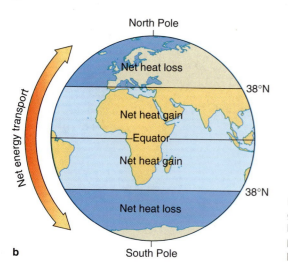

b

Figure 7.3 (**a**) Areas of heat gain and loss on Earth's surface. Since the area of heat gained (orange area) equals the area of heat lost (blue areas), Earth's heat budget is balanced. (**b**) The ocean does not boil away near the equator, nor freeze solid near the poles, because heat is transferred by ocean currents and winds from equatorial to polar regions.

by some mysterious force, and it isn't an optical illusion or some other trick caused by the shape of the globe itself. The observed deflection is caused by the observer's moving frame of reference on the spinning Earth.

The influence of this deflection can be illustrated by performing a mental experiment involving objects—in this case, cities and cannonballs—and then by applying the principle to atmospheric circulation. Let's pick as examples for our experiment the equatorial city of Quito, capital of Ecuador, and Buffalo, New York. Both cities are on almost the same line of longitude (79°W), so Buffalo is almost exactly north of Quito (as **Figure 7.6** shows). Like everything else attached to the rotating Earth, both cities make one trip around the world each 24 hours. The north-south relationship of the two cities never changes—Quito is *always* due south of Buffalo.

A complete trip around the world is 360°, so each city moves eastward at an angular rate of 15° per hour (360°/24

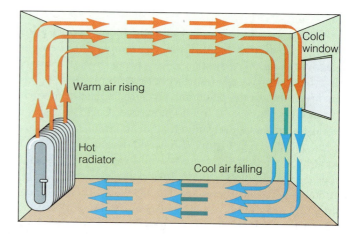

Figure 7.4 A convection current forms in a room when air flows from a hot radiator to a cold window and back. (For a practical oceanic application of this principle, look ahead to Figure 7.12.)

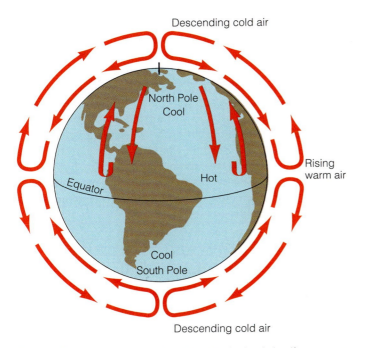

Figure 7.5 A hypothetical model of Earth's air circulation if uneven solar heating were the only factor to be considered. The thickness of the atmosphere is greatly exaggerated in this drawing.

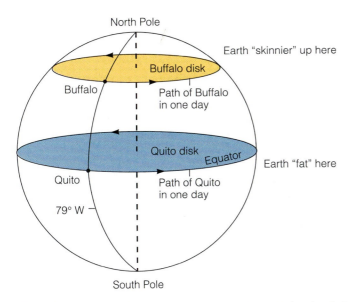

Figure 7.6 Sketch of the thought experiment in the text, showing that Buffalo travels a shorter path on the rotating Earth each day than does Quito.

hours = 15°/hour). Yet, even though their angular rates are the same, the two cities move eastward at different speeds. Quito is on the equator, the "fattest" part of Earth. Buffalo is farther north, at a "skinnier" part. Imagine both cities isolated on flat disks, and imagine Earth's sphere being made of a great number of these disks strung together on a rod connecting North Pole to South Pole. From Figure 7.6, you can see that Buffalo's disk has a smaller circumference than Quito's. Thus,

Buffalo doesn't have as far to go in one day as Quito. That means that Buffalo must move eastward *more slowly* than Quito to maintain its position due north of Quito.

Look at Earth from above the North Pole in **Figure 7.7**. The Quito disk and the Buffalo disk must turn through 15° of longitude each hour (or Earth would rip itself apart), but the city on the equator must move faster to the east to turn its 15° each hour because its disk is larger. Buffalo must move at 1,260 kilometers (783 miles) per hour to go around the world in one day, while Quito must move at 1,668 kilometers (1,036 miles) per hour to do the same.

Now imagine a massive object moving between the two cities. A cannonball shot north from Quito toward Buffalo would carry Quito's eastward component as it goes; that is, regardless of its northward speed, the cannonball is also moving *east* at 1,668 kilometers (1,036 miles) per hour. The fact of being fired northward by the cannon does not change its eastward movement in the least. As the cannonball streaks north, an odd thing happens. The cannonball veers from its northward path, angling slightly to the right (east) (**Figure 7.8**). Actually, this first cannonball is moving just as an observer from space would expect it to, but to those of us on the ground the cannonball "gets ahead of Earth." As cannonball 1 moves north, the ground beneath it is no longer moving eastward at 1,668 kilometers per hour. During the ball's time of flight, Buffalo (on its smaller disk) *has not moved eastward enough to be where the ball will hit.* If the time of flight for the cannonball is one hour, a city 408 kilometers east of Buffalo (1,668 [Quito's speed] − 1,260 [Buffalo's speed] = 408) will have an unexpected surprise. Albany may be in for some excitement!

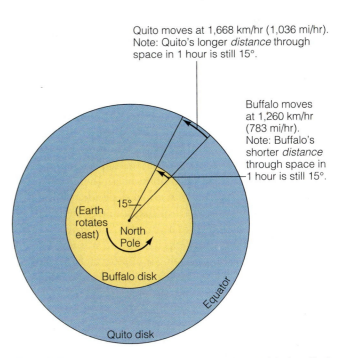

Figure 7.7 A continuation of the thought experiment. A look at Earth from above the North Pole shows that Buffalo and Quito move at different velocities.

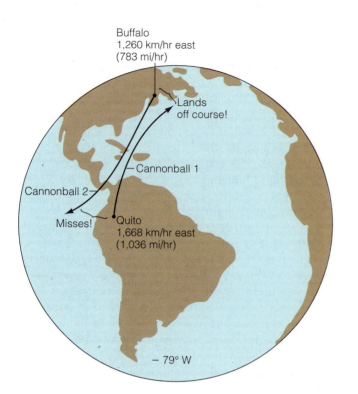

Buffalo
1,260 km/hr east
(783 mi/hr)

Lands
off course!

Cannonball 1

Cannonball 2

Misses!

Quito
1,668 km/hr east
(1,036 mi/hr)

— 79° W

Figure 7.8 The final step in the experiment. As observed from space, cannonball 1 (shot northward) and cannonball 2 (shot southward) move as we might expect; that is, they travel straight away from the cannons and fall to Earth. Observed from the ground, however, cannonball 1 veers slightly east and cannonball 2 veers slightly west of their intended targets. The effect depends on the observer's frame of reference.

If a cannonball were fired south from Buffalo toward Quito, the situation would be reversed. This second cannonball has an eastward component of 1,260 kilometers (783 miles) per hour even while it sits in the muzzle. Once fired and moving southward, cannonball 2 travels over portions of Earth that are moving ever faster in an eastward direction. The ball again appears to veer off course to the right (see Figure 7.8 again), falling into the Pacific to the west of Ecuador. Don't be deceived by the word *appears* in the last sentence. The cannonballs really do veer to the right. Only to an observer in space would they appear to go straight, and points on Earth would appear to move out from underneath them.

The Coriolis effect is a real effect dependent on our rotating frame of reference. Part of that frame of reference involves the direction from which you view the problem. Thus in Figure 7.8, cannonball 2 looks to you as if it is veering left, but to the citizens of Buffalo facing south to watch the cannonball disappear, it curves to the right (west). Coriolis deflection causes moving objects to veer to the left in the Southern Hemisphere because the frame of reference there is reversed. Also, except at the equator (where the Coriolis effect is nonexistent), the Coriolis effect influences the path of objects moving from east to west, or west to east, just the same as it does for objects moving north to south, or south to north.

A quick way to remember the Coriolis effect is that in the Northern Hemisphere, moving objects curve to the right; in the Southern Hemisphere, moving objects curve to the left.

Because the Coriolis effect influences any object with mass—*as long as that object is moving*—it plays a large role in the movements of air and water on Earth. The Coriolis effect is most apparent in mid-latitude situations involving the almost frictionless flow of fluids: between layers of water in the ocean and in the circuits of winds. Does the Coriolis effect influence the directions of cars and airplanes? Yes, but in these cases friction (of tires on pavement, of wings on air) is much greater than the influence of the Coriolis effect; so the deflection is not observable.

The Coriolis Effect and Atmospheric Circulation Cells   7-8

We can now modify our first model of atmospheric circulation (Figure 7.5) into the more correct representation provided in **Figure 7.9.** Yes, air does warm, expand, and rise at the equator; and air does cool, contract, and fall at the poles. But instead of continuing all the way from equator to pole in a continuous loop in each hemisphere, air rising from the equatorial region moves poleward and is gradually deflected eastward; that is, it turns to the *right* in the Northern Hemisphere and to the *left* in the Southern Hemisphere. This eastward deflection is caused by the Coriolis effect. (Note that the Coriolis effect does not *cause* the wind; it only *influences* the wind's direction.)

As air rises at the equator, it loses moisture by precipitation (rainfall) caused by expansion and cooling. This drier air now grows denser in the upper atmosphere as it radiates heat to space and cools. When it has traveled about a third of the way from the equator to the pole—that is, to about 30°N and 30°S latitude—the air becomes dense enough to fall back toward the surface. Most of the descending air turns back toward the equator when it reaches the surface. In the Northern Hemisphere the Coriolis effect again deflects this surface air to the right, and the air blows across the ocean or land from the northeast. (This air is represented by the arrows labeled "NE trade winds" in Figure 7.9.) Though it has been heated by compression during its descent, this air is generally still colder than the surface over which it flows. The air soon warms as it moves equatorward, however, evaporating surface water and becoming humid. The warm, moist, less dense air then begins to rise as it approaches the equator and completes the circuit.

Such a large circuit of air is called an **atmospheric circulation cell.** A pair of these tropical cells exists, one on each side of the equator. They are known as **Hadley cells** in honor of George Hadley, the London lawyer and philosopher who worked out an overall scheme of wind circulation in 1735. Look for them in Figure 7.9.

A more complex pair of circulation cells operates at mid-latitudes in each hemisphere. Some of the air descending at 30° latitude turns poleward rather than equatorward. Before

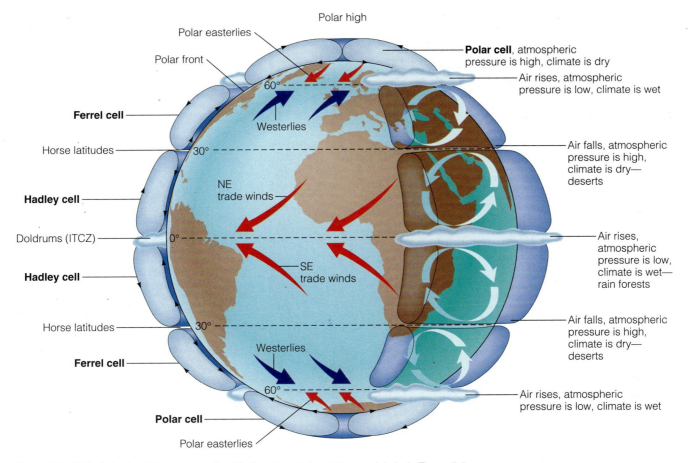

Polar high

Polar easterlies

Polar front

Polar cell, atmospheric
pressure is high, climate is dry

Air rises, atmospheric
pressure is low, climate is wet

Ferrel cell

60°

Westerlies

Horse latitudes

30°

Air falls, atmospheric
pressure is high,
climate is dry—
deserts

Hadley cell

NE
trade winds

Doldrums (ITCZ)

0°

Air rises,
atmospheric
pressure is low,
climate is wet—
rain forests

Hadley cell

SE
trade winds

Horse latitudes

30°

Air falls, atmospheric
pressure is high,
climate is dry—
deserts

Ferrel cell

Westerlies

60°

Air rises, atmospheric
pressure is low, climate is wet

Polar cell

Polar easterlies

Figure 7.9 Global air circulation as described in the six-cell circulation model. As in Figure 7.5, air rises at the equator and falls at the poles. But instead of one great circuit in each hemisphere from equator to pole, there are three circuits in each hemisphere. Note the influence of the Coriolis effect on wind direction.

this air descends to the surface it is joined at high altitude by air returning from the north. As can be seen in Figure 7.9, a loop of air forms between 30° and about 50°–60° of latitude. As before, the air is driven by uneven heating and influenced by the Coriolis effect. Surface wind in this circuit is again deflected to the right, this time flowing from the west to complete the circuit. (This air is represented by the arrows labeled "Westerlies" in Figure 7.9.) The mid-latitude circulation cells of each hemisphere are named **Ferrel cells** after William Ferrel, the American who discovered their inner workings in the mid-nineteenth century. They, too, can be seen in Figure 7.9.

Meanwhile, air that has grown cold over the poles begins blowing toward the equator at the surface, turning to the west as it does so. At between 50° and 60° latitude in each hemisphere, this air has taken up enough heat and moisture to ascend. However, this polar air is denser than the air in the adjacent Ferrel cell and does not mix easily with it. The unstable zone between these two cells generates most mid-latitude weather. At high altitude the ascending air from 50°–60° latitude turns poleward to complete a third circuit. These are the **polar cells.**

Thus, three large atmospheric circulation cells—a Hadley cell, a Ferrel cell, and a polar cell—exist in each hemisphere. Air circulation within each cell is powered by uneven solar heating and influenced by the Coriolis effect.

WIND PATTERNS

7-9

The model of atmospheric circulation described above has many interesting features. At the bands between circulation cells, the air is moving *vertically,* and surface winds are weak and erratic. Such conditions exist at the equator (where air rises and atmospheric pressure is generally low) or at 30° latitude in each hemisphere (where air falls and atmospheric pressure is generally high). Places within these circulation cells where air moves rapidly *horizontally* across the surface from zones of high pressure to zones of low pressure are characterized by strong, dependable winds.

Sailors have a special term for the calm equatorial areas where the surface winds of the two Hadley cells converge: the equatorial low called the **doldrums.** The word has come to be

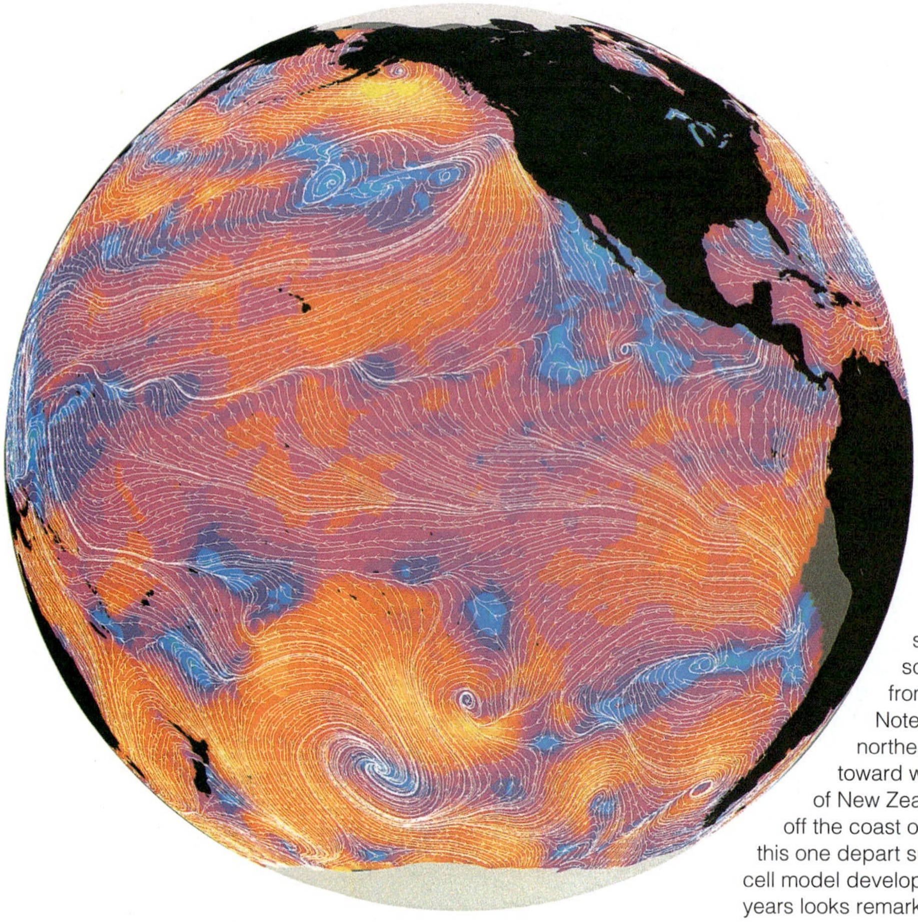

Figure 7.10 Winds over the Pacific Ocean on 20 and 21 September 1996. Wind speed increases as colors change from blue-purple to yellow-orange, with the strongest winds at 20 meters per second (45 miles per hour). Wind direction is shown by the small white arrows. The measurements were made with a NASA radar scatterometer aboard Japan's Advanced Earth Orbiting Satellite, launched 16 August 1996. The scatterometer measures and analyzes the backscatter (reflection) of high frequency radar pulses from small wind-caused ripples on the sea surface. Note the Hawaiian islands in the midst of the persistent northeast trade winds, the vigorous westerlies driving toward western Canada, a large extratropical cyclone east of New Zealand, and the last remnants of a tropical cyclone off the coast of Japan. Although *instantaneous* views such as this one depart substantially from wind flow predicted in the six-cell model developed in Figure 7.9, the *average* wind flow over many years looks remarkably like what we would expect from the model.

associated with a gloomy, listless mood, perhaps reflecting the sultry air and variable breezes found there. Scientists who study the atmosphere call this area the **intertropical convergence zone,** or **ITCZ,** to reflect the influence of wind convergence on conditions near the equator. Strong heating in the ITCZ causes surface air to expand and rise. The humid, rising, expanding air loses moisture as rain, some of which contributes to the success of tropical rain forests.

Sinking air, in contrast, is generally arid. The great deserts of both hemispheres, dry bands centered around 30°, mark the intersection of the Hadley and Ferrel cells. Because evaporation is higher than precipitation in these areas, ocean surface salinity tends to be highest at these latitudes. At sea, these areas of high atmospheric pressure and little surface wind are called the subtropical high, or **horse latitudes.** Spanish ships laden with supplies for the New World were often becalmed there, sometimes for weeks on end. When the mariners ran out of water and feed for their livestock, they were forced to throw the dead horses over the side.

Of much more interest to sailing masters were the bands of dependable surface winds *between* the zones of ascending and descending air. Most constant of these are the persistent **trade winds,** or easterlies, centered at about 15°N and 15°S latitude. The trade winds are the surface winds of the Hadley cells as they move from the horse latitudes to the doldrums. In the Northern Hemisphere they are the northeast trade winds; the southeast trade winds are the Southern Hemisphere coun-

terpart.[1] The **westerlies,** surface winds of the Ferrel cells centered at about 45°N and 45°S latitude, flow between the horse latitudes and the boundaries of the polar cells in each hemisphere. Thus, the westerlies approach from the southwest in the Northern Hemisphere and from the northwest in the Southern Hemisphere. Sailors outbound from Europe to the New World learned to drop south to catch the trade winds and to return home by a more northerly route to take advantage of the westerlies. Trade winds and westerlies are shown in Figure 7.9.

The six-cell model of atmospheric circulation (three cells in each hemisphere) discussed here represents an *average* of air flow through many years over the planet as a whole. Though the model is accurate in a general sense, local details of cell circulation vary because surface conditions are different at different longitudes. The ocean's thermostatic effect is the major factor reducing irregularities in cell circulation over water. **Figure 7.10** is a depiction of winds over the Pacific on two days in September 1996. As you can see, the patterns depart substantially from what we would expect from the six-cell model of Figure 7.9. Most of the difference is caused by the

---

[1] Winds are named by the direction from which they blow. A west wind blows from the west toward the east; a northeast wind blows from the northeast toward the southwest. The trade winds are named in honor of their ability to blow steadily—the early English word *trade* meant "steadily" or "constantly."

geographical distribution of land masses and the different responses of land and ocean to solar heating. But, as noted above, over long periods of time (many years), *average* flow looks remarkably like what we would expect.

Monsoons

7-10

A **monsoon** is a pattern of wind circulation that changes with the season. (The word *monsoon* is derived from *mausim,* the Arabic word for season.) Areas subject to monsoons generally have wet summers and dry winters.

Monsoons are linked to the different specific heats of land and water, and to the annual north-south movement of the ITCZ. In the spring, land heats more rapidly than the adjacent ocean. Air above the land becomes warmer and so rises. Relatively cool air flows from over the ocean to the land to take its place. Continued heating causes this humid air to rise, condense, and form clouds and rain. In autumn the land cools more rapidly than the adjacent ocean. Air cools and sinks over the land, and dry surface winds move seaward. The intensity and location of monsoon activity depends on the position of the ITCZ. Note in **Figure 7.11** that the monsoons follow the ITCZ south in the Northern Hemisphere's winter, and north in its summer.

In Africa and Asia, more than 2 billion people depend on summer monsoon rains for drinking water and agriculture. The most intense summer monsoons occur in Asia. The great landmass of Asia draws vast quantities of warm, moist air from the Indian Ocean (see again Figure 7.11). Southerly winds drive this moisture toward Asia, where it rises and condenses to produce a months-long deluge. Much smaller monsoons occur in North America as warming and rising air over the South and West draws humid air and thunderstorms from the Gulf of Mexico.

Sea Breezes and Land Breezes

7-11

Land breezes and sea breezes are small, daily mini-monsoons. Morning sunlight falls on land and adjacent sea, warming both. The temperature of the water doesn't rise as much as the temperature of the land, however. The warmer inland rocks transfer heat to the air, which expands and rises, creating a zone of low atmospheric pressure over the land. Cooler air from over the sea then moves toward land; this is the **sea breeze** (see **Figure 7.12a**). The situation reverses after sunset, with land losing heat to space and falling rapidly in temperature. After a while, the air over the still-warm ocean will be warmer than the air over the cooling land. This air will then rise, and the breeze direction will reverse, becoming a **land breeze** (see **Figure 7.12b**). Land breezes and sea breezes are common and welcome occurrences in coastal areas.

El Niño and La Niña

7-12

Sometimes cell circulation doesn't seem to play by the rules. In three- to eight-year cycles, atmospheric circulation changes significantly from the patterns shown in Figure 7.9. A reversal

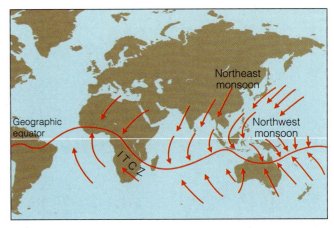

a January

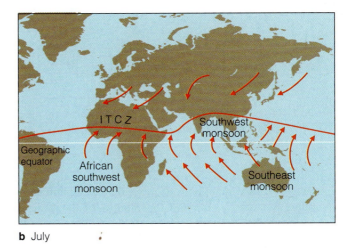

b July

Figure 7.11 Monsoon patterns. During the monsoon circulations of January and July, surface winds are deflected to the right in the Northern Hemisphere and to the left in the Southern Hemisphere.

in the distribution of atmospheric pressure between the eastern and western Pacific causes the trade winds to weaken or reverse. The trade winds normally drag huge quantities of water westward along the ocean's surface near the equator, but without the winds these equatorial currents crawl to a stop. Warm water that has accumulated at the western side of the Pacific can build eastward along the equator toward the coast of Central and South America, greatly changing ocean conditions there.

El Niño and La Niña are primarily ocean current phenomena, so we will study them in Chapter 8's discussion of ocean currents.

STORMS

7-13

Storms are regional atmospheric disturbances characterized by strong winds often accompanied by precipitation. Few natural events underscore human insignificance like a great storm. When powered by stored sunlight, the combination of atmosphere and ocean can do fearful damage.

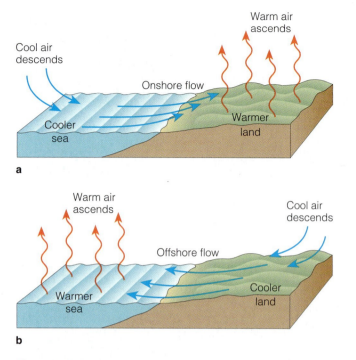

Figure 7.12 The flow of air in sea breezes and land breezes. (**a**) Afternoon onshore sea breeze. (**b**) Nighttime offshore land breeze.

In Bangladesh, on 13 November 1970, a tropical cyclone (a hurricane) with wind speeds of more than 200 kilometers (125 miles) per hour roared up the mouths of the Ganges River, carrying with it masses of seawater up to 12 meters (40 feet) high. Water and wind clawed at the aggregation of small islands, most just above sea level, that makes up this impoverished country. In only 20 minutes at least 300,000 lives were lost, and estimates ranged to 1 million dead! Property damage was absolute. Photographs taken soon after the storm showed a horizon-to-horizon morass of flooded, deep-gashed ground tortured by furious winds. There was almost no trace of human inhabitants, farms, domestic animals, or villages. Another great storm that struck in May 1991 killed another 200,000 people. The economy of the shattered country may never fully recover.

A much different type of storm hammered the U.S. East Coast in March 1993. Mountainous snows from New York to North Carolina (1.3 meters, or 50 inches, at Mount Mitchell), 175-kilometer- (109-mile-) per-hour winds in Florida, and record cold in Alabama ($-17°C$, or $2°F$, in Birmingham) were elements of a four-day storm that spread chaos from Canada to Cuba. At least 238 people died on land; another 48 were lost at sea. At one point more than 100,000 people were trapped in offices, factories, vehicles, and homes; 1.5 million were without electricity. Economic damage exceeded $1 billion.

These two great storms exemplify *tropical cyclones* and *extratropical cyclones* at their worst.[2] As the name implies, tropical cyclones like the hurricane that struck Bangladesh are

_____
[2] Note that the prefix *extra-* means "outside" or "beyond." *Extratropical* refers to the location of the storm, not its intensity.

primarily a tropical phenomenon. Extratropical cyclones—the winter weather disturbances with which residents of the U.S. eastern seaboard and other mid-latitude dwellers are most familiar—are found mainly in the Ferrel cells of each hemisphere.

Both kinds of storms are **cyclones,** huge rotating masses of low-pressure air in which winds converge and ascend. The word *cyclone,* derived from the Greek noun *kyklon* (meaning "an object moving in a circle"), underscores the spinning nature of these disturbances. (Don't confuse a cyclone with a **tornado,** a much smaller funnel of fast-spinning wind associated with severe thunderstorms.)

Air Masses 7-14

Cyclonic storms form between or within air masses. An **air mass** is a large body of air with nearly uniform temperature, humidity, and therefore density throughout. Air pausing over water or land will tend to take on the characteristics of the surface below. Cold, dry land causes the mass of air above to become chilly and dry. Air above a warm ocean surface will become hot and humid. Cold, dry air masses are dense and form zones of high atmospheric pressure. Warm, humid air masses are less dense and form zones of lower atmospheric pressure.

Air masses can move within or between circulation cells. Density differences, however, will prevent the air masses from mixing when they approach one another. Energy is required to mix air masses. Since that energy is not always available, a dense air mass may slide beneath a lighter air mass, lifting the lighter one and causing its air to expand and cool. Water vapor in the rising air may condense. All of these effects contribute to turbulence at the boundaries of the air masses.

The boundary between air masses of different density is called a **front.** The term was coined by a meteorologist who saw a similarity between the zone where air masses meet and the violent battle fronts of the First World War.

Extratropical cyclones form at a front between *two* air masses. Tropical cyclones form from disturbances within *one* warm and humid air mass.

Extratropical Cyclones 7-15

Extratropical cyclones form at the boundary between each hemisphere's polar cell and its Ferrel cell—the **polar front.** These great storms occur mainly in the winter hemisphere when temperature and density differences across the polar front are most pronounced. Remember, the cold wind poleward of the front is generally moving from the east; the warmer air equatorward of the front is generally moving from the west (see again Figure 7.9). The smooth flow of winds past each other at the front may be interrupted by zones of alternating high and low atmospheric pressure that bend the front into a series of waves. Because of the difference in wind direction in the air masses north and south of the polar front, the wave shape will enlarge, and a twist will form along the front. The different densities of the air masses prevent easy mixing; so

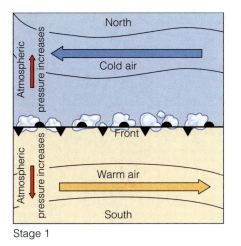

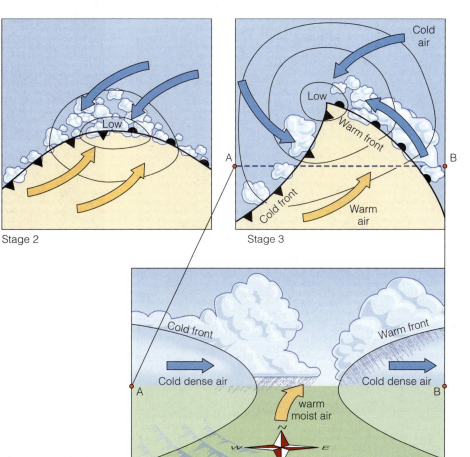

Figure 7.13 The genesis and early development of an extratropical cyclone in the Northern Hemisphere. (The arrows depict air flow.)

the cold, dense air mass will slide beneath the warmer, lighter one. Formation of this twist in the Northern Hemisphere, as seen from above, is shown in **Figure 7.13.** The twisting mass of air becomes an extratropical cyclone.

The twist that generates an extratropical cyclone circulates to the left in the Northern Hemisphere, seemingly in opposition to the Coriolis effect. The reasons for this paradox become clear, however, when we consider the wind directions and the nature of interruption of the airflow between the cells. (In fact, the leftward motion of the cyclone *is* Coriolis-driven because the large-scale airflow pattern at the *edges* of the cells is generated in part by the Coriolis effect.) Wind speed increases as the storm "wraps up" in much the same way that a spinning skater increases rotation speed by pulling in his or her arms close to the body. Air rushing toward the center of the spinning storm rises to form a low-pressure zone at the center. Extratropical cyclones are embedded in the westerly winds and thus move eastward. They are typically 1,000 to 2,500 kilometers (620 to 1,600 miles) in diameter and last from two to five days. **Figure 7.14** provides a beautiful example. The wind and precipitation associated with these fronts are sometimes referred to as **frontal storms.**

Figure 7.14 An extratropical cyclone over Ohio, 25 April 1991.

Figure 7.15 Storm damage from a particularly severe nor'easter, the Ash Wednesday (7 March) storm of 1962. Massive damage such as this, on Fire Island, New York, is most common for buildings near the shoreline.

Figure 7.16 Hurricane Florence over the North Atlantic, photographed from the space shuttle *Discovery* in November 1994. The cloudfree eye of the storm is clearly visible. Note the spiral bands of cloud extending outward from the eye.

North America's most violent extratropical cyclones are the **nor'easters** (northeasters) that sweep the eastern seaboard in winter. The name indicates the direction from which the storm's most powerful winds approach. About 30 times a year, nor'easters moving along the mid-Atlantic and New England coasts generate wind and waves with enough force to erode beaches and offshore barrier islands, disrupt communication and shipping schedules, damage shore and harbor installations, and break power lines. About every hundred years a nor'easter devastates coastal settlements. In spite of a long history of destruction, people continue to build on unstable exposed coasts (**Figure 7.15**).

Tropical Cyclones

 7-16

Tropical cyclones are great masses of warm, humid, rotating air. They occur in all tropical oceans except the equatorial South Atlantic. Large tropical cyclones are called **hurricanes** (*Huracan* = the god of the wind of the Caribbean Taino people) in the North Atlantic and eastern Pacific, *typhoons* (*Tai-fung* = Chinese for great wind) in the western Pacific, *tropical cyclones* in the Indian Ocean, and *willi-willies* in the waters near Australia. To qualify formally as a hurricane or typhoon, the tropical cyclone must have sustained winds of at least 119 kilometers (74 miles) per hour. About a hundred tropical cyclones grow to hurricane status each year. A very

few of these develop into superstorms, with winds near the core that exceed 250 kilometers (155 miles) per hour! (To imagine what winds in such a storm might feel like, picture yourself clinging to the wing of a twin-engine private airplane in flight.) Tropical cyclones containing winds less than hurricane force are called *tropical storms* and *tropical depressions*.

From above, tropical cyclones appear as circular spirals, as **Figure 7.16** shows. They may be 1,000 kilometers (620 miles) in diameter and 15 kilometers (9.3 miles or 50,000 feet) high. The calm center, or *eye* of the storm—a zone some 13 to 16 kilometers (8 to 10 miles) in diameter—is sometimes surrounded by clouds so high and dense that the daytime sky above looks dark. Farther out, churned by furious winds, the rainband clouds condense huge amounts of water vapor into rain. A tropical cyclone is diagrammed in **Figure 7.17**.

Unlike extratropical cyclones, these greatest of storms form within *one* warm, humid air mass between 10° and 25°

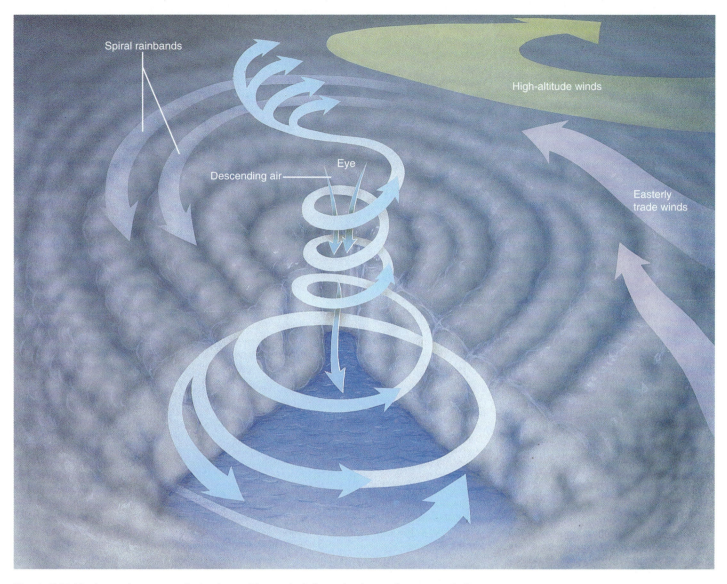

Figure 7.17 The internal structure of a hurricane. The vertical dimension is greatly exaggerated in this drawing.

latitude in both hemispheres. (Though air conditions would be favorable, the Coriolis effect closer to the equator is too weak to initiate rotary motion.) The origins of tropical cyclones are not well understood. A tropical cyclone usually develops from a small tropical depression. Tropical depressions form in easterly waves, areas of lower pressure within the easterly trade winds that are thought to originate over a large, warm landmass. When air containing the disturbance is heated over tropical water with a temperature of about 26°C (79°F) or more, circular winds begin to blow in the vicinity of the wave, and some of the warm humid air is forced upward. Condensation begins, and the storm takes shape. Under ideal conditions the embryo storm reaches hurricane status—that is, with wind speeds in excess of 119 kilometers (74 miles) per hour—in two to three days.

The centers of most tropical cyclones move westward and poleward at 5 to 40 kilometers (3 to 25 miles) per hour. Typical tracks of these storms are shown in **Figure 7.18.** A trio

of Pacific tropical cyclones that formed in 1976 is shown in **Figure 7.19.**

Although its birth process is somewhat mysterious, the source of a storm's power is well understood. Its strength comes from the same seemingly innocuous process that warms a chilled soft-drink can when water from the atmosphere condenses on its surface. As you may recall from Chapter 6, it takes quite a bit of energy to break the bonds holding water molecules together and evaporate water into the atmosphere—water's latent heat of evaporation is very high. That heat energy is released when the water vapor recondenses as liquid. It tends to warm your drink very quickly, and the more humid the air, the faster the condensing and warming. In tropical cyclones, the condensation energy generates air movement (wind), not more heat. Fortunately, only 2% to 4% of this energy of condensation is converted into motional energy!

A tropical cyclone is an ideal machine for "cashing in" water vapor's latent heat of evaporation. Warm, humid air forms

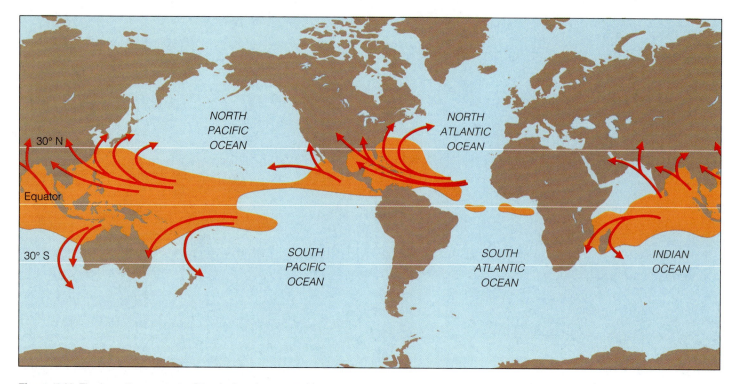

Figure 7.18 The breeding grounds of tropical cyclones are shown as orange shaded areas. The storms follow curving paths: First they move westward with the trade winds. Then they either die over land or turn eastward until they lose power over the cooler ocean of mid-latitudes. Cyclones are not spawned over the South Atlantic or the southeast Pacific because their waters are too chilly; nor are they born in the still air—the doldrums—within 5° of the equator.

in great quantity only over a warm ocean. As hot, humid tropical air rises and expands, it cools and is unable to retain the moisture it held when warm. Rainfall begins. The rainfall rate in some parts of the storm routinely exceeds 2.5 centimeters (1 inch) per hour, and 20 billion metric tons of water can fall from a large tropical cyclone in a day! Tremendous energy is released as this moisture changes from water vapor to liquid. In one day, a large tropical cyclone generates about 2.4 trillion kilowatt-hours of power, equivalent to the electrical energy needs of the entire United States for a year! Thus, solar energy ultimately powers the storm in a cycle of heat absorption, evaporation, condensation, and conversion of heat energy to kinetic energy. This energy is available as long as the storm stays over warm water and has a ready source of hot, humid air.

Three aspects of a tropical cyclone can cause property damage and loss of life: wind, rain, and storm surge. The destructive force of winds of 250 kilometers (155 miles) per hour—or more—is self-evident (and is described in **Box 7.1**). Rapid rainfall can cause severe flooding when the storm moves onto land. But the greatest danger lies in **storm surge,** a mass of water driven by the storm. The low

Figure 7.19 Three tropical cyclones lined up between Hawaii and Central America in 1976. On the left is Hurricane Kate; in the center is a tropical depression that never reached hurricane force; and on the right is Hurricane Liza, which within hours smashed into La Paz, Mexico, and caused millions of dollars in damage.

The most lethal Atlantic hurricane of modern times began as a chain of rolling thunderclouds drifting westward across the coast of central Africa. By the morning of 22 October 1998, the clouds had become organized into a tropical depression. Earth's rotation had given the clouds a gentle twist, and warm oceanic air rose into their core. Like water spinning down a bathtub drain, humid air was sucked into the growing vortex. The moisture condensed to form rain, liberating heat energy that caused the storm to grow into the thirteenth hurricane of the season: Mitch.

Three days passed. Mitch tracked uncertainly toward the west, paralleling the north coast of Honduras. On two occasions, shearing winds threatened to tear the storm apart; each time it survived and strengthened. Then, on the morning of 26 October, Mitch metamorphosed into a towering black mountain of wind and rain, a rare category-5 storm (**Figure a**). Its spreading violence sucked the sea surface into a 6-meter (20-foot) dome tens of kilometers across. The wind speed near its center rose to 290 kilometers (180 miles) per hour, driving ocean waves 15 meters (50 feet) high against the coast. In the predawn hours of 28 October, the leading edge of the storm touched the shore. During the next three days Mitch stalked Honduras and Nicaragua—a tearing, clawing force that entered every window and door, that tore the roofs off shacks and children from their parents' grasp, that hurled cars and structures about like toy blocks. The division between ocean and land blurred beneath an onrushing wall of windblown seawater. At the height of the storm the hot, wet air glowed yellow-green and was ear-poppingly thin.

As **Figure b** shows, devastation was all but absolute. More than 9,000 people died, 5,500 were missing, and 570,000 homeless. The steep terrain of Honduras and Nicaragua is covered with poorly consolidated volcanic soil, and mudflows and landslides buried at least ten communities. The banana crop, economic mainstay of the region, was all but wiped out, the area's fragile infrastructure decimated. Fifty years of progress disappeared in 50 hours.

The damage done by Hurricane Mitch in Nicaragua alone has been estimated at $1.36 billion, or 67% of that country's gross domestic product (GDP). If a natural disaster in the United States caused damage equivalent to 67% of our GDP, the cost would be a staggering $4.3 trillion, equivalent to 143 hurricane landfalls the magnitude of Andrew (Florida, 1992) or 108 Northridge (California, 1994) earthquakes. Taken as a whole, Mitch was one of the Western Hemisphere's greatest twentieth-century natural disasters, a sobering demonstration of uneven solar heating and the power of water's latent heat of fusion.

7-17

a Hurricane Mitch rakes the Central American Atlantic coast, 28 October 1998.

b Residents of the Honduran capital of Tegucigalpa dig themselves out after floods caused by Hurricane Mitch. Officials say the hurricane damage set their country's infrastructure back 50 years.

atmospheric pressure at the storm's center produces a dome of seawater that can reach a height of 1 meter (3.3 feet) in the open sea. The water height increases when waves and strong hurricane winds ramp the water mass ashore. If a high tide coincides with the arrival of all this water at a coast, or if the coastline converges (as is the case at the mouths of the Ganges in Bangladesh), rapid and catastrophic flooding will occur. Storm surges of up to 12 meters (40 feet) were reported in Bangladesh in 1970. Much of the $3 billion in damage done by Hurricane Andrew to south Florida in August 1992 was caused by a 4-meter (13-foot) storm surge arriving near high tide. You'll learn more about storm surges in Chapter 9's discussion of large waves.

You may have noticed that tropical cyclones turn leftward (or counterclockwise when viewed from above) in the Northern Hemisphere and rightward (clockwise) in the Southern Hemisphere. Does this mean that the Coriolis effect does not apply to tropical cyclones? No. Their spin is caused by the Coriolis deflection of winds approaching the center of a low-pressure area from great distances. In the Northern Hemisphere, there is rightward deflection of the *approaching air*. The edge spin given by this incoming air causes the storm to spin to the left in the Northern Hemisphere.

Tropical cyclones last from three hours to three weeks; most have lives of five to ten days. They eventually run down when they move over land or over water too cool to supply the humid air that sustains them. The friction of a land encounter rapidly drains a tropical cyclone of its energy, and a position above ocean water cooler than 24°C (75°F) is a sure harbinger of the storm's demise. When deprived of energy, the storm "unwinds" and becomes a mass of unstable humid air pouring rain, lightning, and even tornadoes from its clouds. Tropical cyclones can be dangerous to the end: Torrential rain streaming from the remnants of Hurricane Agnes in 1976 caused more than $2 billion in damage, mostly to Pennsylvania. Chesapeake and Delaware Bays were flooded with fresh water and sediments, destroying much of the shellfish industry there.

Tropical cyclones are nature's escape valves, flinging solar energy poleward from the tropics. They are beautiful, dangerous examples of the energy represented by water's latent heat of fusion.

QUESTIONS FROM STUDENTS

1. Earth's orbit brings it closer to the sun in the Northern Hemisphere's winter than in its summer. Yet it's warmer in summer. Why?

Earth's orbit around the sun is elliptical, not circular. The whole Earth receives about 7% more solar energy through the half of the orbit during which we are closer to the sun than through the other half. The time of greater energy input comes during our winter, but the entire Northern Hemisphere is tilted toward the sun during the summer, which results in much more light reaching it in the summer. Three times more energy enters the Northern Hemisphere each day at midsummer than at midwinter.

2. Does the ocean affect weather at the centers of continents?

Absolutely. In a sense, *all* large-scale weather on Earth is oceanically controlled. The ocean acts as a solar collector and heat sink, storing and releasing heat. Most great storms (tropical and extratropical cyclones alike) form over the ocean and then sweep over land.

3. Are any of the results of large-scale atmospheric circulation apparent to the casual observer?

Yes. The view of atmospheric circulation developed in this chapter explains some phenomena you may have experienced. For instance, flying from Los Angeles to New York takes about 40 minutes less than flying from New York to Los Angeles, because of westerly headwinds (indicated in Figure 7.9).

Because of these same prevailing westerlies, most storms travel over the United States from west to east. Weather prediction is based on observations and samples taken from the air masses as they move. Forecasting is often easier in the East and Midwest than in the West because more data are available from an air mass when it's over land.

Temperatures are milder (not as hot in summer, not as cold in winter) on the West Coast than on the East Coast, again because of the prevailing westerlies. The wind blows over water (which moderates temperatures) toward the West Coast, but over land toward the East Coast.

4. Does the Coriolis effect really make explorers wander to the left in the snows of Antarctica, or tree trunks grow in rightward spirals in Canadian forests, or water swirl counterclockwise down a washbasin drain in Las Vegas?

The Coriolis effect depends on the speed, mass, and latitude of the moving object. Small, lightweight objects moving slowly are subject to many forces and conditions (such as wind currents, natural variations in basin shape, and friction) that overwhelm Coriolis acceleration. For example, think of how *very* small the difference in eastward speed of the northern edge of a washbasin is in comparison to the southern edge. Any small irregularity in the washbasin's shape will be hundreds of times more important than the Coriolis effect in determining whether the water will exit in a rightward spin or a leftward spin! Explorers and trees aren't massive enough and don't move quickly enough to be affected by the Coriolis effect.

But if the moving object is at mid-latitude, is heavy, and is moving quickly, it *will* be deflected to the right of its intended path. Jet airliners, artillery shells, and weather systems are all noticeably influenced by the Coriolis effect.

5. Has anything similar to Earth's weather patterns been seen on other planets?

Yes, indeed. Saturn, Jupiter, and Neptune have tremendous cyclonic storms. A few on Jupiter are large enough to be seen from Earth through small telescopes. Tracks of tornadoes have recently been identified on Mars, and a huge cyclonic storm was photographed there in 1999 (**Figure 7.20**). Venus

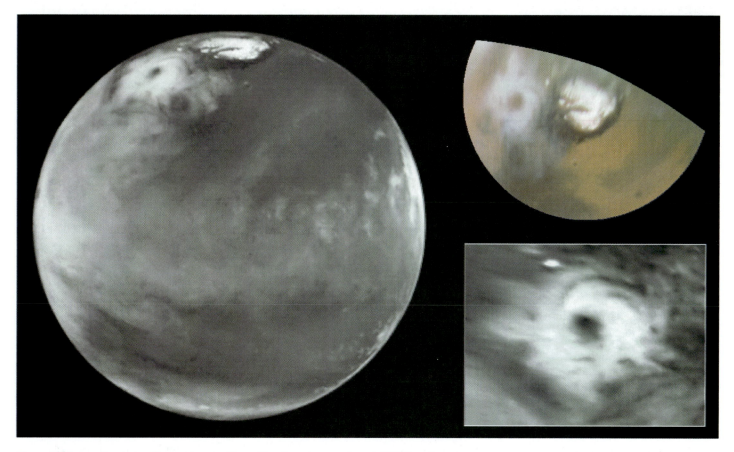

Figure 7.20 A vast extratropical cyclone on Mars. The storm is more than 1,600 kilometers (1,000 miles) across.

has huge cloud banks suggestive of polar fronts and extratropical cyclones. The Coriolis effect and uneven solar heating are common to all planets with atmospheres in this solar system, so we shouldn't be surprised by similarities.

areas of unstable air occurring between or within air masses. Extratropical cyclones originate at the boundary between air masses; tropical cyclones, the most powerful of Earth's atmospheric storms, occur within a single humid air mass. The immense energy of tropical cyclones is derived from water's latent heat of evaporation.

CHAPTER SUMMARY

The water, gases, and energy at Earth's surface are shared between the atmosphere and the ocean. The two bodies are in continuous contact, and conditions in one are certain to influence conditions in the other. The interaction of ocean and atmosphere moderates surface temperatures, shapes Earth's weather and climate, and creates most of the sea's waves and currents.

The atmosphere responds to uneven solar heating by flowing in three great circulating cells over each hemisphere. This circulation of air is responsible for about two-thirds of the heat transfer from tropical to polar regions. The flow of air within these cells is influenced by the rotation of Earth. To observers on the surface, Earth's rotation causes moving air (or any moving mass) in the Northern Hemisphere to curve to the right of its initial path, and in the Southern Hemisphere to the left. The apparent curvature of path is known as the Coriolis effect.

Uneven flow of air within cells is one cause of the atmospheric changes we call *weather*. Large storms are spinning

TERMS AND CONCEPTS TO REMEMBER

air mass
atmospheric circulation
   cell
convection current
Coriolis effect
Coriolis, Gaspard
   Gustave de
cyclone
doldrums
extratropical cyclone
Ferrel cell
front
frontal storm
Hadley cell
horse latitudes
hurricane

intertropical convergence
   zone (ITCZ)
land breeze
monsoon
nor'easter (northeaster)
polar cell
polar front
precipitation
sea breeze
storm
storm surge
tornado
trade winds
tropical cyclone
water vapor
westerlies

STUDY QUESTIONS

1. What factors contribute to the uneven heating of Earth by the sun?

2. How does the atmosphere respond to uneven solar heating? How does the rotation of Earth affect the resultant circulation?

3. Describe the atmospheric circulation cells in the Northern Hemisphere. At what latitudes does air move vertically? Horizontally? What are the trade winds? The westerlies? Where are deserts located? Why? What do you think ocean surface salinity is like in these desert bands?

4. How do the two kinds of large storms differ? How are they similar? What causes an extratropical cyclone? What happens in one?

5. What triggers a tropical cyclone? From what is its great power derived? What causes the greatest loss of life and property when a tropical cyclone reaches land?

6. If the Coriolis effect causes the rightward deflection of moving objects in the Northern Hemisphere, why does air rotate to the left around zones of low pressure in that hemisphere?

FOR FURTHER STUDY

Ahrens, A. C. 1998. *Essentials of Meteorology: An Invitation to the Atmosphere.* Pacific Grove, CA: Brooks/Cole. Excellent general text.

Broecker, W. S. 1995. "Chaotic Climate." *Scientific American,* November, 62–8. Global temperatures have been known to change substantially in only a decade or two.

Dolan, R., and H. Lins. 1987. "Beaches and Barrier Islands." *Scientific American,* July, 68–77. Discusses the folly of building close to shore.

Emanuel, K. A. 1988. "Toward a General Theory of Hurricanes." *American Scientist* 76 (no. 4): 370–9. The world's fiercest storms are produced by the most benign climates. Fine review article with some controversial views on the sources of a tropical cyclone's power.

Friedman, R. M. 1989. *Appropriating the Weather: Vilhelm Bjerknes and the Construction of a Modern Meteorology.* New York: Cornell University Press. An excellent recent history of Bjerknes's contributions.

Karl, T. R., N. Nicholls, and J. Gregory. 1997. "The Coming Climate." *Scientific American,* May, 79–83. Meteorological records and computer models permit insights into some of the broad weather patterns that might be expected in a warmer world.

Piccard, B. 1999. "Around at Last!" *National Geographic,* September, 30–51. Balloonists celebrate being first to travel around the globe.

Webster, P. J. 1981. "Monsoons." *Scientific American,* August, 108–18.

Webster, P. J., and J. A. Curry. 1998. "The Oceans and Weather." *Scientific American,* Fall special issue on "The Oceans," 38–43. By driving the formation of storms and guiding wind circulation, the ocean makes itself felt far inland.

Whipple, A. B. C. 1982. *Storm.* Alexandria, VA: Time–Life Books. Fine general overview of storms. Excellent photos, diagrams, and charts, coupled with lively writing.

Williams, J., F. Doehring, and I. Duedall. 1993. "Heavy Weather in Florida." *Oceanus* 36 (no. 1): 19–26. A discussion of Florida hurricanes and tropical storms over a period of 122 years.

For additional readings, go to InfoTrac College Edition, your online research library at:

http://infotrac.thomsonlearning.com/

Ocean Circulation 8

STREAMING ALONG

In July 1969 six researchers sealed themselves into a specially designed submarine for a 2,640-kilometer (1,650-mile) drift in the Gulf Stream, the great ocean current that flows northward off the east coast of the United States. Their vehicle, named the *Ben Franklin* in honor of the man who first took a scientific interest in the current, was designed by Dr. Jacques Piccard, the builder of the bathyscaphe *Trieste* and father of balloonist Bertrand Piccard (see openers for Chapters 4 and 7).

The 15-meter (50-foot) *Ben Franklin* was towed into the Gulf Stream off West Palm Beach, Florida, and positioned at a depth of 150 meters (500 feet). The month-long program of observations was designed to track the Gulf Stream and measure physical conditions within it. The scientists aboard were in regular contact with two surface ships that followed the drifting submarine and monitored its exact position. They kept a nearly continuous record of water temperature, salinity, and density.

The team made surprising discoveries right from the start. First, the submarine drifted more rapidly than expected. The speed of the surface current often exceeds 8 kilometers (about 5 miles) per hour, but oceanographers believed the current would move more slowly at greater depths. It didn't; the surface ships drifted at about the same rate as the submarine. And the drift was not always smooth. Underwater hills looming from the seabed, many of them previously uncharted, disrupted the

The submarine *Ben Franklin* enters New York harbor in August 1969 after drifting for 31 days in the Gulf Stream.

flow of the submerged current and caused the submarine to rise and fall alarmingly. Gulf Stream eddies also caused problems. Thirteen days into the mission, a huge eddy spun the submarine 80 kilometers (50 miles) out of the Gulf Stream core, forcing it to surface in order to be towed back into the center of the stream. These great eddies were found to reach much deeper than previously believed (we now know that they can reach all the way to the ocean bottom).

Another surprise was the general lack of marine life. The crew did sight several species of sharks, a huge jellyfish with tentacles 10 centimeters (4 inches) thick, and an aggressive broadbilled swordfish that jousted with their vessel before swimming away unharmed; but this was less life than they had expected. The biology program did make some strides, however. Tape recordings of underwater sounds of biological origin were studied on board, and some of these recordings were later used in the first detailed analyses of whale vocalizations. Dr. Piccard photographed some unusual bioluminescent organisms, a few of which were new to science. The crew also completed medical and psychological investigations on each other and analyzed the performance of the vehicle's life support systems. These studies were used in later submarine and spacecraft design.

When the *Ben Franklin* finally bobbed to the surface 510 kilometers (317 miles) south of Halifax, Nova Scotia, researchers were confident of their biggest finding: They hadn't known as much about the Gulf Stream as they thought they did. 🔲 **8-1**

KEY CONCEPTS

1. Ocean water circulates in currents caused mainly by wind friction and differences in water mass density.

2. Surface currents affect the uppermost 10% of the world ocean. Some surface currents are rapid and riverlike, with well-defined boundaries; others are slow and diffuse. The largest surface currents are organized into huge circuits known as gyres.

3. Circulation of the 90% of ocean water beneath the surface zone is driven by the force of gravity, as dense water sinks and less dense water rises. Since density is largely a function of temperature and salinity, the movement of deep water due to density differences is called thermohaline circulation.

4. El Niño, an anomaly in surface circulation, occurs when the trade winds falter, allowing warm water to build eastward across the Pacific at the equator.

CHAPTER AT A GLANCE

About 10% of the water in the world ocean is involved in **surface currents,** water flowing horizontally in the uppermost 400 meters (1,300 feet) of the ocean's surface, driven mainly by wind friction. Most surface currents move water above the pycnocline, the zone of rapid density change with depth.

Water within and below the pycnocline also circulates, but the power for this slower, deeper circulation comes from the action of gravity on adjacent water masses of different densities. Since density is largely a function of temperature and salinity, circulation due to density differences is called *thermohaline circulation.* We'll investigate thermohaline circulation after a discussion of surface currents.

SURFACE CURRENTS

8-2

The primary force responsible for surface currents is wind. As you read in Chapter 7, surface winds form global patterns within latitude bands (see Figures 7.9 and **8.1**). Most of Earth's surface wind energy is concentrated in each hemisphere's trade winds (easterlies) and westerlies. Waves on the sea surface transfer some of the energy from the moving air to the water by friction. This tug of wind on the ocean surface begins a mass flow of water. The water flowing beneath the wind forms a surface current.

The moving water will "pile up" in the direction the wind is blowing. Water pressure will be higher on the "piled-up" side, and the force of gravity will act to pull the water down the slope—against the *pressure gradient*—in the direction from which it came. But the Coriolis effect intervenes. Because of the Coriolis effect, Northern Hemisphere surface currents

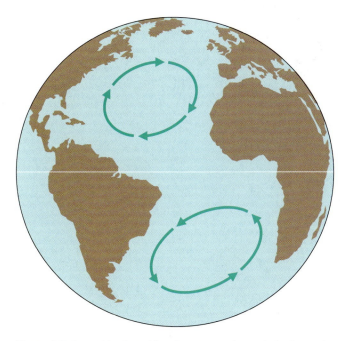

Figure 8.2 A combination of four forces—surface winds, the sun's heat, the Coriolis effect, and gravity—circulates the ocean surface clockwise in the Northern Hemisphere and counterclockwise in the Southern Hemisphere, forming gyres.

flow to the *right* of the wind direction. Southern Hemisphere currents flow to the *left*. Continents and basin topography often block continuous flow and help to deflect the moving water into a circular pattern. This flow around the periphery of an ocean basin is called a **gyre** (*gyros* = a circle). Two gyres are shown in **Figure 8.2.**

Flow Within a Gyre

8-3

Figure 8.3 shows the North Atlantic gyre in more detail. Though it flows continuously without obvious places where one current ceases and another begins, oceanographers subdivide the North Atlantic gyre into four interconnected currents because each has distinct flow characteristics and temperatures. (Gyres in other ocean basins are similarly divided.) Notice that the east-west currents in the North Atlantic gyre flow to the right of the driving winds; once initiated, water flow in these currents continues in a roughly east-west direction. Where their flow is blocked by continents, the currents turn clockwise to complete the circuit.

Why does water flow around the *periphery* of the ocean basin instead of spiraling to the center? After all, the Coriolis effect influences any moving mass *as long as it moves,* so water in a gyre might be expected to curve to the center of the North Atlantic and stop. To understand this aspect of current movement, imagine the forces acting on the surface water at 45° north latitude (point A in **Figure 8.4**). Here the westerlies blow from the southwest, so initially the water will move toward the northeast. The rightward Coriolis deflection then causes the water to flow almost due east. A particle at 15° north latitude (point B) responds to the push of the trade winds from

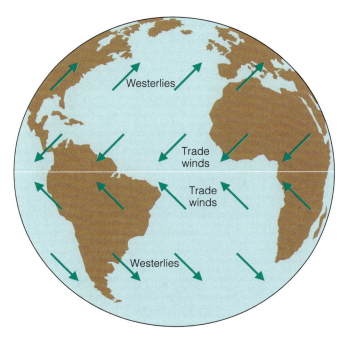

Figure 8.1 Winds, driven by uneven solar heating and Earth's spin, drive the movement of the ocean's surface currents. The prime movers are the powerful westerlies and the persistent trade winds (easterlies).

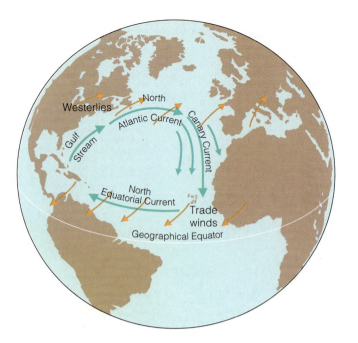

Figure 8.3 The North Atlantic gyre, a series of four interconnecting currents with different flow characteristics and temperatures.

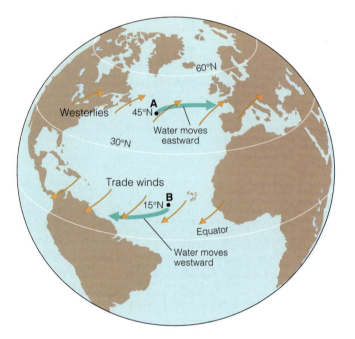

Figure 8.4 Surface water blown by the winds at point A will veer to the right of its initial path and continue eastward. Water at point B veers right and continues westward.

the northeast, however, and with Coriolis deflection it will flow almost due west.

When driven by the wind, the topmost layer of ocean water in the Northern Hemisphere flows at about 45° to the right of the wind direction, a flow consistent with the arrows leading away from points A and B in Figure 8.4. But what about the water in the next layer down? It can't "feel" the wind at the surface; it "feels" only the movement of the water immediately above. This deeper layer of water moves *at an angle to the right*

of the overlying water. The same thing happens in the layer below that, and the next layer, and so on, to a depth of about 100 meters (330 feet) at mid-latitudes. Each layer slides horizontally over the one beneath it like cards in a deck, with each lower card moving at an angle slightly to the right of the one above. Because of frictional losses, each lower layer also moves more slowly than the layer above. The resulting situation, portrayed in **Figure 8.5,** is known as an *Ekman spiral* after the Swedish oceanographer who worked out the mathematics involved. The term is somewhat misleading; the water itself does not spiral downward in a whirlpool-like motion. Rather, the spiral is a way of conceptualizing the horizontal movements in a layered water column, each layer moving in a slightly different horizontal direction. An unexpected result of the Ekman spiral is that at some depth (known as the friction depth), water will be flowing in the opposite direction from the surface current!

The *net* motion of the water down to about 100 meters, after allowance for the summed effects of the Ekman spiral (the sum of all the arrows indicating water direction in the affected layers), is known as **Ekman transport.** In theory, the direction of Ekman transport is 90° to the *right* of wind direction in the Northern Hemisphere and 90° to the *left* in the Southern Hemisphere.

Armed with this information, we can look in more detail at the area around point B in Figure 8.4, which is enlarged in **Figure 8.6.** In nature, Ekman transport in gyres is less than 90°; in most cases the deflection barely reaches 45°. This deviation from theory occurs because of an interaction between the Coriolis effect and the pressure gradient. Some flowing Atlantic water has turned to the right to form a hill of water—it followed the rightward dotted-line arrow in Figure 8.6. Why does the water now go straight west from point B without turning? Because, as **Figure 8.7a** shows, to turn further *right* the water would have to move uphill against the pressure gradient (and in defiance of gravity), but to turn *left* in response to the pressure gradient would defy the Coriolis effect. So the water continues westward and then clockwise around the whole North Atlantic gyre, dynamically balanced between the downhill urge of the pressure gradient and the uphill tendency of Coriolis deflection.

Yes, there really *is* a hill near the middle of the North Atlantic, centered in the area of the Sargasso Sea (**Figure 8.7b**). This hill is formed of surface water gathered at the ocean's center of circulation. It is not a steep mountain of water—its maximum height is an unspectacular 2 meters (6.5 feet)—but rather a gradual rise and fall from coastline to open ocean and back to opposite coastline. Its slope is so gradual you wouldn't notice it on a transatlantic crossing.

The hill is maintained by wind energy. If the winds did not continuously inject new energy into currents, friction within the fluid mass and with the surrounding ocean basins would slow the flowing water, gradually converting its motion into heat. The balance of wind energy and friction, and of the Coriolis effect and the pressure gradient (through the effect of gravity), propels the currents of the gyre and holds them along the outside edges of the ocean basin.

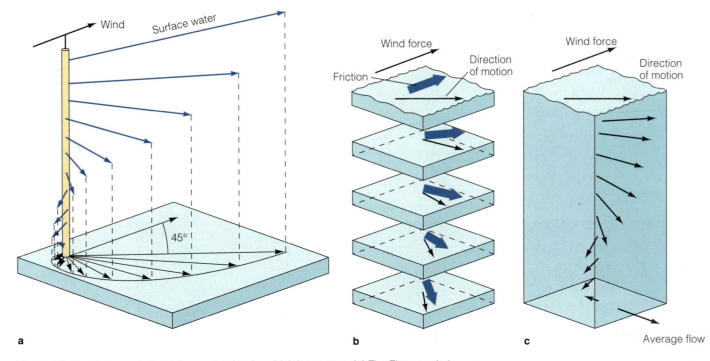

a

b

c   Average flow

Figure 8.5 The Ekman spiral and the mechanism by which it operates. (**a**) The Ekman spiral model. (**b**) A body of water can be thought of as a set of layers. The top layer is driven forward by the wind, and each layer below is moved by friction. Each succeeding layer moves at a slower speed than and at an angle to the layer immediately above it—to the right in the Northern Hemisphere, to the left in the Southern Hemisphere—until friction becomes negligible. (**c**) Though the direction of movement varies for each layer in the stack, the theoretical net flow of water in the Northern Hemisphere is 90° to the right of the prevailing wind force. The length of the arrows is proportional to the speed of the current in each layer.

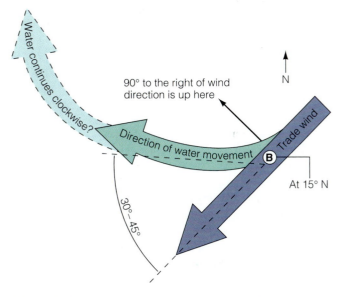

Figure 8.6 The movement of water away from point B in Figure 8.4 is influenced by the rightward tendency of the Coriolis effect and the gravity-powered movement of water down the pressure gradient.

Geostrophic Gyres

 8-4

Gyres in balance between the pressure gradient and the Coriolis effect are called **geostrophic gyres** (*geos* = Earth, *strophe* = turning), and their currents are *geostrophic currents*. Because of the patterns of driving winds and the present positions of continents, the geostrophic gyres are largely independent of each other in each hemisphere.

There are six great current circuits in the world ocean, two in the Northern Hemisphere and four in the Southern Hemisphere. They are shown in **Figure 8.8.** Five are geostrophic gyres: the North Atlantic gyre, the South Atlantic gyre, the North Pacific gyre, the South Pacific gyre, and the Indian Ocean gyre. Though it is a closed circuit, the sixth and largest current is technically not a gyre because it does not flow around the periphery of an ocean basin. The **West Wind Drift,** or **Antarctic Circumpolar Current,** as this exception is called, flows endlessly eastward around Antarctica, driven by powerful, nearly ceaseless westerly winds. This greatest of all surface ocean currents is never deflected by a continent.

We might expect the two gyres in the North and South Pacific (and the two gyres in the North and South Atlantic) to converge exactly at the geographical equator. However, as Figure 8.8 shows, the junction of equatorial currents lies a few

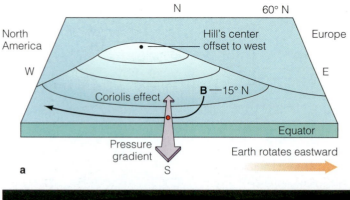

a

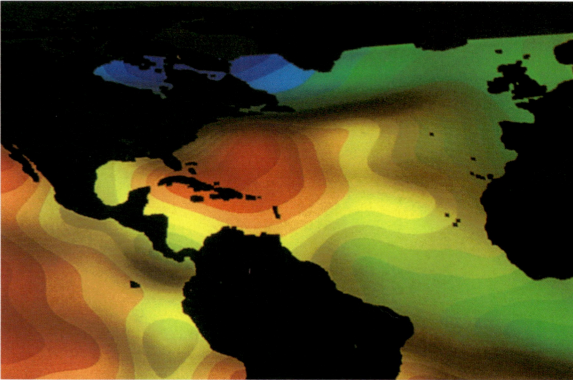

b

Figure 8.7 (**a**) The surface of the North Atlantic is raised through wind motion and Ekman transport to form a low hill. The eastward rotation of Earth offsets the center of the hill to the west. Water from point B (see also Figures 8.4 and 8.6) turns westward and flows along the side of this hill. The westward-moving water is balanced between the Coriolis effect (which would turn the water to the right) and flow down the pressure gradient, driven by gravity (which would turn it to the left). Thus, water in a gyre moves along the outside edge of an ocean basin. (**b**) The average height of the surface of the North Atlantic is shown in color in this image derived from data taken in 1992 by the *TOPEX/Poseidon* satellite. Red indicates the highest surface, green and blue the lowest. Note that the measured position of the hill is offset to the west as seen in (**a**). The gradually sloping hill is only 2 meters (6.5 feet) high and would not be apparent to anyone crossing the ocean.

degrees north of the geographical equator, at the **meteorological equator** (the imaginary line that divides Earth into two equally warm hemispheres). The meteorological equator and the intertropical convergence zone (the band at which the trade winds converge) are displaced about 6° northward mainly because of the heat accumulated in the Northern Hemisphere's greater tropical land surface area. Ocean circulation, like atmospheric circulation, is balanced around the meteorological equator.

Currents Within Gyres   8-5

Because of the different factors that drive and shape them, the currents constituting geostrophic gyres have different characteristics. Geostrophic currents may be classified by their position within the gyre as western boundary currents, eastern boundary currents, or transverse currents.

WESTERN BOUNDARY CURRENTS   The fastest and deepest geostrophic currents are found at the western boundaries of ocean basins (that is, off the *east* coast of continents). These narrow, fast, deep currents move warm water poleward in each of the gyres. There are five large **western boundary currents**: the Gulf Stream (in the North Atlantic), the Japan or Kuroshio Current (in the North Pacific), the Brazil Current (in the South Atlantic), the Agulhas Current (in the Indian Ocean), and the East Australian Current (in the South Pacific).

The **Gulf Stream** is the largest of the western boundary currents. Studies of the Gulf Stream, such as those undertaken by the drift submarine *Ben Franklin* (see chapter opener), have revealed that off Miami, the Gulf Stream moves at an average speed of 2 meters per second (5 miles per hour) to a depth of over 450 meters (1,500 feet). Water in the Gulf Stream can move more than 160 kilometers (100 miles) in a day. Its average width is about 70 kilometers (43 miles).

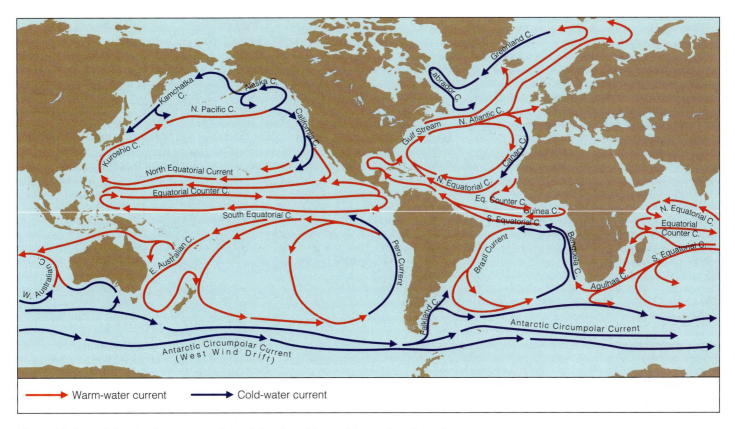

Figure 8.8 A chart showing the names and usual direction of the world ocean's major surface currents.

➤ Warm-water current ➤ Cold-water current

The volume of water transported in western boundary currents is extraordinary. The unit used to express volume transport in ocean currents is the **sverdrup** (**sv**), named in honor of Harald Sverdrup, one of this century's pioneering oceanographers. A sverdrup equals 1 million cubic meters per second.[1] The Gulf Stream flow is at least 55 sv (55 million cubic meters per second), about 300 times the usual flow of the Amazon, the greatest of rivers. In **Figure 8.9,** the surface currents of the North Atlantic gyre are shown with their volume transport (in sverdrups) indicated.

Water in a current, especially a western boundary current, can move for surprisingly long distances within well-defined boundaries. In the Gulf Stream the current-as-river analogy can be startlingly apt: The western edge of the current is often clearly visible. Water within the current is usually warm, clear, and blue, often depleted of nutrients and incapable of supporting much life. By contrast, water over the continental slope adjacent to the current is often cold, green, and teeming with life.

Long, straight edges are the exception rather than the rule in western boundary currents, however. Unlike rivers, ocean currents lack well-defined banks, and friction with adjacent water can cause a current to form waves along its edges. Western boundary currents meander as they flow poleward. The looping meanders sometimes connect to form turbulent rings,

or **eddies,** that trap cold or warm water in their centers and then separate from the main flow. For example, *cold-core eddies* form in the Gulf Stream as it meanders eastward upon leaving the coast of North America off Cape Hatteras (see **Figure 8.10**). *Warm-core eddies* can form north of the Gulf Stream when the warm current loops into the cold water lying to the north. When these loops are cut off, they become

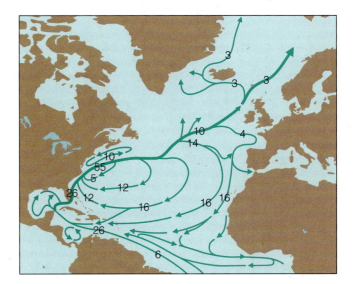

Figure 8.9 The general surface circulation of the North Atlantic. The numbers indicate flow rates in sverdrups (1 sv = 1 million cubic meters of water per second).

---

[1] One million cubic meters is about half the volume of the Louisiana Superdome.

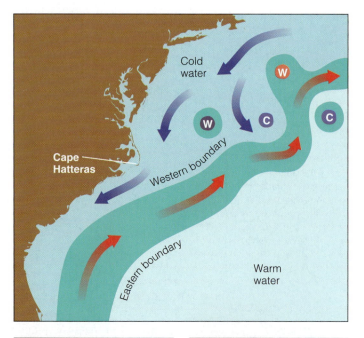

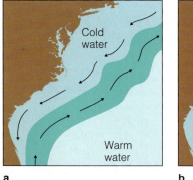

a

b

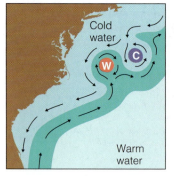

c

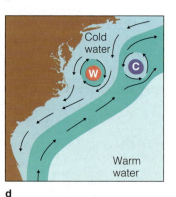

d

Figure 8.10 Eddy formation. The western boundary of the Gulf Stream is usually distinct, marked by abrupt changes in water temperature, speed, and direction. Meanders (eddies) form at this boundary as the Gulf Stream leaves the U.S. coast at Cape Hatteras (**a**). The meanders can pinch off (**b**) and eventually become isolated cells of warm water between the Gulf Stream and the coast (**c**). Likewise, cold cells can pinch off and become entrained in the Gulf Stream itself (**d**). (C = cold water, W = warm water; blue = cold, red = warm.) Figure 8.11 shows the Gulf Stream from space with meanders and eddies clearly visible.

freestanding spinning masses of water. Warm-core eddies rotate clockwise, and cold-core eddies rotate counterclockwise.

The slowly rotating eddies move away from the current and are distributed across the North Atlantic. Some may be 1,000 kilometers (620 miles) in diameter and retain their identity for more than three years. In the mid-latitudes as much as one-fourth of the surface of the North Atlantic may consist of old, slow-moving, cold-core eddy remnants! Both cold and warm eddies are visible in the satellite image of **Figure 8.11**. Their influence reaches to the seafloor. Warm- and cold-core eddies are probably responsible for the slowly moving *abyssal storms* that leave often-observed ripple marks in deep sediments.

Figure 8.11 The Gulf Stream viewed from space. This image is a composite of temperature data returned from NOAA polar-orbiting meteorological satellites during the first week of April 1984. The composite image is printed with an artificial color scale: Reds and oranges are a warm 24–28°C (76–84°F); yellows and greens are 17–23°C (63–74°F); blues are 10–16°C (50–61°F); and purples are a cold 2–9°C (36–48°F). The Gulf Stream appears like a red (warm) river as it moves from the southern tip of Florida ① north along the east coast. Moving offshore at Cape Hatteras ②, it begins to meander, with some meanders pinching off to form warm-core ③ and cold-core ④ eddies. As it moves northeastward, the water cools dramatically, releasing heat to the atmosphere and mixing with the cooler surrounding waters. By the time it reaches the middle of the North Atlantic, it has cooled so much that its surface temperature can no longer be distinguished from that of the surrounding waters.

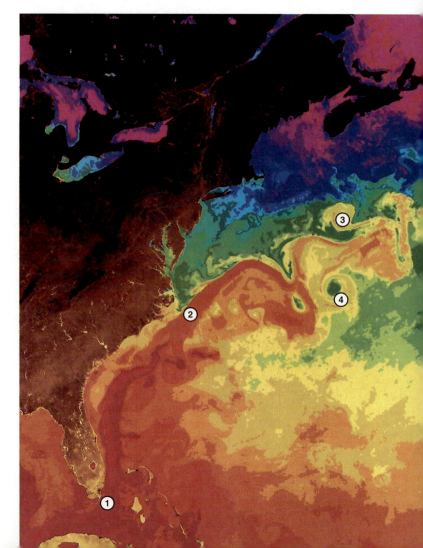

EASTERN BOUNDARY CURRENTS   There are five **eastern boundary currents** found at the eastern edge of ocean basins (that is, off the *west* coast of continents): the Canary Current (in the North Atlantic), the Benguela Current (in the South Atlantic), the California Current (in the North Pacific), the West Australian Current (in the Indian Ocean), and the Peru or Humboldt Current (in the South Pacific).

Eastern boundary currents are the opposite of their western boundary counterparts in nearly every way: They carry cold water equatorward; they are shallow and broad, sometimes more than 1,000 kilometers (620 miles) across; their boundaries are not well defined; and eddies tend not to form. Their total flow is less than that of their western counterparts. The Canary Current in the North Atlantic carries only 16 sv of water at about 2 kilometers (1.2 miles) per hour. The current is so shallow and broad that sailors may not notice it. Contrast the flow rates of the North Atlantic's western and eastern boundary currents in Figure 8.9. **Table 8.1** summarizes the major differences between boundary currents in the Northern Hemisphere.

WESTWARD INTENSIFICATION   Why should western boundary currents be concentrated and eastern boundary currents be diffuse? One reason is the converging flow of the trade winds on either side of the equator. Water moved by the trades approaches the meteorological equator and is shepherded west, where it piles up at the western edge of the basin before turning swiftly poleward. This concentration of water produces the poleward-moving western boundary currents. This poleward-moving water is traveling fast, and because the Coriolis effect increases with velocity, the pressure gradient must also be stronger on the steeper (western) side of the hill described in Figure 8.7. The heightened rightward tendency of the Coriolis effect is then balanced by the stronger, steeper pressure gradient, resulting in even more rapid poleward flow. (In contrast, the westerly winds of each hemisphere do not converge, and water driven by them is not swept along a line of convergence. Coriolis deflection can therefore move some of the eastward-moving water equatorward before the basin's eastern boundary is reached.)

A second reason is the rotation of Earth itself. The hill of Figure 8.7 is offset to the west because of Earth's eastward rotation, so water must squeeze closer to the ocean basin's western edge to pass around the hill at the western boundary. The combined effect on current flow is known as **westward intensification,** a phenomenon clearly visible in Figures 8.7 and 8.9.

TRANSVERSE CURRENTS   As we have seen, most of the power for ocean currents is derived from the trade winds at the fringes of the tropics, and from the mid-latitude westerlies. The stress of winds on the ocean in these bands gives rise to the **transverse currents**—currents that flow from east to west and west to east, thus linking the eastern and western boundary currents.

The trade wind–driven North and South Equatorial Currents in the Atlantic and Pacific are moderately shallow and broad, but each transports about 30 sv westward. Because of the thrust of the trades, Atlantic water at Panama is usually 20 centimeters (8 inches) higher, on average, than water across the isthmus in the Pacific. The Pacific's greater expanse of water at the equator and stronger trade winds develop more powerful westward-flowing equatorial currents, and the height differential between the western and eastern Pacific is thought to approach 1 meter (3.3 feet)!

Westerly winds drive the eastward-flowing transverse currents of the mid-latitudes. Because they are not shepherded by the trade winds, eastward-flowing currents are wider and flow more slowly than their equatorial counterparts. The North Pacific and North Atlantic Currents are Northern Hemisphere examples.

As can be seen in Figure 8.8, the westward flow of the transverse currents near the equator proceeds unimpeded for great distances, but the eastward flow of transverse currents at middle and high latitudes in the northern ocean basins is interrupted by continents and island arcs. In the far south,

Table 8.1   Boundary Currents in the Northern Hemisphere

Type of Current, Example	General Features	Speed	Transport (millions of cubic meters per second)	Special Features
Western Boundary Currents Gulf Stream, Kuroshio (Japan) Current	**Warm** Narrow, <100 km. Deep—substantial transport to depths of 2 km.	Swift, hundreds of kilometers per day	Large, usually 50 sv or greater	Sharp boundary with coastal circulation system; little or no coastal upwelling; waters tend to be depleted in nutrients, unproductive; waters derived from trade wind belts.
Eastern Boundary Currents California Current, Canary Current	**Cold** Broad, ~1,000 km. Shallow, <500 m.	Slow, tens of kilometers per day	Small, typically 10–15 sv.	Diffuse boundaries separating from coastal currents; coastal upwelling common; waters derived from mid-latitudes

however, eastward flow is almost completely free. Intense westerly winds over the southern ocean drive the greatest of all ocean currents, the unobstructed Antarctic Circumpolar Current (or West Wind Drift). This current carries more water than any other—at least 100 sv west-to-east in the Drake Passage between the tip of South America and the adjacent Palmer Peninsula of Antarctica.

Gyres: A Final Word 8-6

Although we have stressed individual currents in our discussion, remember that gyres consist of currents that blend into one another. Flow is continuous without obvious places where one current ceases and another begins. The balance of wind energy, friction, Coriolis effect, and pressure gradient propels gyres and holds them along the outside of ocean basins.

EFFECTS OF SURFACE CURRENTS ON CLIMATE 8-7

Surface currents distribute tropical heat worldwide. Warm water flows to higher latitudes, transfers heat to the air and cools, moves back to low latitudes, absorbs heat again, and the cycle repeats. The greatest amount of heat transfer occurs at the mid-latitudes, where about 10 million billion calories of heat are transferred *each second*—more than a million times the power consumed by all the world's human population in the same length of time! This combination of water flow and heat transfer from and to water influences climate and weather in several ways.

In winter, for example, Scotland and England are bathed in eastward-moving air only recently in contact with the relatively warm North Atlantic Current and therefore have a maritime climate. Edinburgh and London are warmed in part by the energy of tropical sunlight transported to their high latitudes by the Gulf Stream (see once again Figure 8.8).

At lower latitudes on an ocean's eastern boundary the situation is often reversed. Mark Twain is supposed to have said that the coldest winter he ever spent was a summer in San Francisco. Summer months in that West Coast city are cool, foggy, and mild, while Washington, D.C., on nearly the same line of latitude (but on the western boundary of an ocean basin), is known for its August heat and humidity. Why the difference? Look at Figure 8.8 and follow the currents responsible. The California Current, carrying cold water from the north, comes close to the coast at San Francisco. As shown in **Figure 8.12,** air normally flows clockwise in summer around an offshore zone of high atmospheric pressure. Wind approaching the California coast loses heat to the cold sea and comes ashore to chill San Francisco. Summer air often flows around a similar high off the East Coast (the Bermuda High). Winds approaching Washington, D.C. therefore blow from the south and east. Heat and moisture from the Gulf Stream contribute to the capital's oppressive summers. (In winter, on the other

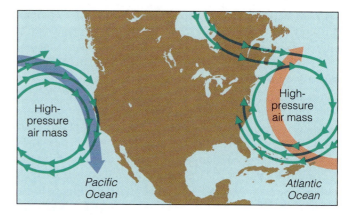

Figure 8.12 General summer air circulation patterns of the east and west coasts of the United States. Warm ocean currents are shown in red, cold currents in blue. Air is chilled as it approaches the west coast and warmed as it approaches the east coast.

hand, Washington D.C. is colder than San Francisco because westerly winds approaching Washington are chilled by the cold continent they cross.)

UPWELLING AND DOWNWELLING 8-8

The wind-driven *horizontal* movement of water can sometimes induce *vertical* movement in the surface water. This movement is called *wind-induced vertical circulation.* Upward movement of water is known as **upwelling**; the process brings deep, cold, usually nutrient-laden water toward the surface. Downward movement is called **downwelling.**

Equatorial Upwelling 8-9

Because the meteorological equator usually lies about 5° north of the geographical equator, the South Equatorial Currents of the Atlantic and Pacific straddle the geographical equator. Though the Coriolis effect is weak near the geographical equator (and absent *at* the geographical equator), water moving in the currents on either side of the geographical equator is deflected slightly poleward and replaced by deeper water (**Figure 8.13**). Thus, **equatorial upwelling** occurs in these westward-flowing equatorial surface currents. Upwelling is an important process because this water from within and below the pycnocline is often rich in the nutrients needed by marine organisms for growth. The long, thin band of upwelling and biological productivity extending along the equator westward from South America is clearly visible in Figure 8.15b. The layers of ooze on the equatorial Pacific seabed (Figure 5.11) are testimony to the biological productivity of surface water there. By contrast, generally poor conditions for growth prevail in most of the open tropical ocean—because strong layering isolates deep, nutrient-rich water from the sunlit ocean surface.

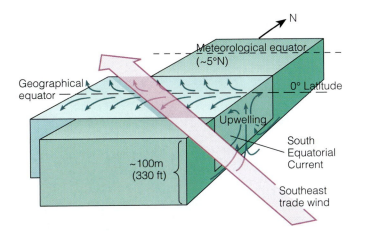

Figure 8.13 Equatorial upwelling. The South Equatorial Current, especially in the Pacific, straddles the geographical equator (see again Figure 8.8). Water north of the equator veers to the right (northward), and water to the south veers to the left (southward). Surface water therefore diverges, causing upwelling. Most of the upwelled water comes from the area above the equatorial undercurrent, at depths of 100 meters or less.

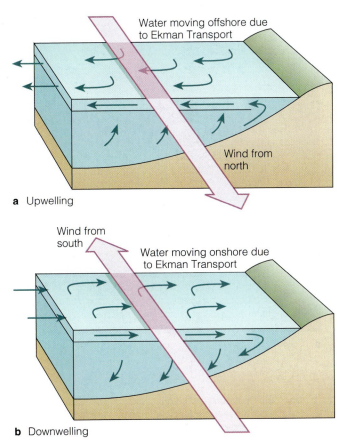

a Upwelling

b Downwelling

Figure 8.14 Coastal upwelling and downwelling. (**a**) In the Northern Hemisphere, coastal upwelling can be caused by winds from the north blowing along the west coast of a continent. Water moved offshore by Ekman transport is replaced by cold, deep, nutrient-laden water. (**b**) A prolonged southerly wind along a Northern Hemisphere west coast can result in downwelling.

Coastal Upwelling and Downwelling

 8-10

Wind blowing parallel to shore or offshore can cause **coastal upwelling.** The friction of wind blowing along the ocean surface causes the water to begin moving, the Coriolis effect deflects it to the right (in the Northern Hemisphere), and the resultant Ekman transport moves it offshore. As shown in **Figure 8.14a,** coastal upwelling occurs when this surface water is replaced by water rising along the shore. Again, because the new surface water is often rich in nutrients, prolonged wind can result in increased biological productivity. Coastal upwelling along the coast of South America is also visible in Figure 8.15b.

Upwelling can also influence weather. Wind blowing from the north along the California coast causes offshore movement of surface water and subsequent coastal upwelling. The overlying air becomes chilled, contributing to San Francisco's famous fog banks and cool summers. Wind-induced upwelling is also common in the Peru Current, along the west coast of Antarctica's Palmer Peninsula, in parts of the Mediterranean, and near some large Pacific islands.

Water driven toward a coastline will be forced downward, returning seaward along the continental shelf. This downwelling (**Figure 8.14b**) helps supply the deeper ocean with dissolved gases and nutrients, and it assists in the distribution of living organisms. Unlike upwelling, downwelling has no direct effect on the climate or productivity of the adjacent coast.

EL NIÑO AND LA NIÑA

 8-11

Surface winds across most of the tropical Pacific normally move from east to west (review Figure 7.9). The trade winds blow from the normally high-pressure area over the eastern Pacific (near Central and South America) to the normally stable low-pressure area over the western Pacific (north of Australia). However, for reasons that are still unclear, these pressure areas change places at irregular intervals of roughly three to eight years: High pressure builds in the western Pacific, and low pressure dominates the eastern Pacific. Winds across the tropical Pacific then reverse direction and blow from west to east—the trade winds weaken or reverse. This change in atmospheric pressure (and thus in wind direction) is called the **Southern Oscillation.** Ten of these oscillations have occurred since 1950.

The trade winds normally drag huge quantities of water westward along the ocean's surface on each side of the equator, but as the winds weaken these equatorial currents crawl to a stop. Warm water that has accumulated at the western side of the Pacific—the warmest water in the world ocean—can then build to the east along the equator toward the coast of Central and South America. The eastward-moving warm water usually arrives near the South American coast around Christmastime. In the 1890s it was reported that Peruvian fishermen were using the expression *Corriente del Niño* ("current of the Christ Child") to describe the flow; hence the current's name,

Figure 8.15 A non-El Niño year. (**a**) Normally the air and surface water flow westward, the thermocline rises, and upwelling of cold water occurs along the west coast of Central and South America. (**b**) This map from satellite data shows the temperature of the equatorial Pacific on 31 May 1988. The warmest water is indicated by the dark red, and progressively cooler water by yellow and green. Note the coastal upwelling along the coast at the lower right of the map, and the tongue of recently upwelled water extending westward along the equator from the South American coast. An El Niño year.

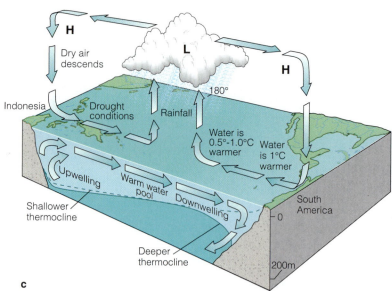

c

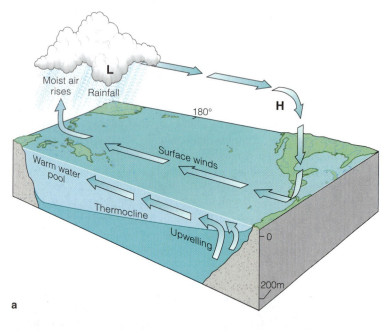

a

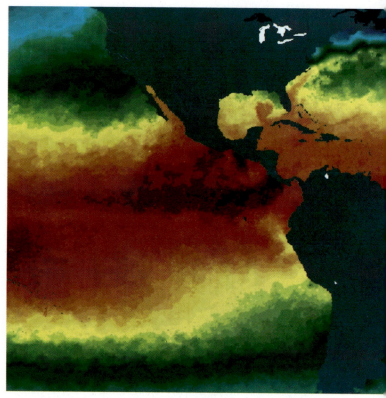

d

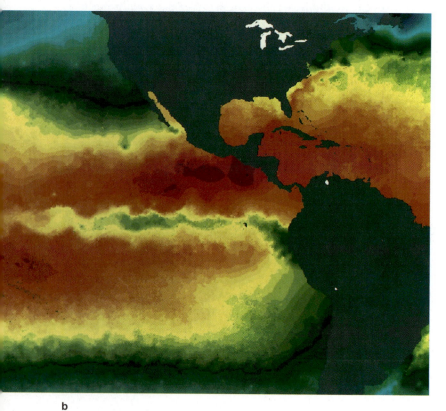

b

(**c**) When the Southern Oscillation develops, the trade winds diminish and then reverse, leading to an eastward movement of warm water along the equator. The surface waters of the central and eastern Pacific become warmer, and storms over land may increase. (**d**) Sea-surface temperatures on 13 May 1992, a time of El Niño conditions. The thermocline was deeper than normal, and equatorial upwelling was suppressed. Note the absence of coastal upwelling along the coast and the lack of a tongue of recently upwelled water extending westward along the equator.

El Niño. The phenomena of the Southern Oscillation and El Niño are coupled, so the terms are often combined to form the acronym **ENSO,** for El Niño/Southern Oscillation. An ENSO event typically lasts about a year, but some have persisted for more than three years. The effects are felt not only in the Pacific; all ocean areas at trade wind latitudes in both hemispheres can be affected.

Normally, a current of cold water, rich in upwelled nutrients, flows north and west away from the South American continent (**Figures 8.15a** and **b**). When the propelling trade winds falter during an ENSO event, warm equatorial water that would normally flow westward in the equatorial Pacific backs up to flow east (**Figures 8.15c** and **d**). The normal northward flow of the cold Peru Current is interrupted or overridden by the warm water. Upwelling within the nutrient-laden Peru Current is responsible for the great biological productivity of the ocean off the coasts of Peru and Chile. Although upwelling may continue during an ENSO event, the source of the upwelled water is nutrient-depleted water in the thickened surface layer approaching from the west. When the Peru Current slows and its upwelled water lacks nutrients, fish and seabirds dependent on the abundant life it contains die or migrate elsewhere. Peruvian fishermen are never cheered by this Christmas gift!

During major ENSO events, sea level rises in the eastern Pacific, sometimes by as much as 20 centimeters (8 inches) in the Galápagos. Water temperature also increases by up to 7°C (13°F). The warmer water causes more evaporation, and the area of low atmospheric pressure over the eastern Pacific intensifies. Humid air rising in this zone, centered some 2,000 kilometers (1,200 miles) west of Peru, causes high precipitation in normally dry areas. The increased evaporation intensifies coastal storms, and rainfall inland may be much higher than normal. Marine and terrestrial habitats and organisms can be affected by these changes.

The two most severe ENSO events of the twentieth century occurred in 1982–83 and 1997–98 (**Figure 8.16**). In both cases, effects associated with El Nio were spectacular over much of the Pacific and some parts of the Atlantic and Indian Oceans. In February 1998, 40 people were killed and 10,000 buildings damaged by a "wall" of tornadoes advancing over the southeastern United States. This record-breaking tornado event was spawned by the collision of warm, moist air that had lingered over the warm Pacific and a polar front that dropped from the north. In the eastern Pacific, heavy rains through the 1997–98 winter in Peru left at least 250,000 people homeless, destroyed 16,000 dwellings, and closed every port in the country for at least a month. Hawaii, however, was left with record

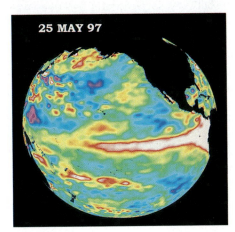

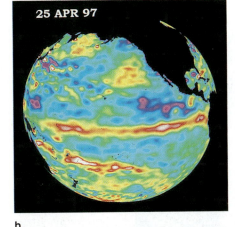

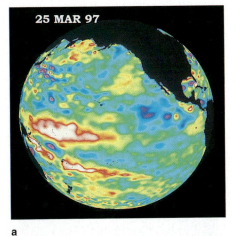

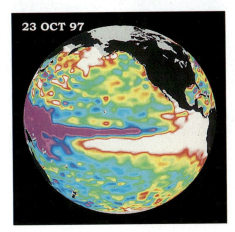

Figure 8.16 Development of the 1997–98 El Niño, observed by the *TOPEX/Poseidon* satellite. (**a**) March 1997. The slackening of the trade winds and westerly wind bursts allow warm water to move away from its usual location in the western Pacific Ocean. Red and white colors indicate sea level above average height. (**b**) April 1997. About a month after it began to move, the leading edge of the warm water reaches South America. (**c**) May 1997. Warm water piles up against the South American continent. The white area of sea level is 13–30 centimeters (5–12 inches) above normal height, and 1.6–3°C (3–5°F) warmer. (**d**) October 1997. By October, sea level is up to 30 centimeters (12 inches) lower than normal near Australia. The bulge of warm water has spread northward along the coast of North America from the equator to Alaska. Fisheries in Peru are severely affected—the warm water prevents upwelling of cold, nutrient-rich water necessary for the support of large fish populations.

drought, and some parts of southwestern Africa and Papua New Guinea received so little rain that crops failed completely and whole villages were abandoned due to starvation. Most of the United States escaped serious consequences—indeed, the midwestern states, Pacific Northwest, and eastern seaboard enjoyed a relatively mild fall, winter, and spring. But California's trials were widely reported—rainfall in most of the state exceeded twice normal amounts, and landslides, avalanches, and other weather-related disasters crowded the evening news. Conditions did not return to near-normal until the late spring of 1998. Estimates of worldwide 1997–98 ENSO-related damage exceed 23,000 deaths and $33 billion.

Normal circulation sometimes returns with surprising vigor, producing strong currents, powerful upwelling, and chilly and stormy conditions along the South American coast. These contrasting colder-than-normal events are given a contrasting name: **La Niña** (the girl). As conditions to the east cool off, the ocean to the west (north of Australia) warms rapidly. The renewed thrust of the trades piles this water upon itself, depressing the upper curve of the thermocline to more than 100 meters (328 feet). In contrast, the thermocline during a La Niña event in the eastern equatorial Pacific rests at about 25 meters (82 feet). A vigorous La Niña followed the 1997–98 El Niño and persisted for more than a year and a half (**Figure 8.17**).

Studies of the ocean and the atmosphere in 1982–83 and 1997–98 have given researchers new insight into the behavior

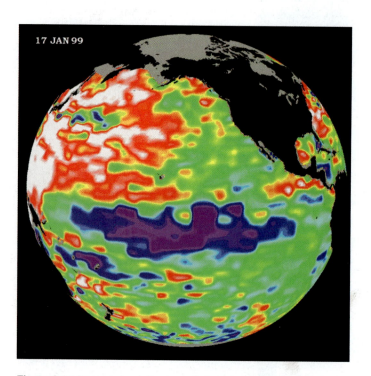

17 JAN 99

Figure 8.17 Normal circulation sometimes returns with surprising vigor after an El Niño event, producing strong currents, powerful upwelling, and chilly and stormy conditions along the South American coast. This image was prepared from data for 17 January 1999. Note the mass of cold surface water and relatively low sea level (purple). Such cold water tends to deflect winds around it, changing the course of weather systems locally and the nature of weather patterns globally.

and effects of the Southern Oscillation. Some researchers believe that the 1982–83 event was triggered by the violent 1982 eruption of El Chichón, a Mexican volcano, which injected huge quantities of obscuring dust and sulfur-rich gases into the atmosphere. No similar trigger occurred before the 1997–98 ENSO, however, and researchers have recently suggested that the Southern Oscillation may be triggered by pulses of heat entering the Pacific at crustal spreading zones near the equator. Some of this heat is moved by currents to the ocean surface. The warming of surface water may cause air above the ocean to warm, atmospheric pressure to fall, and the east-to-west trade winds to slacken. Though the exact cause or causes of the Southern Oscillation are not yet understood, subtle changes in the atmosphere permit meteorologists to predict a severe El Niño nearly a year in advance of its most serious effects.

THERMOHALINE CIRCULATION 8-12

The surface currents we have discussed affect the uppermost layer of the world ocean (about 10% of its volume), but horizontal and vertical currents also exist below the pycnocline in the ocean's deeper waters. The slow circulation of water at great depths is driven by density differences rather than by wind energy. Because density is largely a function of water temperature and salinity, the movement of water due to differences in density is called **thermohaline circulation** (*therme* = heat, *halos* = salt). Virtually the entire ocean is involved in slow thermohaline circulation, a process responsible for most of the vertical movement of ocean water, and the circulation of the global ocean as a whole.

Water Masses 8-13

As you may recall from Chapter 6, the ocean is density stratified, with the densest water near the seafloor and the least dense near the surface. Each water mass has specific temperature and salinity characteristics. Density stratification is most pronounced at temperate and tropical latitudes because the temperature difference between surface water and deep water is greater there than near the poles.

The water masses possess distinct, identifiable properties. Like air masses, water masses don't often mix easily when they meet, due to their differing densities; instead, they usually flow above or beneath each other. Water masses can be remarkably persistent and will retain their identity for great distances and long periods of time. Oceanographers name water masses according to their relative position.

In temperate and tropical latitudes, there are five common water masses:

- *Surface water*, to a depth of about 200 meters (660 feet)
- *Central water*, to the bottom of the main thermocline (which varies with latitude)

- *Intermediate water,* to about 1,500 meters (5,000 feet)
- *Deep water,* water below intermediate water but not in contact with the bottom, to a depth of about 4,000 meters (13,000 feet)
- *Bottom water,* water in contact with the seafloor

Surface currents move in the relatively warm upper environment of surface and central water. The boundary between central water and intermediate water is the most abrupt and pronounced.

No matter at what depth they are located, the characteristics of each water mass are usually determined by conditions of heating, cooling, evaporation, and dilution that occurred at the ocean surface when the mass was formed. The densest (and deepest) masses were formed by surface conditions that caused the water to become very cold and salty. Water masses near the surface can be warmer and less saline; they may have formed in warm areas where precipitation exceeded evaporation. Water masses at intermediate depths are intermediate in density.

In spite of this differentiation, the relatively cold water masses lying beneath the thermocline exhibit smaller variations in salinity and temperature than the water in the currents that move across the ocean's surface.

Formation and Downwelling of Deep Water

 8-14

Antarctic Bottom Water, the most distinctive of all deep-water masses, is characterized by a salinity of 34.65‰, a temperature of $-0.5°C$ (30°F), and a density of 1.0279 grams per cubic centimeter. This water is noted for its extreme density (the densest in the world ocean), for the great amount of it produced near Antarctic coasts, and for its ability to migrate north along the seafloor.

Most Antarctic Bottom Water forms in the Weddell Sea during winter. Sea ice can incorporate only about 15% of seawater's salt, and the salt remaining in the unfrozen water beneath the ice forms a frigid brine. Between 20 and 50 million cubic meters of this brine forms every second! The water's great density causes it to sink toward the continental shelf, where it mixes with nearly equal parts of water from the southern Antarctic Circumpolar Current.

The mixture settles along the edge of Antarctica's continental shelf, descends along the slope, and spreads along the deep-sea bed, creeping north in slow sheets. Antarctic Bottom Water flows many times more slowly than the water in surface currents: In the Pacific it may take a thousand years to reach the equator. Six hundred years later it may be as far away as the Aleutian Islands at 50°N! Antarctic Bottom Water also flows into the Atlantic Ocean basin, where it flows north at a faster rate than in the Pacific. Antarctic Bottom Water has been identified as high as 40° *north* latitude on the Atlantic floor, a journey that has taken some 750 years.

Some dense bottom water also forms in the northern polar ocean, but the topography of the Arctic Ocean basin prevents most of it from escaping, except in the deep channels formed in the submarine ridges separating Scotland, Iceland, and Greenland. These channels allow the cold, dense water formed in the Arctic to flow into the North Atlantic, forming **North Atlantic Deep Water.** Very little deep Arctic Ocean water enters the Pacific.

Other distinct deep-water masses exist. Their positions in relation to each other are always determined by their relative densities. North Atlantic Deep Water forms when the relatively warm and salty North Atlantic Ocean cools as cold winds from northern Canada sweep over it. Exposed to the chilled air, water at the latitude of Iceland releases heat, cools from 10°C to 2°C (50°F to 36°F), and sinks. (Transferred to the air, this bonus heat helps to moderate European winters.) As can be seen in **Figure 8.18,** similar water forms at the Antarctic Circumpolar Current in the South Atlantic. Pacific Deep Water also forms in the Antarctic Circumpolar Current, and along the east coast of the Kamchatka Peninsula. Atlantic or Pacific Deep Water is less dense than Antarctic Bottom Water and so floats above it, out of contact with the ocean floor.

A deep-water mass also forms in the enclosed Mediterranean Sea, where surface water is made more saline by the excess of evaporation over freshwater input. About 300,000 cubic kilometers (72,000 cubic miles) more water evaporates annually from the Mediterranean than is replaced by river runoff or precipitation. In the cool winter months, Mediterranean water with a salinity of about 38‰ flows past the lip of Gibraltar and spreads into the Atlantic as Mediterranean Deep Water. Mediterranean Deep Water underlies much of the central water mass in the Atlantic, and some of this water can be traced as far south as the basins of the Antarctic.

Researchers can determine the age of deep water by analyzing its dissolved oxygen content. Ocean water picks up free oxygen only at or near the surface by contact with the atmosphere or through the action of photosynthetic plants. After leaving the surface, the water gradually loses its oxygen through the respiration of organisms or by chemical reactions with sediments, rocks, or dissolved components. The dissolved oxygen content of a water mass is therefore a rough index of its age—the length of time since the water left the surface. Researchers have also found that water masses slowly mix with the surrounding water as they flow away from their sources, losing their individual identity as they grow older.

Thermohaline Circulation Patterns

8-15

The great quantities of dense water sinking at ocean basin edges must be offset by equal quantities of water rising elsewhere. **Figure 8.19** shows an idealized model of thermohaline flow. Note that water sinks relatively rapidly in a small area where the ocean is very cold, and it rises much more gradually across a very large area in the warmer temperate and tropical zones. It then slowly returns poleward near the surface to repeat the cycle. The continual diffuse upwelling of deep water maintains the existence of the permanent thermocline found everywhere at low and mid-latitudes. This slow upward movement is estimated to be about 1 centimeter (½ inch) per day over most of the ocean. If this rise were to stop, downward

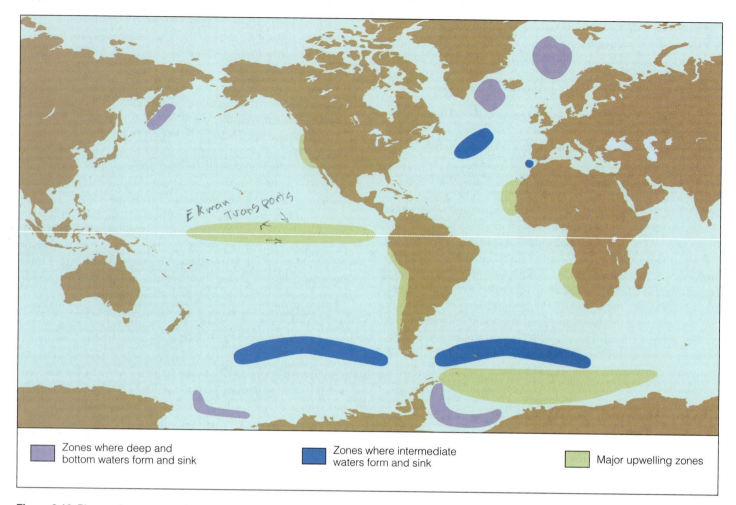

Figure 8.18 Places where most surface water sinks and where most deep or intermediate water rises.

Legend:

- Zones where deep and bottom waters form and sink
- Zones where intermediate waters form and sink
- Major upwelling zones

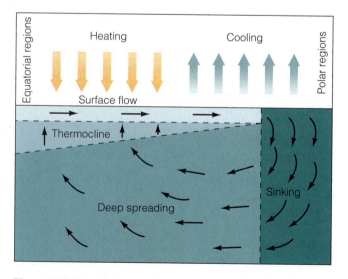

Figure 8.19 The classical model of a pure thermohaline circulation, caused by heating in lower latitudes and cooling in higher latitudes. The thermocline at middle and low latitudes is "held up" by the slow upward movement of cold water. Compare this figure to Figure 7.4.

movement of heat would cause the thermocline to descend and would reduce its steepness. In a sense, the thermocline is "held up" by the continual slow upward movement of cold water.

Most features of this ideal circulation pattern exist in nature. **Figure 8.20** shows deep circulation in the Atlantic. The water masses, each of distinct density and sandwiched in layers, are slowly propelled by gravity. Masses butt against one another in **convergence zones,** and the heavier water can slide beneath the lighter water. Hundreds of years may pass before water masses complete a circuit or blend to lose their identities. Antarctic Bottom Water in the Pacific retains its character for up to 1,600 years! The residence time of most deep water is less, however; it takes about 200 to 300 years to rise to the surface. (By contrast, a bit of surface water in the North Atlantic gyre may take only a little more than a year to complete a circuit.)

Not all thermohaline circulation is so sedate. Ripple marks in sediments, scour lines, and the erosion of rocky outcrops on deep-ocean floors provide evidence that relatively strong, localized bottom currents exist (see Figure 5.3). Some of these currents may move as rapidly as 60 centimeters (24 inches) per second. These relatively fast currents are strongly influenced

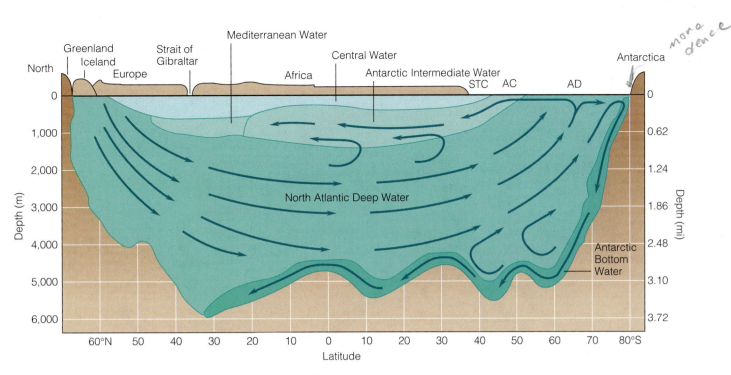

handwritten: more dence

Figure 8.20 The water layers and deep circulation of the Atlantic Ocean. Arrows indicate the direction of water movement. STC indicates the position of the Subtropical Convergence, AC the Antarctic Convergence, and AD the Antarctic Divergence. The surface layer is too thin to show clearly at this scale. The vertical scale is greatly exaggerated.

by bottom topography, and they are sometimes called *contour currents* because their dense water flows around (rather than over) seafloor projections. Bottom currents generally move equatorward at or near the western boundaries of ocean basins (below the western boundary surface currents). The subtle density differences between deep-water masses are not capable of moving water at the speed of the wind-driven surface currents. Water in some bottom currents may move only 1 to 2 meters (3 to 7 feet) per day. Even at that slow speed, the Coriolis effect modifies their pattern of flow.

Thermohaline Flow and Surface Flow: The Global Heat Connection

8-16

As we have seen, swift and narrow currents along the western margins of ocean basins carry warm tropical surface waters toward the poles. In a few places, as shown in Figure 8.18, the water loses heat to the atmosphere and sinks to become deep water and bottom water. This sinking is most pronounced in the North Atlantic. The cold, dense water moves at great depths toward the Southern Hemisphere and eventually wells up into the surface layers of the Indian and Pacific Oceans. Almost a thousand years are required for this water to make a complete circuit.

The transport of tropical water to the polar regions is part of a global conveyor belt for heat. A simplified outline of the

global circuit, the result of three decades of concentrated effort to understand deep circulation, is shown in **Figure 8.21.** This slow circulation straddles the hemispheres and is superimposed upon the more rapid flow of water in surface gyres. Recent analysis of this global circuit suggests that some of the heat warming the European coast enters the ocean in the vicinity of Indonesia and Australia, travels to the Indian Ocean, and enters the Gulf Stream by way of the Agulhas Current rounding the southern tip of Africa. The surface water that leaves the Pacific is driven in part by excess rainfall and river runoff throughout the Pacific basin. The slow, steady, three-dimensional flow of water in the conveyor belt distributes dissolved gases and solids, mixes nutrients, and transports the juvenile stages of organisms between ocean basins.

STUDYING CURRENTS

8-17

Surface currents can be traced with drift bottles or drift cards. These tools are especially useful in determining coastal circulation, but they provide no information on the path the drift bottle or card may have taken between its release and collection points. Researchers wishing to know the precise track taken by a drifting object can deploy more elaborate drift devices, such as drogues that can be tracked continuously by radio direction finders or radar. Surface currents can also be

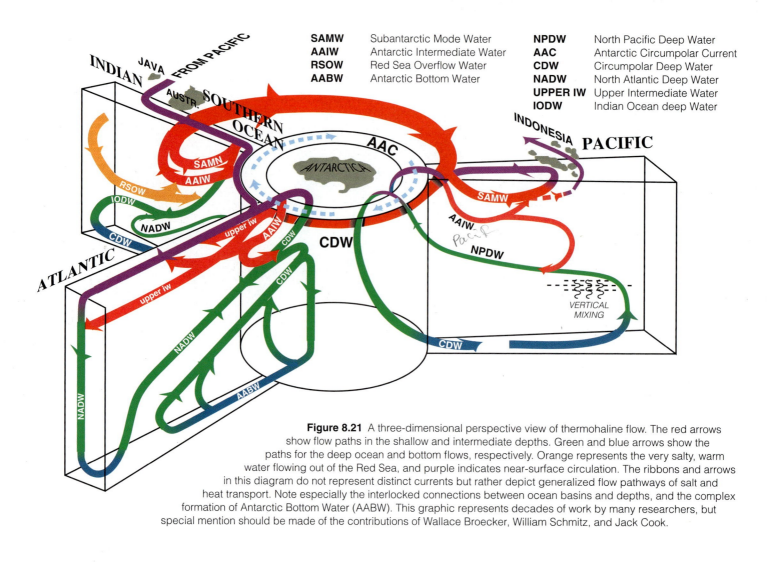

SAMW Subantarctic Mode Water
AAIW Antarctic Intermediate Water
RSOW Red Sea Overflow Water
AABW Antarctic Bottom Water

NPDW North Pacific Deep Water
AAC Antarctic Circumpolar Current
CDW Circumpolar Deep Water
NADW North Atlantic Deep Water
UPPER IW Upper Intermediate Water
IODW Indian Ocean deep Water

Figure 8.21 A three-dimensional perspective view of thermohaline flow. The red arrows show flow paths in the shallow and intermediate depths. Green and blue arrows show the paths for the deep ocean and bottom flows, respectively. Orange represents the very salty, warm water flowing out of the Red Sea, and purple indicates near-surface circulation. The ribbons and arrows in this diagram do not represent distinct currents but rather depict generalized flow pathways of salt and heat transport. Note especially the interlocked connections between ocean basins and depths, and the complex formation of Antarctic Bottom Water (AABW). This graphic represents decades of work by many researchers, but special mention should be made of the contributions of Wallace Broecker, William Schmitz, and Jack Cook.

tracked by noting the difference between the daily expected and observed positions of ships at sea.

Bottle, card, or drogue studies are almost always carefully planned and executed, but not all surface drift releases have been intentional. In May 1990, a violent storm struck the containership *Hansa Carrier* enroute between Korea and Seattle. Twenty-one boxcar-sized cargo containers were lost overboard, among them containers holding 30,910 pairs of Nike athletic shoes. About six months later, shoes from the broken containers began washing up on beaches from British Columbia to Oregon. Because the shoes were not tied together in pairs, beachcombers placed advertisements in local newspapers and held swap meets to exchange the shoes (which were in excellent condition despite having been exposed to the ocean). Oceanographers noticed the ads and asked the media to request that individuals let them know where and when they found shoes. By knowing the place where the shoes were lost and the places where they were found, researchers have been able to refine their computer models of the North Pacific gyre. Some of the shoes have completed a full circuit of the North Pacific. A similar spill occurred on 10 January 1992,

when a storm-beset freighter lost a container filled with 29,000 rubber ducks, turtles, and other bathtub toys in the North Pacific (**Figure 8.22a**). At last report, 400 of the toys had been recovered from 500 miles of Alaskan shoreline.

Deeper currents can also be surveyed by free-floating devices. Sophisticated devices developed in the 1970s and 1980s depend on the sofar layer (see Figure 6.21) to transmit sound. Low-frequency tones are broadcast from autonomous submerged probes to moored listening stations (see **Figure 8.22b**). These sofar probes have accumulated more than 240 float-years of data in the North Atlantic, at depths of from 700 to 2,000 meters (2,300 to 6,600 feet). One of these drifting probes has been sending data for nine years!

Another device (**Figure 8.22c**) glides smoothly up and down through the water column powered by gravity and buoyancy. Energy to pump ballast overboard, allowing the little glider to rise, is provided by a simple heat engine powered by the difference in temperature between ocean surface and great depths. To fall again, the glider pumps seawater aboard. Named "Slocums" after Joshua Slocum, the first person to circumnavigate the globe alone, such gliders could travel with

a **b** **c**

Figure 8.22 Methods for measuring currents. (**a**) A rubber duck of the kind lost overboard in large numbers in 1992 in an unintentional drift experiment (see text). (**b**) A sofar float being launched from the Woods Hole Oceanographic Institution's research ship *Oceanus*. The probe will drop to a depth of 3,500 meters (11,500 feet) and produce a low-frequency tone once each day for tracking. (**c**) A Slocum glider—a probe that uses energy from gravity, buoyancy, heat, and batteries to power long-range exploration of water masses.

the currents and map the ocean's thermohaline depth profiles for years, transmitting data to satellites on their occasional visits to the surface.

Yet another method, developed for the study of thermohaline circulation, senses the presence in seawater of tritium, a radioactive isotope of hydrogen. A small amount of tritium is produced in the upper atmosphere when hydrogen is bombarded by cosmic rays, but most of it was produced by hydrogen bomb testing in the 1960s. Most tritium combines with oxygen to become radioactive water, and some of this water enters the ocean by precipitation or river runoff. At high latitudes this water sinks and, labeled by its tritium content, can be traced by sampling. The speed of deep currents has been measured by analysis of their tritium content.

Study of currents is the very heart of physical oceanography. Their global effects on climate, vast masses of water, complex flow, and possible influence on human migrations make their study of particular importance.

QUESTIONS FROM STUDENTS

1. If the Gulf Stream warms Britain during the winter and keeps Baltic ports free of ice, why doesn't it moderate New England winters? After all, Boston is closer to the warm core of the Gulf Stream than London is.

Yes, but remember the direction of prevailing winds in winter. Winter winds at Boston's latitude are generally from the west, so any warmth is simply blown out to sea. On the other side of the Atlantic the same winds blow toward London. It does get cold in London, but generally winters in London are much milder than those in Boston.

2. Could currents be used as a source of electrical power? With the Gulf Stream so close to Florida, it seems that some way could be devised to take advantage of all that water flow to turn a turbine.

It's been considered. The total energy of the Gulf Stream flowing off Miami has been estimated at 25,000 megawatts! A Woods Hole Oceanographic Institution team has proposed a honeycomblike array of turbines for the layer between 30 and 130 meters (100 and 430 feet) across 20 kilometers (13 miles) of the current. They estimate a power output of around 1,000 megawatts, equal to the generation potential of two large nuclear power plants. Engineering difficulties would be considerable, however.

SURFING

More than 2 million Americans surf regularly. Rushing down the face of a growing, breaking wave is exhilarating! People willingly endure cold, boredom, and some danger for a ride lasting only a few seconds. If you're a good swimmer, give it a try. You will need to paddle your board or swim vigorously to match your speed to that of the advancing wave crest. As the wave rises to break, your forward speed (and sense of timing) will place you on the leading edge of the crest, accelerating downward and forward. The technique takes time to master, but the feeling is worth the effort.

Although riding a surfboard can be great fun, I have long believed that the most rewarding surfing is body surfing. The buoyant forces on which board surfing depends are diminished in body surfing, but forces within the wave propel the body surfer with greater acceleration. A body surfer's intimate contact with the water and close proximity to the wave surface greatly increase the sensation of movement.

The ultimate trick is to surf the wave completely submerged, as a dolphin does. You push powerfully off the bottom (or swim rapidly toward the surface) as the wave crest approaches, inserting yourself into the freshly breaking wave from below. The rapid flow of water within the toppling wave will ripple your skin, and the sudden acceleration can thrust you from the face of the wave like a wet bar of soap shooting from a fist. Altogether a *wonderful* oceanic experience! 9-1

Body surfing. Many surfers prefer body surfing to board surfing because of the body surfer's proximity to the water and the sensation of great speed and acceleration.

1. Waves transmit energy, not water mass, across the ocean's surface.

2. The speed of ocean waves usually depends on their wavelength, with long waves moving fastest.

3. In order of wavelength from short to long (and therefore from slowest to fastest), ocean waves are generated by very small disturbances (capillary waves), wind (wind waves), rocking of water in enclosed spaces (seiches), seismic and volcanic activity or other sudden displacements (tsunami), and gravitational attraction (tides).

4. The behavior of a wave depends largely on the relation between the wave's size and the depth of water through which it is moving.

CHAPTER AT A GLANCE

To most people an ocean wave in deep water appears to be a massive moving object—a ridge of water traveling across the sea surface. An ocean wave is one of several kinds of **waves,** all of which are disturbances caused by the movement of energy from a source through some medium (solid, liquid, or gas). As the energy of the disturbance travels, the medium through which it passes moves in specific ways. Sometimes this movement is visible to us as crests in the medium. The traveling crests produce the appearance of movement we see in a wave. In an ocean wave, a ribbon of *energy* is moving at the speed of the wave, but *water* is not.

Picture a resting sea gull as it bobs on the wavy ocean surface far from shore. The gull moves in *circles*—up and forward as the tops of the waves move to his position, down and backward as the tops move past. Each circle is equal in diameter to the wave's height. As can be seen in **Figure 9.1,** energy in waves flows past the resting bird, but the gull and its patch of water move only a very short distance forward in each up-and-forward, down-and-back wave cycle. The water on which the bird rests does not move continuously across the sea surface as the wave illusion suggests.[1]

The transfer of energy from water particle to water particle in these circular paths, or **orbits,** transmits wave energy across the ocean surface and causes the wave form to move. This kind of wave is known as an **orbital wave**—a wave in which particles of the medium (water) move in closed circles as the wave passes. Orbital ocean waves occur at the boundary between two fluid media (between air and water) and between layers of water of different densities. Because the waveform moves forward, these waves are a type of **progressive wave.**

The progressive wave that moved the gull was probably caused by wind. Other forces can generate much greater progressive waves in which water molecules move through much larger circular or elliptical orbits. Some of these waves are so large that they do not appear to us as waves at all, but rather as the slow sloshing of water in a harbor or bay, as dangerous flooding surges of water, or as rhythmic and predictable ocean tides.

Ocean waves have distinct parts. The **wave crest** is the highest part of the wave above average water level; the **wave trough** is the valley between wave crests below average water level. **Wave height** is the vertical distance between a wave crest and the adjacent trough, while **wavelength** is the horizontal distance between two successive crests (or troughs). The relationship between these parts is shown in **Figure 9.2.** The time it takes for a wave to move a distance of one wavelength is known as the **wave period. Wave frequency** is the number of waves passing a fixed point per second.

---

[1] To clarify the important idea of wave-as-illusion, imagine yourself at a sports stadium where spectators are doing "the wave." Your role in wave propagation is simple—you stand up and sit down in precise synchronization with your neighbors. Though you move only a few feet vertically, the wave of which you were a part circles the arena at high speed. You and all the other participants stay in place, but the wave moves faster than anyone can run.

Figure 9.1 A floating sea gull demonstrates that wave forms travel but the water itself does not. In this sequence, a wave moves from left to right as the gull (and the water in which it is resting) revolves in a circle, moving slightly to the left up the front of an approaching wave, then to the crest, then sliding to the right down the back of the wave.

The circular motion of water particles at the surface of a wave continues underwater. As **Figure 9.3** shows, the diameter of the orbits through which water particles move diminishes rapidly with depth. For all practical purposes wave motion is negligible below a depth of half the wavelength, where the circles are only 1/23 the diameter of those at the surface. This means divers in 20 meters of water would not notice the passage of a wind wave of 30-meter wavelength and might barely

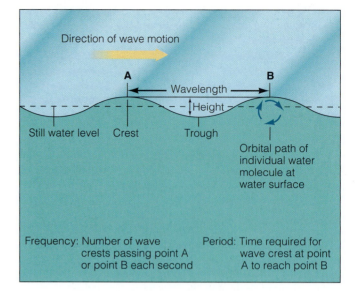

Figure 9.2 The anatomy of a progressive wave.

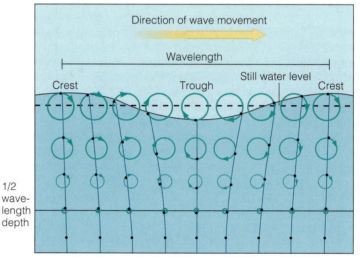

Figure 9.3 The orbital motion of water particles in a wave, which extends to a depth of about one-half of the wavelength.

notice the wave if they were at a depth of 15 meters. Since most ocean waves have moderate wavelengths, the circular disturbance of the ocean that propagates these waves affects only the uppermost layer of water. Note that the movement of water in circles doesn't resemble interlocking mechanical gears. Instead, there is a coordinated, uniform circular movement of water molecules in one direction as the waves pass.

CLASSIFYING WAVES

9-2

Ocean waves are classified by the *disturbing force* that creates them, the *restoring force* that tries to flatten them, and their *wavelength*. (Wave height is not often used for classification because it varies greatly depending on water depth, interference between waves, and other factors.)

Disturbing Force

 9-3

Energy that causes ocean waves to form is called a **disturbing force.** Wind blowing across the ocean surface provides the disturbing force for *wind waves.* Arrival of a storm surge or seismic sea wave in an enclosed harbor or bay, or a sudden change in atmospheric pressure, are the disturbing forces for a resonant rocking of water known as a *seiche.* Landslides, volcanic eruptions, or faulting of the seafloor associated with earthquakes are the disturbing forces for *tsunami.* The disturbing forces for *tides* are changes in the magnitude and direction of gravitational forces among Earth, moon, and sun, combined with Earth's rotation.

Restoring Force

9-4

Restoring force is the dominant force trying to return the water surface to flatness after a wave has formed in it. If the restoring force of a wave were quickly and fully successful, a disturbed sea surface would immediately become smooth and the energy of the embryo wave would be dissipated as heat. But that isn't what happens. Waves continue after they form because the restoring force overcompensates and causes oscillation. The situation is analogous to a weight bobbing at the bottom of a very flexible spring, constantly moving up and down past its normal resting point.

The dominant restoring force for very small water waves—those with wavelengths of less than 1.73 centimeters (0.68 inch)—is surface tension (cohesion), the property that enables individual water molecules to stick to each other by means of hydrogen bonds (see Figure 6.2). These **capillary waves** are transmitted across a puddle because cohesion tugs the tiny wave troughs and crests toward flatness.

Capillary waves are the first waves to form when the wind blows. These small ripples are important in transferring energy from air to water to drive ocean currents but are of little consequence in the overall picture of ocean waves because they are tiny and carry very little energy.

All waves with wavelengths greater than 1.73 centimeters depend mostly on gravity to provide the restoring force. Gravity pulls the crests downward, but the momentum of the water causes the crests to overshoot and become troughs. The repetitive nature of this movement, like the spring weight moving up and down, gives rise to the circular orbits of individual water molecules in an ocean wave. Since the circular motion of water molecules in a wave is nearly friction-free, gravity waves

can travel across thousands of miles of ocean surface without dissipating, eventually to break on a distant shore.

Wavelength

9-5

Wavelength is an important measure of wave size. **Table 9.1** lists the causes and typical wavelengths of the four types of ocean waves: wind waves, seiches, tsunami, and tides. **Figure 9.4** shows the relation between the disturbing and restoring forces, the period, and the relative amount of energy present in the ocean's surface for each wave type. Note that more energy is stored in wind waves than in any other wave type.

DEEP-WATER WAVES, SHALLOW-WATER WAVES

9-6

Most of the characteristics of ocean waves depend on the relationship between their wavelength and water depth. Wavelength determines the *size* of the orbits of water molecules within a wave, but water depth determines the *shape* of the orbits. The paths of water molecules in a wind wave are circular only when the wave is traveling in deep water—that is, water deeper than half its wavelength. Waves moving through water deeper than half their wavelength are known as **deep-water waves.** For example, a wind wave with a 20-meter wavelength will act as a deep-water wave if it is passing through water more than 10 meters deep (**Figure 9.5a**). In this case, the wave cannot "feel" the bottom, because too little wave energy is contained in the small circles below that depth.

The situation is different for wind-generated waves close to shore. The orbits of water molecules in waves moving through shallow water are flattened by the proximity of the bottom. Water just above the seafloor cannot move in a circular path, only forward and backward. Waves in water shallower than 1/20 their original wavelength are known as **shallow-water waves** (**Figure 9.5b**). A wave with a 20-meter wavelength will act as a shallow-water wave if the water is less than 1 meter deep.

Of the four wave types listed in Table 9.1, only wind waves can ever be deep-water waves. To understand why, remember that most of the ocean floor is deeper than 125 meters (400 feet), half the wavelength of very large wind waves. The wavelengths of the larger waves are *much* longer—the wavelength of seismic sea waves usually exceeds 100 kilometers

Table 9.1	Wavelengths and Disturbing Forces of the Four Types of Ocean Waves	
Wave Type	**Typical Wavelength**	**Disturbing Force**
Wind wave	60–150 m (200–500 ft)	Wind over ocean
Seiche	Large, variable; a fraction of basin size	Changes in atmospheric pressure, storm surge, tsunami
Tsunami	200 km (125 mi)	Faulting of seafloor, volcanic eruption, landslide
Tide	½ circumference of Earth	Gravitational attraction, rotation of Earth

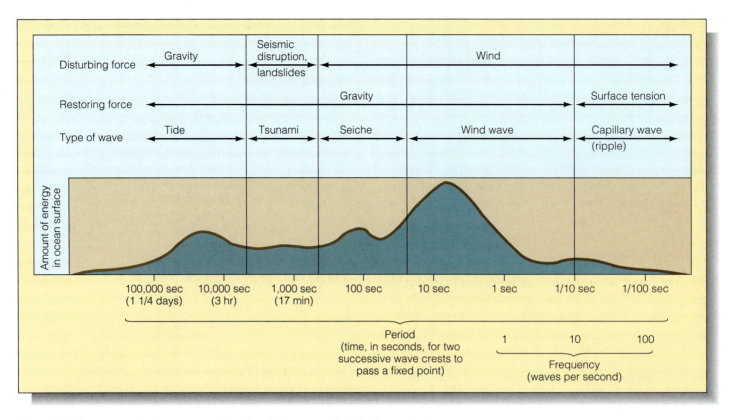

Disturbing force	Gravity		Seismic disruption, landslides	Wind			
Restoring force				Gravity		Surface tension	
Type of wave	Tide		Tsunami	Seiche	Wind wave	Capillary wave (ripple)	

Amount of energy in ocean surface

| 100,000 sec (1 1/4 days) | 10,000 sec (3 hr) | 1,000 sec (17 min) | 100 sec | 10 sec | 1 sec | 1/10 sec | 1/100 sec |

Period
(time, in seconds, for two
successive wave crests to
pass a fixed point)

1   10   100

Frequency
(waves per second)

Figure 9.4 Wave energy in the ocean as a function of the wave period. As the graph shows, most wave energy is typically concentrated in wind waves. However, large tsunami, rare events in the ocean, can transmit more energy than all wind waves for a brief time.

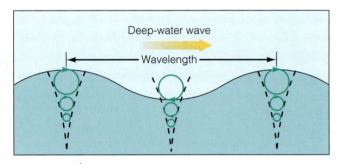

a Depth $\geq \frac{1}{2}$ wavelength

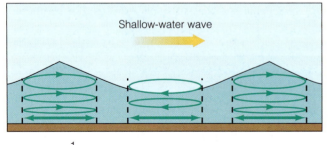

b Depth $\leq \frac{1}{20}$ wavelength

Figure 9.5 Progressive waves. (**a**) A deep-water wave; (**b**) a shallow-water wave. These diagrams are not to scale.

(62 miles). No part of the ocean is 50 kilometers (31 miles) deep; so seiches, seismic sea waves, and tides are always in water that—to them—is shallow. Their huge orbital circles flatten against a distant bottom always less than half a wavelength away.

In general, the longer the wavelength of a wave, the faster the wave energy will move through the water. For *deep-water* waves this relationship is shown in the formula

$$S = L/T$$

in which S represents speed, L is wavelength, and T is time, or period (in seconds). Wavelength is difficult to determine at sea, but period is comparatively easy to find—for example, by an observer timing the movement of waves past the bow of a stopped ship. If period (T) is known, speed (S) can be calculated from the relation

$$S \text{ (in meters/sec)} = 1.56(T)$$

or, if you prefer,

$$S \text{ (in feet/sec)} = 5(T).$$

Figure 9.6 shows the relationship between wavelengths of deep-water waves and their speed and period.

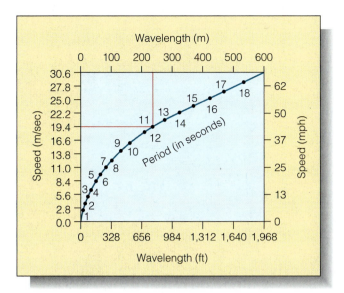

Figure 9.6 The theoretical relationship among speed, wavelength, and period in deep-water waves. Speed is equal to wavelength divided by period. If one characteristic of a wave can be measured, the other two can be calculated. The easiest to measure exactly is period. In the example shown in red, the speed of a wave with a wavelength of 228 meters and a period of 12 seconds is 19 meters per second.

The action of shallow-water waves is described by a different equation that can be written as

$$S = \sqrt{gd}, \text{ or } S = 3.1\sqrt{d}$$

where S is speed (in meters per second), g the acceleration due to gravity (an average of 9.8 meters per second per second), and d the depth of the water in meters. The *period* of a wave remains unchanged regardless of the depth of the water through which it is moving, but as deep-water waves enter the shallows and feel bottom their velocity is reduced and their crests "bunch up," so their wavelength shortens.

Following is a comparison of two very different kinds of ocean waves (a comparison of apples and oranges), but notice the general relationship between wavelength and wave velocity: The longer the wavelength, the greater the velocity.

Wind waves (deep-water waves):
- Period to about 20 *seconds*
- Wavelength to perhaps 600 meters (2,000 feet) in extreme cases
- Speed to perhaps 112 kilometers (70 miles) per hour in extreme cases

Seismic sea waves (shallow-water waves):
- Period to perhaps 20 *minutes*
- Wavelength typically 200 kilometers (125 miles)
- Speeds of 760 kilometers (470 miles) per hour

Remember that *energy*—not the water mass itself—is moving through the water at the astonishing speed of 760 kilometers (470 miles) per hour (the speed of a jet airliner!) in seismic sea waves.

Wind waves are gravity waves formed by the transfer of wind energy into water. Most wind waves are less than 3 meters (10 feet) high. Wavelengths of from 60 to 150 meters (200 to 500 feet) are most common in the open ocean.

Wind waves grow from capillary waves, tiny waves nearly always present on the ocean. Capillary waves interrupt the smooth sea surface and cause some of the wind's energy to be transferred into the water to drive the capillary wave crest forward. The wind may eddy briefly behind the tiny crest, creating a slight partial vacuum there. Atmospheric pressure pushes the trailing crest forward (downwind) toward the trough, adding still more energy to the water surface. The process is shown in **Figure 9.7**. The increasing energy in the water surface expands the circular orbits of water particles in the direction of the wind, enlarging the small wave's size. The capillary wave becomes a wind wave when its wavelength exceeds 1.73 centimeters (0.68 inch), the wavelength at which gravity supersedes capillary action as the dominant restoring force.

If the wind wave remains in water deeper than half its wavelength, and the wind continues to blow, the wave becomes larger. Its crest is thrust higher into faster wind, extracting even more energy from the moving air. The circular orbits of water particles within the wave grow larger with more energy input; height, wavelength, and period increase proportionally. The irregular peaked waves in the area of wind wave formation are called **sea**; the chaotic surface is formed by simultaneous wind waves of many wavelengths, periods, and heights.

When the wind slows or ceases, as it does away from a storm, the wave crests become rounded and regular. These mature wind waves of uniform wavelength outside their original area of generation are called **swell.** Swell form smooth undulations in the ocean surface (**Figure 9.8**). Because waves with the longest wavelengths move fastest, swell are sorted by wavelength with longer waves moving away from the storm

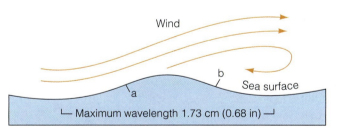

Figure 9.7 Wind forces acting on a capillary wave. Capillary waves interrupt the smooth sea surface, deflect surface wind upward, slow it, and cause some of the wind's energy to be transferred into the water to drive the capillary wave crest forward (point a). The wind may eddy briefly downwind of the tiny crest, creating a slight partial vacuum there. Atmospheric pressure pushes the trailing crest forward (downwind) toward the trough (point b), adding still more energy to the water surface. The increasing energy in the water surface expands the circular orbits of water particles in the direction of the wind, enlarging the small wave's size. The capillary wave becomes a wind wave when its wavelength exceeds 1.73 centimeters (0.68 inch), the wavelength at which gravity supersedes capillary action as the dominant restoring force.

Figure 9.8 Swell—mature, regular wind waves—off the Oregon coast. The small waves super-imposed on the swell are the result of local wind conditions.

most rapidly. Observers at a distance would first encounter large quick-moving waves of long wavelength, then middle-sized waves, then slow, small ones.

Factors Affecting Wind Wave Development

9-8

Three factors affect the growth of wind waves. Wind must be moving faster than the wave crests for energy transfer from air to sea to continue, so the mean speed of the wind, or **wind strength,** is clearly important to wind wave development. A second factor is the length of time the wind blows, or **wind duration**; high winds that blow only a short time will not generate large waves. The third is the uninterrupted distance over which the wind blows without significant change in direction— the **fetch** (**Figure 9.9**).

A strong wind must blow continuously for nearly three days for the largest waves to develop fully. A **fully developed sea** is the maximum wave size theoretically possible for a wind of a specific strength, duration, and fetch. Longer exposure to wind at that speed will not increase the size of the waves.

The greatest potential for large waves occurs beneath the strong and nearly continuous winds of the Antarctic Circum-

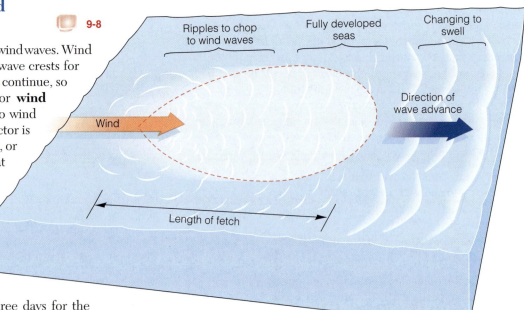

Figure 9.9 The fetch, the uninterrupted distance over which the wind blows without significant change in direction. Wave size increases with increased wind speed, duration, and fetch. A strong wind must blow continuously in one direction for nearly three days for the largest waves to develop fully.

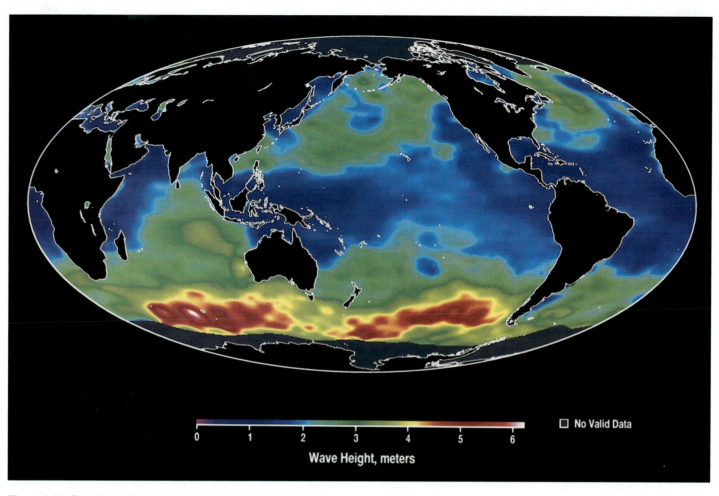

Figure 9.10 Global wave height acquired by a radar altimeter aboard the *TOPEX/Poseidon* satellite in October 1992. In this image, the highest waves occur in the southern ocean, where waves were more than 6 meters (19.8 feet) high (represented in white). The lowest waves (indicated by dark blue) are found in the tropical and subtropical ocean, where wind speed is lowest.

polar Current. The early nineteenth-century French explorer of the South Seas Jules Dumont d'Urville encountered waves with heights estimated "in excess" of 30 meters (100 feet) in Antarctic waters. Satellite observations (like those of **Figure 9.10**) have shown that wave heights to 11 meters (36 feet) are fairly common in this area.

In zones of high winds, a less than fully developed sea can also attract attention. Though wind speed within cyclonic storms is often very strong, the circular motion of air doesn't allow long fetches, and fully developed seas rarely occur beneath them. Officers standing deck watches during storms rarely quibble with theoretical maximum height versus observed height, however. Wind waves can be overwhelming even if they are not fully developed.

Wind Wave Height and Wavelength

9-9

Wave height is not directly represented in the deep-water wave formula $S = L/T$. During their formation, moderately sized wind waves in the open ocean exhibit a maximum 1:7 ratio of wave height to wavelength (see **Figure 9.11**); this ratio

is the **wave steepness.** Waves 7 meters long will not be more than 1 meter high, and waves with a 70-meter wavelength will not exceed 10 meters of height. The angle at their crest will not exceed 120°. A peaked appearance usually indicates the continuing injection of wind energy. If a wave gets any higher than the 1:7 ratio for its wavelength, it will break and excess energy from the wind will be dissipated as turbulence—hence the *whitecaps* or *combers* associated with a fully developed sea.

The highest wave ever measured was sighted on the night of 7 February 1933 by Lieutenant Frederick Marggraff, a watch officer aboard the U.S. Navy tanker *Ramapo*. USS

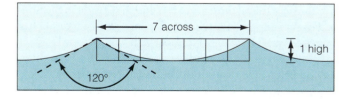

Figure 9.11 A wind wave of moderate size shown during its formation. The ratio of height to wavelength, called wave steepness, is 1:7; the crest angle does not exceed 120°.

Figure 9.12 How the great wave observed from the USS *Ramapo* was measured. An officer on the bridge was looking toward the stern and saw the crow's nest in his line of sight to the crest of the wave, which had just come in line with the horizon. Wave height was later calculated based on the geometry of the situation.

Ramapo was steaming from Manila to San Diego through a furious storm—a storm made more intense by the coalescence of three low-pressure centers. For days a steady wind had blown at 107 kilometers per hour (67 miles per hour), and gusts to 126 kilometers per hour (78 miles per hour) often lashed the decks. But the wind blew persistently from one direction, and though the monstrous waves it generated dwarfed the tanker, they were surprisingly orderly in form.

"The conditions for observing the seas from the ship were ideal," wrote *Ramapo*'s executive officer. "We were running directly down the wind with the sea. There were no cross seas and therefore no peaks along wave crests. There was practically no rolling, and the pitching motion was easy because of the fact that the sides of the waves were much longer than the ship. The moon was out astern and facilitated observations during the night. The sky was partly cloudy." At about three in the morning, Mr. Marggraff observed a train of tremendous waves looming in the moonlight. As the trough of the first wave approached the ship, he noted that its distant crest was on a level with the crow's nest on the mainmast. At that instant the ship's stern sank into the bottom of the onrushing trough. The next two waves were about the same size. Not surprisingly, such immense waves made an indelible impression on all who witnessed them.

How big were these waves? When the executive officer had some time to spare, he did some calculations. **Figure 9.12** illustrates the attitude of the ship when the largest waves were measured. The height of the waves was determined by using a set of the ship's plans, a calculation of the height of the observer above the sea surface, the draft of the ship, and a sight to the horizon. The largest wave for which a dependable observation had been made was 34 meters (112 feet) high, still a record!

But is that as big as wind waves can get? This is an interesting question, and one not dispassionately answered. Calculating the wavelength from wave period (T in $S = L/T$), historical researchers reported that mariners have sighted waves with wavelengths of 451 meters (1,481 feet) off the west coast of Ireland, 583 meters (1,914 feet) off the Cape of Good Hope,

and 829 meters (2,719 feet) in the equatorial Atlantic. (That last monster would have had a period of 25 seconds!) The nineteenth-century French admiral J. Mottez reported a wave with a wavelength of 790 meters (2,600 feet), a period of 23 seconds, and a speed of 123 kilometers (76 miles) per hour in the equatorial Atlantic west of Africa. The wavelength of the *Ramapo* wave was calculated at 360 meters (1,180 feet), so these sightings are perhaps exaggerated. Of course, much smaller waves can also add a great deal of excitement to a deck officer's day (see **Figure 9.13**)!

Interference 9-10

The real situation of wind waves at sea is not as simple as has been suggested here. The ideal vision of one set of waves moving in one direction at one speed across an otherwise smooth surface is almost never observed in the ocean.

Figure 9.13 A U.S. Navy destroyer battles large waves generated by a tropical storm in the Pacific. Despite the battering, the ship is holding a steady enough course to take on fuel and stores from the aircraft carrier barely visible on the right.

Independent wave trains exist simultaneously in the ocean most of the time. Since long waves outrun shorter ones, wind waves from different storm systems can overtake and interfere with one another. One wave doesn't crawl over the others when they meet—instead they add to (or subtract from) one another. Such interaction is known as **interference**. In **Figure 9.14a** a wave of one wavelength is represented as a blue line, a slightly longer wavelength wave as a green line. In the sea surface where these waves coincide (shown in **Figure 9.14b**), you can see the alternation between addition (large crests and troughs) and subtraction (almost no waves at all). The cancellation effect of subtraction is termed **destructive interference,** not because of harm to lives or property but because wave interference destroys or cancels the waves in-

volved. **Constructive interference** is the additive formation of large crests or deep troughs. This pattern of constructive and destructive interference explains why the ocean surface will be relatively calm for a time, rise to a few big waves, and then return to calm. Interference also explains why in some instances every ninth wave might be unusually large. As wavelengths change, though, every seventh wave might be large, or every fifth, or twelfth. Contrary to folklore, there is no set ratio.

Interference can have sudden unpleasant consequences on the open sea. In or near a large storm, wind waves at many wavelengths and heights may approach a single spot from different directions. If such a rare confluence of crests occurred at your position, a huge wave crest would suddenly erupt from a moderate sea to threaten your ship. The freak wave—called a **rogue wave**—would be much larger than any noticed before or after and would be higher than the theoretical maximum wave capable of being sustained in a fully developed sea. In such conditions one wave in about 1,175 is more than three times average height, and one in every 300,000 is more than four times average height!

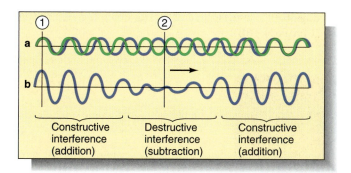

Figure 9.14 Interference. Two overlapping waves of different wavelength (**a**) generate both constructive and destructive interference (**b**).

Wind Waves Approaching Shore 9-11

Most wind waves eventually find their way to a shore and break, dissipating all their order and energy. The process begins with the transition of our now familiar deep-water wave to a shallow-water wave in water less than half a wavelength deep. **Figure 9.15** outlines the events leading to the break:

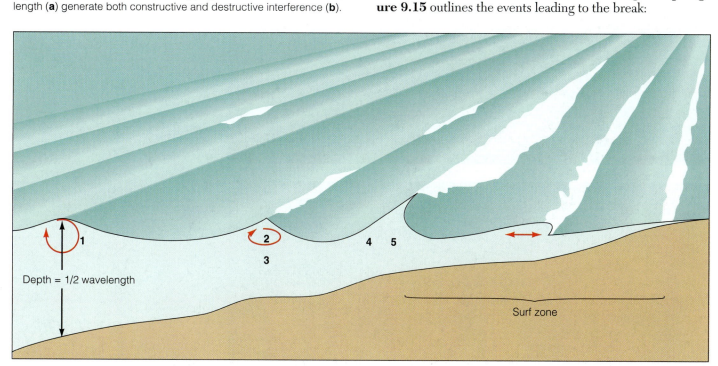

Figure 9.15 How a wave train breaks against the shore. (1) The swell "feels" bottom when the water is shallower than half the wavelength. (2) The wave crests become peaked because the wave's energy is packed into less water depth. (3) Constraint of circular wave motion by interaction with the ocean floor slows the wave, while waves behind it maintain their original rate. Therefore, wavelength shortens but period remains unchanged. (4) The wave approaches the critical 1:7 ratio of wave height to wavelength. (5) The wave breaks when the ratio of wave height to water depth is about 3:4. The movement of water particles is shown in red. Note the transition from a deep-water wave to a shallow-water wave.

1. The wave train moves toward shore. When the water is less than half the wavelength in depth, the wave "feels" bottom.
2. The circular motion of water molecules in the wave is interrupted. Circles near the bottom flatten to ellipses. The wave's energy must now be packed into less water depth, so the wave crests become peaked rather than rounded.
3. Interaction with the bottom slows the wave. Waves behind it continue toward shore at the original rate. Wavelength therefore decreases but period remains unchanged.
4. The wave becomes too high for its wavelength, approaching the critical 1:7 ratio.
5. As the water becomes even shallower, the part of the wave below average sea level slows because of the restricting effect of the ocean floor on wave motion. When the wave was in deep water, molecules at the top of the crest were supported by the molecules ahead (thus transferring energy forward). This is now impossible because the *water* is moving faster than the *wave*. As the crest moves ahead of its supporting base, the wave breaks. The break occurs at about a 3:4 ratio of wave height to water depth. (That is, a 3-meter wave will break in 4 meters of water.) The turbulent mass of agitated water rushing shoreward during and after the break is known as **surf.** The **surf zone** is the region between the breaking waves and the shore.

Waves break against the shore in different ways, depending in part on the slope of the bottom. The break can be violent and toppling, leaving an air-filled channel (or "tube") between the falling crest and the foot of the wave. These **plunging waves** (**Figure 9.16a**) form when waves approach a steeply sloping bottom. A more gradually sloping bottom generates a milder **spilling wave** as the crest slides down the face of the wave (**Figure 9.16b**).

a

b

Figure 9.16 Types of breaking waves. (**a**) Plunging waves form when the bottom slopes steeply toward shore. (**b**) Spilling waves form when the bottom slopes gradually.

Figure 9.17 *Neptune's Horses* by Walter Crane, a graphic representation of the power of breaking waves. The energy of a wave is proportional to the square of its height. Researchers report that each linear meter of a wave 2 meters above average sea level represents an energy flow of about 25 kilowatts (34 horsepower), enough to light 250 100-watt light bulbs; a wave twice as high would contain four times as much energy. A single wave 1.2 meters high striking the west coast of the United States may release as much as 50 million horsepower.

Slope alone does not determine the position and nature of the breaking wave. The contour and composition of the bottom can also be important. Gradually shoaling bottoms can sap waves of their strength because of prolonged interaction against the bottom of the lowest elliptical water orbits. Energy may be lost even more rapidly if the bottom is covered with loose gravel or irregular growths of coral. Masses of moving seaweeds or jostling chunks of sea ice can also extract energy from a wave. In a few rare cases the shore is configured in such a way that waves don't break at all—the waves have lost virtually all their energy by the time their remnants arrive at the beach.

Wave Refraction

9-12

What happens when a wave line approaches the shore at an angle, as it almost always does (**Figure 9.18**)? The line does not break simultaneously, because different parts of it are in different depths of water. The part of the wave line in shallow water slows down, but the attached segment still in deeper water continues at original speed, so the wave line bends (or refracts). The bend can be as much as 90° from the original direction of the wave train. This slowing and bending of waves in shallow water is called **wave refraction.** The refracted waves break in a line almost parallel to the shore.

Figure 9.18 Wave refraction. (**a**) Diagram showing the elements that produce refraction. (**b**) Wave refraction around Maili Point, Oahu, Hawaii. Note how the wave crests bend almost 90° as they move around the point.

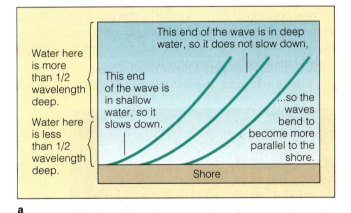

a

b

Figure 9.23 A tsunami in progress. The largest wave of the 1 April 1946 tsunami rushes ashore in Hilo. Terrified residents run for their lives; more than 150 died. V-shaped Hilo Bay is an especially dangerous place during a tsunami because its funnel-like outline concentrates the energy of the waves. Fourteen years after the 1946 disaster, a seismic disturbance in the subduction zone off western South America generated a tsunami that drowned 61 people in Hilo. Wave-cut scars on cliffs north of the town suggest that visits by large tsunami are not rare occurrences.

Figure 9.24 Massive property damage in the fishing village of Aonae on the island of Okushiri, Japan, was caused by a tsunami generated by the Hokkaido Nansei-Oki earthquake in 1993. Maximum wave height was 31 meters (101 feet); 239 people died. This event has become the best-documented tsunami disaster in history.

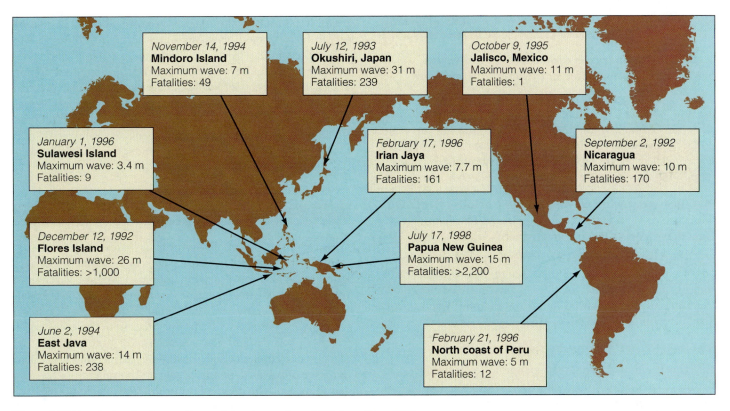

November 14, 1994
Mindoro Island
Maximum wave: 7 m
Fatalities: 49

July 12, 1993
Okushiri, Japan
Maximum wave: 31 m
Fatalities: 239

October 9, 1995
Jalisco, Mexico
Maximum wave: 11 m
Fatalities: 1

January 1, 1996
Sulawesi Island
Maximum wave: 3.4 m
Fatalities: 9

February 17, 1996
Irian Jaya
Maximum wave: 7.7 m
Fatalities: 161

September 2, 1992
Nicaragua
Maximum wave: 10 m
Fatalities: 170

December 12, 1992
Flores Island
Maximum wave: 26 m
Fatalities: >1,000

July 17, 1998
Papua New Guinea
Maximum wave: 15 m
Fatalities: >2,200

June 2, 1994
East Java
Maximum wave: 14 m
Fatalities: 238

February 21, 1996
North coast of Peru
Maximum wave: 5 m
Fatalities: 12

Figure 9.25 Ten destructive tsunami have claimed more than 4,000 lives since 1990.

a

b

Figure 9.26 (**a**) Tsunami hazard warning sign in a town in central coastal Oregon. The configuration of the coast and the slope and instability of the bottom in this area make tsunami particularly dangerous. (**b**) An unfortunate effect of the tsunami warning network. Sightseers flock to Oahu's Makapu'u Beach in Hawaii, awaiting a tsunami generated by a 6.5 earthquake centered in the Aleutian Trench on 7 May 1986.

in Alaska (see Chapter 3). A 3.7-meter (12-foot) wave, probably the fourth crest to reach the coast, swept into Crescent City, California, about six hours after the quake. Though more than 300 buildings were destroyed or damaged, five gasoline storage tanks set ablaze, and 27 blocks of the city demolished, there were relatively few casualties.

It has been more than 30 years since a tsunami caused substantial damage in the United States. Some public safety experts suggest we have become complacent about the risks associated with these destructive waves. As can be seen in **Figure 9.26,** some coastal communities remind residents and visitors of the danger.

1. If so much energy is expended as wind waves break, why doesn't water in the surf zone get hot?

The energy of wind waves is dissipated mostly as heat in the surf, but since water has such a high specific heat (discussed in Chapter 6), the injection of heat into the surf zone doesn't significantly increase water temperature. The surf zone is also an area of vigorous mixing, so any heat released is quickly distributed throughout a large volume of water.

2. What about wind waves and the Coriolis effect? Do wind waves turn right in the Northern Hemisphere as they move across the ocean surface?

The Coriolis effect has no influence on waves with periods of less than about five minutes. Large shallow-water waves such as tsunami, seiches, and tides involve the mass movement of water and are influenced by Earth's rotation. Wind waves, however, with a period rarely exceeding 20 seconds, are not.

3. I love to surf. Where should I plan to live?

The Pacific has the largest potential fetch distances, so the best chances for large wind waves are there. The biggest wind waves are made in the West Wind Drift, but temperatures there are much too cold for comfortable surfing. Besides, the sea state in areas of continuous high wind is chaotic, and surfing requires orderly waves. It is best to let the wave trains sort out. Hawaii is in the middle of the Pacific, so wind waves from polar storms in either hemisphere strike its shores only after much sorting by wavelength. Order is assured. The water is warm, too. I vote for Hawaii.

4. One of my friends has discovered I'm studying waves. He demands to know about the world's largest wave. What's the record, anyway?

If your friend classifies the largest wave by wavelength, the answer is easy: Tides take the prize, with a wavelength that is ideally half of Earth's circumference. Tides also win the speed race: Under ideal circumstances, their crests can move at speeds of up to 1,600 kilometers (1,000 miles) per hour.

If your friend classifies the largest wave by height, things become more complicated. As you read in this chapter, the crew of *Ramapo* measured a wave 34 meters (112 feet) high in 1933. In general, the highest wind waves probably develop in the Antarctic Circumpolar Current, the nasty area of windy ocean ringing Antarctica. Multiple storms there, some with huge fetch distances, can cause very large waves and opportunity for constructive interference between them. In 1916 during an agonizing small-boat trip from an island on which their expedition was marooned, Frederic Worsley, captain of Ernest Shackleton's ship *Endurance*, encountered occasional tremendous waves greater than any he had encountered in a lifetime at sea. He estimated a few were more than 30 meters (100 feet) high!

Some tsunami may be higher, however. Landslides in Lituya Bay, Alaska, have generated swash waves in the 50-meter (160-foot) range. And a wave of about 530 meters (1,740 feet) formed briefly as water surged up the opposite side of the bay on 9 July 1958. (It was more a titanic splash than a classic tsunami; one writer has likened the effect to dropping a sack of concrete into a half-filled bathtub.) The greatest height recently recorded for a tsunami in the ocean was 84 meters (278 feet) in 1971 off Ishigaki, Japan. But don't forget the 91-meter (300-foot) tsunami thought to have occurred 66 million years ago on the coast of what is now Texas after an asteroid struck the sea surface near the Yucatán Peninsula.

CHAPTER SUMMARY

Waves transmit energy, not water mass, across the ocean's surface. The speed of ocean waves usually depends on their wavelength, with long waves moving fastest. In order of wavelength from short to long (and therefore from slowest to fastest), ocean waves are generated by very small disturbances (capillary waves), wind (wind waves), rocking of water in enclosed spaces (seiches), seismic and volcanic activity or other sudden displacements (tsunami), and gravitational attraction (tides). The behavior of waves depends largely on the relation between a wave's size and the depth of water through which it is moving. Waves can refract and reflect, break, and interfere with one another.

Unlike wind waves, waves of very long wavelengths are always in "shallow water" (water less than half their wavelength deep). These long waves travel at high speeds. Some of the waves can be destructive, but their ability to cause damage is fortunately not proportional to their wavelengths. The waves with the longest wavelengths of all—the tides—are covered in the next chapter.

TERMS AND CONCEPTS TO REMEMBER

capillary wave	spilling wave
constructive interference	standing wave
deep-water wave	storm surge
destructive interference	surf
disturbing force	surf zone
fetch	swell
fully developed sea	tsunami
interference	wave
orbit	wave crest
orbital wave	wave frequency
plunging wave	wave height
progressive wave	wave period
restoring force	wave refraction
rogue wave	wave steepness
$S = L/T$	wave trough
$S = \sqrt{gd}$	wavelength
sea	wind duration
seiche	wind strength
seismic sea wave	wind wave
shallow-water wave	

STUDY QUESTIONS

1. How is an ocean wave different from a wave in a spring or a rope? How is it similar? How does it relate to a "stadium wave"—a waveform made by sports fans in a circular arena?

2. Draw a deep-water ocean wave and label its parts. Show the orbits of water particles. Include a definition of wave period. How would you measure wave frequency?

3. What factors influence the growth of a wind wave? What is a fully developed sea? Where would we regularly expect to find the largest waves? Are waves in fully developed seas always huge?

4. What happens when a wind wave breaks? What factors affect the break? How are plunging waves different from spilling waves?

5. Though they move across the deepest ocean basins, seiches and tsunami are referred to as "shallow-water waves." How can this be?

6. What causes tsunami? Are all seismic sea waves tsunami? Are all tsunami seismic sea waves? How fast do tsunami travel? Do they move in the same way or at the same speed in a confining bay as they would in the open ocean?

7. Are tsunami ever dangerous if encountered in the open sea? What happens when they reach shore?

FOR FURTHER STUDY

Bascom, W. 1980. *Waves and Beaches.* Rev. ed. New York: Anchor/Doubleday. Perhaps the best work for the lay reader on the subject. Excellent photographs and diagrams, lively and clear writing.

Dudley, W. C., and M. Lee. 1988. *Tsunami!* Honolulu: University of Hawaii Press. Combines firsthand accounts of survivors with scientific information on tsunami.

Folger, T. 1994. "Waves of Destruction." *Discover,* May, 66–73. Information on tsunami, including the September 1992 Nicaraguan event.

Giese, G. S., and D. C. Chapman. 1993. "Coastal Seiches." *Oceanus* 36 (no. 1): 38–46. Internal waves approaching the shore may be responsible for some occasionally destructive seiches.

González, F. I. 1999. "Tsunami!" *Scientific American,* May, 56–65.

Kampion, D. 1989. *The Book of Waves.* Santa Barbara, CA: Arpel Graphics. The definitive coffee-table book of waves. Magnificent photography.

Maxtone-Graham, J. 1972. *The Only Way to Cross.* New York: Collier/Macmillan. Rogue waves and other stories. One of the best books on the era of the fabled transatlantic liners. Very highly recommended for a wonderful writing style.

McCredie, S. 1994. "When Nightmare Waves Appear Out of Nowhere to Smash the Land." *Smithsonian,* March, 28–39. Disturbing photos of tsunami, including the July 1993 event in Japan.

Noll, G., and A. Gabbard. 1989. *Da Bull: Life over the Edge.* Bozeman, MT: Bangtail Press. Surfer Greg Noll rides Hawaii's largest wave. Historical view of the golden age of board surfing.

Peterson, I. 1996. "Rough Math: Focusing on Rogue Waves at Sea." *Science News* 150 (no. 21): 325. Ocean currents or large fields of random eddies can sporadically concentrate a steady ocean swell to create unusually large waves.

For additional readings, go to InfoTrac College Edition, your online research library at:

http://infotrac.thomsonlearning.com/

MAELSTROM!

According to Homer's account in *The Odyssey*, across Ulysses' homeward path lay a lethal whirlpool named Charybdis:

> We sailed up the narrow strait. On one side was shining Charybdis, who made her terrible ebb and flow of the sea's water. When she vomited it up, like a caldron over a strong fire, the whole sea would boil up in turbulence, and the foam flying spattered the pinnacles of the rocks in either direction; but when in turn again she sucked down the sea's salt water, the turbulence showed all the inner sea, and the rock around it groaned terribly, and the ground showed at the sea's bottom, black with sand. My crew turned sallow with fright, staring into this abyss from which we expected our immediate death.

For millennia, the prospect of navigating past Charybdis terrified sailors in the Strait of Messina, which separates Italy and Sicily. Its activity was greatly reduced in 1908 when the great Calabria–Messina earthquake and tsunami altered the geology of the area.

Edgar Allan Poe wrote about a whirlpool called the Maelstrom, lying among the southern Lofoten Islands off Norway's west coast. The rapid spinning of water in the Maelstrom raises its outer edge and depresses the central core. Here is Poe's description, from his 1841 story "A Descent into the Maelstrom":

> The edge of the whirl was represented by a broad belt of gleaming spray; but no particle of this slipped into the mouth of the terrific funnel, whose interior, as far as the eye could fathom it, was a smooth, shining, and jet-black wall of water, inclining to the horizontal at an angle of some forty-five degrees, speeding dizzily round and round with swaying and sweltering motion, and sending forth to the winds an appalling voice, half shriek, half roar, such as not even the mighty cataract of Niagara ever lifts up in its agony to Heaven.

Do these whirlpools genuinely deserve the dread they inspired, or is their reputation largely a figment of artistic imagination (**Figure a**)?

Whirlpools arise in shallow, restricted straits connecting two large bodies of deep water. They are caused by the tides. If the large bodies of water move

a Maelstrom: fiction. A representation by the early twentieth-century English illustrator Arthur Rackham.

to different tidal cycles, high tide in one will not occur at the same time as high tide in the other. Turbulence develops when one mass of water tries to pass the other during changes in tide. Under ideal conditions, bottom contours and the rush of opposing tidal currents can cause seawater in the confined area to spin vigorously. The larger the opposing bodies of water, and the smaller the passage through which the confined water must pass, the greater the vortex caused by the tidal currents.

Maelstrøm is a Norwegian word derived from the Dutch *malen* (to grind in a circle, as a millstone grinds grain) and *strøm* (stream). Today's maelstrom is a heaving mass of active tidal water connecting Vest Fjord and the Norwegian Sea between the islands of Moskenesøya and Moskenes. Look for the area on a chart at 67°48′N, 12°50′E, about 165 kilometers (100 miles) north of the Arctic Circle. Currents in the passage can reach a speed of 13 kilometers (8 miles) per hour when the tides change, and strong local winds make the passage even more dangerous. The shallow "saddle" separating the large bodies of water to the east and west is only about 20 meters (70 feet) deep, and water rushing over its rocky bottom is driven to spin.

But is the chaos as horrifying as Homer, Poe, Jules Verne, and others have suggested? Well, no. Fishing boats in this rich fishing region carefully avoid the area during times of maximum tidal difference, and no sailor likes to risk his or her life in turbulent confined waters during times of high winds. But as can be seen in **Figure b,** the sucking, roaring, cavernous maw of the maelstrom is more fiction than fact.   10-1

KEY CONCEPTS

1. Tides are huge shallow-water waves—the largest waves in the ocean. They are caused by a combination of the gravitational force of the moon and sun and the motion of Earth.

2. The *equilibrium theory* of tides deals primarily with the position and attraction of Earth, moon, and sun. The *dynamic theory* takes into account the speed of the long-wavelength tide wave in water of varying depth, the presence of interfering continents, and the circular movement or rhythmic back-and-forth rocking of water in ocean basins.

3. The rising and falling of the tides can be used to generate electrical power and are important in many physical and biological coastal processes.

CHAPTER AT A GLANCE

TIDES AND THE FORCES THAT GENERATE THEM

10-2

Tides are periodic, short-term changes in the height of the ocean surface at a particular place, caused by a combination of the gravitational force of the moon and the sun and the motion of Earth. With a wavelength that can equal half of Earth's circumference, tides are the longest of all waves. Unlike the other waves we have met, these huge shallow-water waves are never free of the forces that cause them and thus are called *forced* waves. (After they are formed, wind waves, seiches, and tsunami are *free* waves; that is, they are no longer being acted upon by the force that created them and do not require a maintaining force to keep them in motion.)

The Greek navigator and explorer Pytheas first wrote of the connection between the position of the moon and the height of a tide around 300 B.C., but full understanding of tides had to await Isaac Newton's analysis of gravitation.

Among many other things, Isaac Newton's brilliant 1687 book *Philosophiae Naturalis Principia Mathematica* (*Mathematical Principles of Natural Philosophy*) described the motions of planets, moons, and all other bodies in gravitational fields. A central finding: The pull of gravity between two bodies is proportional to the masses of the bodies but inversely proportional to the square of the distance between them. This means that heavy bodies attract each other more strongly than light bodies do, and that gravitational attraction quickly weak-

ens as the distance grows larger. This can be expressed mathematically as

$$F = G\left(\frac{m_1 m_2}{r^2}\right)$$

where F is the gravitational attraction, G is the universal gravitational constant, m_1 and m_2 are the masses of the two bodies, and r the distance between their centers.

While the main cause of tides is the gravity of the moon and the sun acting on the ocean, the forces that actually generate the tides vary inversely with the *cube* of the distance from Earth's center to the center of the tide-generating object (the moon or the sun). Distance is thus even more important in this relationship, which can be expressed as

$$T = G\left(\frac{m_1 m_2}{r^3}\right)$$

where T is the tide-generating force, G is the universal gravitational constant, m_1 and m_2 are the masses of the two bodies, and r the distance between their centers. The sun is about 27 million times more massive than the moon, but the sun is about 387 times farther away than the moon; so the sun's influence on the tides is only about half that of the moon's.

As we will see, Newton's gravitational model of tides—the *equilibrium theory*—deals primarily with the position and at-

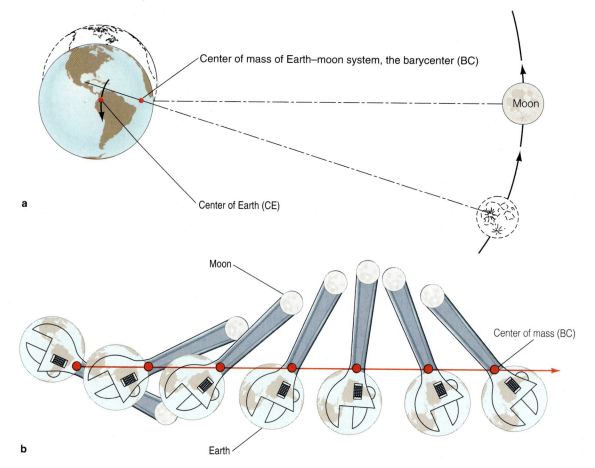

a
Center of mass of Earth–moon system, the barycenter (BC)

Moon

Center of Earth (CE)

b
Moon

Center of mass (BC)

Earth

Figure 10.1 (**a**) The Earth–moon system does not revolve around the center of Earth, but rather around its common center of mass, the barycenter. The moon revolves around Earth once a month, and through each month the center of Earth describes a circle in space around the barycenter. (The drawing is not to scale.) (**b**) The rotation of a thrown wrench is analogous to the motion of the Earth–moon system. Imagine that Earth's mass is located at the heavy head of the wrench and the moon is located at the end of the handle to visualize the motion of the system through space.

traction of Earth, moon, and sun, and does not allow for the influence of ocean depth or the positions of continental land masses on tides. The equilibrium theory would accurately describe tides on a planet uniformly covered by water. A modification proposed by Pierre-Simon Laplace about a century later—the *dynamic theory*—takes into account the speed of the long-wavelength tide wave in relatively shallow water, the presence of interfering continents, and the circular movement or rhythmic back-and-forth rocking of water in ocean basins. We will explore the idealized situation of the equilibrium theory before moving to the real-world dynamic view.

THE EQUILIBRIUM THEORY OF TIDES

10-3

The **equilibrium theory of tides** explains many characteristics of ocean tides by examining the balance and effects of the forces that enable a planet to stay in a stable orbit around the sun, or the moon to orbit Earth. The equilibrium theory assumes that the ocean conforms instantly to the forces affecting the position of its surface; that is, the ocean surface is presumed always to be in equilibrium (balance) with the forces acting on it.

The Moon and Tractive Forces

10-4

We begin our examination of these forces by looking at the moon's effect on the ocean surface. Gravity tends to pull Earth and the moon toward each other, but inertia—the tendency of moving objects to continue in a straight line—keeps them apart. (In this context, we sometimes call inertia *centrifugal force*, the "force" that keeps water against the bottom of a bucket when you swing it overhead in a circle.) Earth and the moon don't smash into each other (or fly apart) because they are in a stable orbit; their mutual gravitational attraction is exactly offset by their inertia. Contrary to what you might think, the moon does not revolve around the center of Earth. Rather, the Earth–moon *system* revolves once a month (every 27.3 days) around the system's center of mass. Because Earth's mass is 81 times that of the moon, this common center of mass is located not in space, but 1,700 kilometers (1,060 miles) *inside* Earth. This center of mass, the **barycenter** (*barys* = heavy), is shown in **Figure 10.1.**

Note in **Figure 10.1a** that the center of Earth is beginning to describe a circle around the barycenter. You can see that circle completed at the center of Earth in **Figure 10.2.** Because Earth and the moon revolve as a system, all points on and in Earth move in circles of the same radii; the circles described by four random points are shown in Figure 10.2.

Now imagine the forces working on those points. In **Figure 10.3** we see the outward-flinging force of inertia. All the outward-directed force arrows are equal and parallel to each other. In **Figure 10.4** we see the inpulling force of gravity, directed toward the center of the moon. Unlike the centrifugal arrows of Figure 10.3, the arrows of Figure 10.4 are *not* parallel to one another. The inward pull of gravity and the outward-

moving tendency of inertia don't always act in exactly the same balanced way on each particle of Earth and the moon. In **Figure 10.5,** notice that points 1 and 2 are closer to the moon; so gravitational attraction at those points slightly exceeds the outward-moving tendency of inertia. Water there tends to be attracted toward the moon and so is pulled along the ocean surface toward a spot beneath the moon. At points 3 and 4, slightly farther from the moon, inertia exceeds gravitational attraction. Water at those points tends to be flung away from the moon and so moves along the ocean surface toward a spot opposite the moon.

The unbalanced forces doing the pulling and flinging are known as *tractive forces*. Together, the tractive forces cause

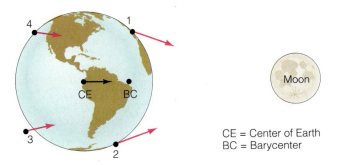

Earth

Figure 10.2 Because Earth and the moon revolve as a system, all points on and in Earth move monthly in a circle identical in size to the circle described by Earth's center. The movement of Earth's center and four points on Earth's surface are shown here.

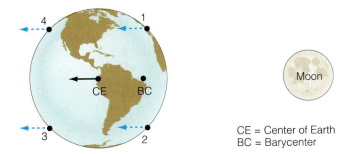

CE = Center of Earth
BC = Barycenter

Figure 10.3 The outward-flinging force of inertia (often called *centrifugal force*) at five points on and in Earth. Note that all the outward-directed force arrows are equal and parallel to each other.

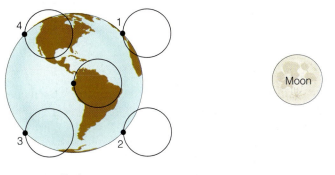

CE = Center of Earth
BC = Barycenter

Figure 10.4 The inpulling force of gravity, directed toward the center of the moon. Unlike the centrifugal arrows of Figure 10.3, these arrows are *not* parallel to one another.

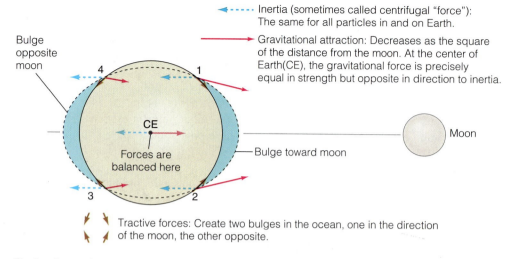

Inertia (sometimes called centrifugal "force"): The same for all particles in and on Earth.

Gravitational attraction: Decreases as the square of the distance from the moon. At the center of Earth(CE), the gravitational force is precisely equal in strength but opposite in direction to inertia.

Bulge opposite moon

CE

Forces are balanced here

Bulge toward moon

Moon

Tractive forces: Create two bulges in the ocean, one in the direction of the moon, the other opposite.

The two forces that can move the ocean are balanced only at the center of Earth (point CE). Elsewhere the net imbalance is a small force that causes ocean water to converge into two equal "bulges," as shown.

Figure 10.5 The actions of gravity and inertia on particles at five different points on Earth. At points 1 and 2, the gravitation attraction of the moon slightly exceeds the outward-moving tendency of inertia, and the imbalance of forces causes water to move along Earth's surface to converge at a point *toward* the moon. At points 3 and 4, inertia exceeds gravitational force; so water moves along Earth's surface to converge at a point *opposite* the moon. Forces are balanced only at the center of Earth (point CE).

two small bulges in the ocean, one in the direction of the moon, the other in the opposite direction. Note that there is no point on Earth's surface where the force of the moon's gravity exactly equals the outward-moving tendency of inertia. Only at point CE—the center of Earth—are the inward pull of gravity and the outward-moving tendency of inertia exactly equal and opposite. The solid Earth cannot move much in response to these forces, but the fluid atmosphere and ocean can. We don't notice the changes in the height of the atmosphere, but changes in water level are visible to coastal observers. In **Figure 10.6,** tractive forces pull water toward a point beneath the moon and another point opposite the moon.

How do these bulges cause the rhythmic rise and fall of the tides? In the idealized equilibrium model we are discussing, the bulges tend to stay aligned with the moon as Earth spins around its axis. **Figure 10.7** shows the situation in Figure 10.6 as it would look from above the North Pole. As Earth turns

eastward, an island on the equator is seen to move in and out of these bulges through one rotation (one day). The bulges are the crests of the planet-sized waves that cause **high tides. Low tides** correspond to the troughs, the area between bulges. Starting at 0000 (midnight) we see the island in shallow water at low tide. Around six hours later, at 0613 (6:13 A.M.), the island is submerged in the lunar bulge at high tide. At 1226 (about noon) the island is within the tide wave trough at low tide. At 1838 (6:38 P.M.) the island is again submerged, this time in the opposite crest caused by inertia. About an hour after midnight (0050) on the next day, the island is back in shallow water, where it began.

The wave crests and troughs that cause high and low tides are actually very small: A 2-meter (7-foot) rise or fall in sea level is insignificant in comparison to the ocean's great size. Earth rotates beneath the bulges (tide wave crests) at about 1,600 kilometers (1,000 miles) per hour at the equator. The

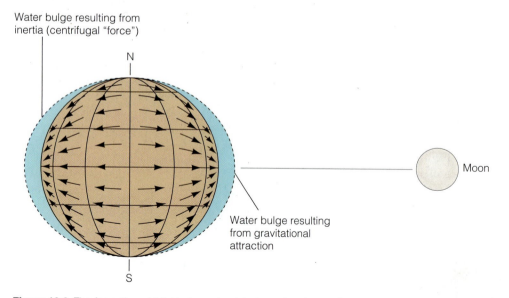

Water bulge resulting from inertia (centrifugal "force")

N

Moon

Water bulge resulting from gravitational attraction

S

Figure 10.6 The formation of tidal bulges at points toward and away from the moon. (This and the similar diagrams that follow are not drawn to scale.)

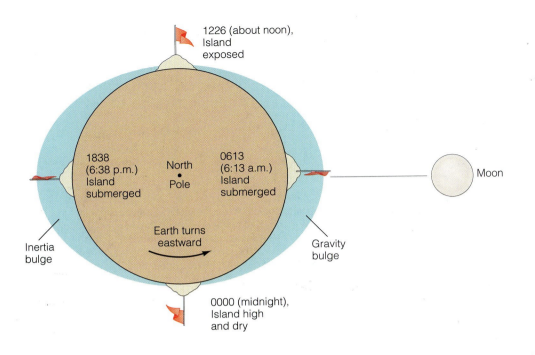

1226 (about noon), Island exposed

1838 (6:38 p.m.) Island submerged

North Pole

0613 (6:13 a.m.) Island submerged

Earth turns eastward

Inertia bulge

Gravity bulge

0000 (midnight), Island high and dry

Moon

Figure 10.7 How Earth's rotation beneath the tidal bulges produces high and low tides. Notice that the tidal cycle is 24 hours 50 minutes long, because the moon rises 50 minutes later each day.

bulges appear to move across the ocean surface at this speed in an attempt to keep up with the moon. Theoretically, the wavelength of these tide waves is as long as 20,000 kilometers (12,500 miles)! The bulges tend to stay aligned with the moon as Earth spins around its axis. *The key to understanding the equilibrium theory of tides is to see Earth turning beneath these bulges.*

There are complications, of course. For one, the **lunar tides** (tides caused by the gravitational and inertial interaction of the moon and Earth as just described) complete their cycle in a tidal day (also called a lunar day). A complete tidal day is 24 hours 50 minutes long, because the moon, which exerts the greatest tidal influence, rises 50 minutes later each day (**Figure 10.8**). Thus, the highest tide also arrives 50 minutes later each day.

Another complication arises from the fact that the moon does not stay right over the equator; each month, it moves

from a position as high as 28½° above Earth's equator to 28½° below. When the moon is above the equator, the bulges are offset accordingly (as in **Figure 10.9**). When the moon is 28½° north of the equator, an island north of the equator will pass through the bulge on one side of Earth but miss the bulge on the other side. During one day the island passes through a very high tide, a low tide, a lower high tide, and another low tide. This is shown in **Figure 10.10**.

The Sun's Role 10-5

The sun's gravity also attracts particles on Earth. Remember, closeness counts for much in determining the strength of gravitational attraction. As we saw earlier, the sun is about 27 million times more massive than the moon but also about 387 times farther away than the moon; so the sun's influence on the tides is only 46% that of the moon's. The sun's tractive

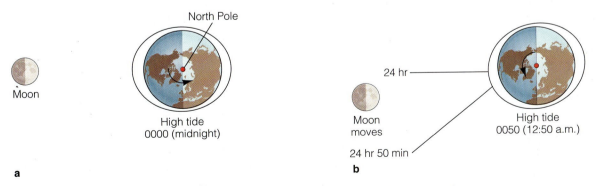

North Pole

Moon

High tide 0000 (midnight)

a

24 hr

Moon moves

24 hr 50 min

High tide 0050 (12:50 a.m.)

b

Figure 10.8 (**a**) A high tide at midnight. (**b**)The high tide one day later. Earth rotates once a day, and the Earth–moon system revolves around its center of mass once a month. Both turn in the same direction (eastward). During one complete Earth rotation (one day), the moon moves eastward 12.2°, so Earth must rotate for an additional 50 minutes for the moon to reach the same place in the sky on consecutive days. Because Earth must turn an additional 50 minutes for the same tidal alignment, lunar tides usually arrive 50 minutes later each day.

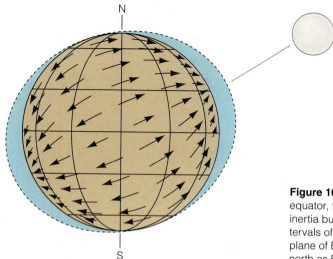

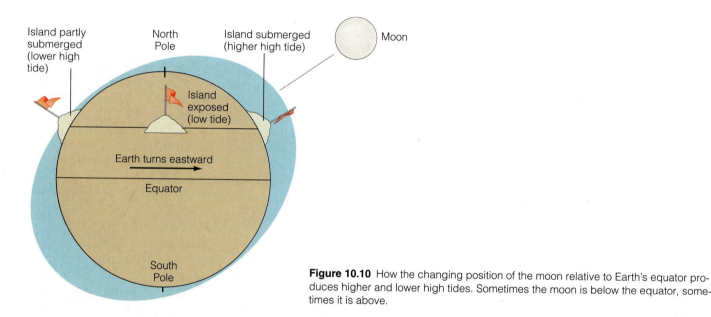

Figure 10.9 The bulges follow the moon. When the moon's position is north of the equator, the bulge toward the moon is also located north of the equator. The opposite inertia bulge is correspondingly below the equator. (Compare with Figure 10.6.) At intervals of about 18½ years the plane of the moon's orbit is inclined 28½° relative to the plane of Earth's equator, so the moon can sometimes appear directly overhead as far north as Florida or southern Texas. Lunar tides are extreme at such times.

Figure 10.10 How the changing position of the moon relative to Earth's equator produces higher and lower high tides. Sometimes the moon is below the equator, sometimes it is above.

forces develop in the same way as the moon's, and the smaller solar bulges tend to follow the sun through the day. These are the **solar tides,** caused by the gravitational and inertial interaction of the sun and Earth.

Like the moon, the sun also appears to move above and below the equator (23½° north to 23½° south, as you may recall from Chapter 7), so the position of the solar bulges varies like that of the lunar bulges. Earth revolves around the sun only once a year, however; so the position of the solar bulges above or below the equator changes much more slowly than does the position of the lunar bulges. (Figure 7.2b, used to explain the cause of the seasons, shows this well.)

Sun and Moon Together
 10-6

The ocean responds simultaneously to inertia and to the gravitational force of both sun and moon. If Earth, moon, and sun are all in a line (as shown in **Figure 10.11a**), the lunar and so-

lar tides will be additive, resulting in higher high tides and lower low tides. But if moon, Earth, and sun form a right angle (as shown in **Figure 10.11b**), the solar tide will tend to diminish the lunar tide. Because the moon's contribution is more than twice that of the sun, the solar tide will not completely cancel the lunar tide.

The large tides caused by the linear alignment of sun, Earth, and moon are called **spring tides** (*springen* = to move quickly). During spring tides, high tides are very high and low tides very low. These tides occur at two-week intervals corresponding to the new and full moons. (Please note that spring tides don't happen only in the spring of the year.) **Neap tides** (*naepa* = hardly disturbed) occur when moon, Earth, and sun form a right angle. During neap tides, high tides are not very high and low tides not very low. Neap tides also occur at two-week intervals, with the neap tide arriving a week after the spring tide. **Figure 10.12** plots tides at two coastal sites through spring and neap cycles.

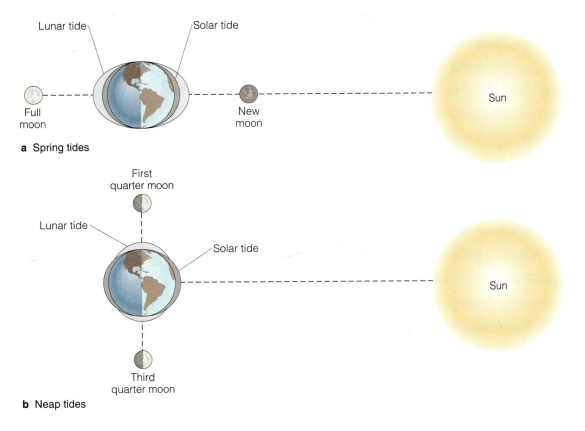

a Spring tides

b Neap tides

Figure 10.11 Relative positions of sun, moon, and Earth during spring and neap tides. (**a**) At the new and full moons, the solar and lunar tides reinforce each other, making spring tides, the highest high and lowest low tides. (**b**) At the first and third quarter moons, sun, Earth, and moon form a right angle, creating neap tides, the lowest high and the highest low tides.

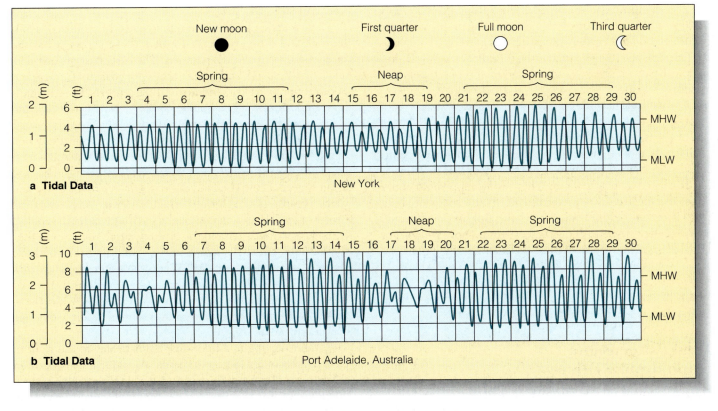

a Tidal Data — New York

b Tidal Data — Port Adelaide, Australia

Figure 10.12 Tidal records for a typical month at (**a**) New York and (**b**) Port Adelaide, Australia. Note the relationship of spring and neap tides to the phases of the moon. MHW = mean high water; MLW = mean low water.

Because their orbits are ellipses, not perfect circles, the moon and the sun are closer to Earth at some times than at others. The difference between **apogee** (the moon's most distant point) and **perigee** (its closest) is 30,600 kilometers (19,015 miles). Because the tidal force is inversely proportional to the cube of the distance between the bodies, the closer moon raises a noticeably higher tidal crest. The difference between **perihelion** (Earth's closest approach to the sun) and **aphelion** (greatest distance) is 3.7 million kilometers (2.3 million miles). If the moon and the sun are over nearly the same latitude, and if Earth is also close to the sun, extreme spring tides will result. Interestingly, spring tides will have greater ranges in the Northern Hemisphere winter than in the Northern Hemisphere summer because Earth is closest to the sun during the northern winter.

THE DYNAMIC THEORY OF THE TIDES

🖥 10-7

Newton knew his explanation was incomplete. For one thing, the maximum theoretical range of a lunar tidal bulge is only 55 centimeters (about 22 inches), and of a solar tide only 24 centimeters (about 10 inches), both considerably smaller than the 2-meter (7-foot) average tidal range we observe in the world ocean. This is because the ocean surface never comes completely to the equilibrium position at any instant. The moon and the sun change their positions so rapidly that the water cannot keep up.

The **dynamic theory of tides,** first proposed in 1775 by Laplace, added a fundamental understanding of the problems of fluid motion to Newton's breakthrough in celestial mechanics. The dynamic theory explains the differences between predictions based on Newton's model and the observed behavior of tides.

Remember that tides are a form of *wave.* The crests of these waves—the tidal bulges—are separated by a distance of half of Earth's circumference (see again Figure 10.6). In the equilibrium model, the crests would remain stationary, pointing steadily toward (or away) from the moon (or the sun) as Earth turned beneath them. They would appear to move across the idealized water-covered Earth at a speed of about 1,600 kilometers (1,000 miles) per hour. But how deep would the ocean have to be to allow these waves to move freely? For a tidal crest (or wave) to move at 1,600 kilometers per hour, the ocean would have to be 22 kilometers (13.7 miles) deep. As you may recall, the average depth of the ocean is only 3.8 kilometers (2.4 miles). So tidal crests (tidal bulges) move as forced waves, their velocity determined by ocean depth.

Tidal Patterns and Amphidromic Points

🖥 10-8

This behavior of tidal waves as *shallow-water waves* is only one variation from the ideal that the dynamic theory explains. The continents also get in the way. As Earth turns, land masses obstruct the tidal crests, diverting, slowing, and otherwise com-

plicating their movements. This interference produces different patterns in the arrival of tidal crests at different places. Imagine, for example, a continent directly facing the moon. There would be no oceanic bulge, and the shores of the continent would experience high tide. A few hours later the moon would be over the ocean. When the continent was not aligned with the moon but the ocean was, the tidal bulge would reform and the continent's edges would experience low tide.

The shape of the basin itself has a strong influence on the patterns and heights of tides. As we have seen, water in large basins can rock rhythmically back and forth in seiches. Though they are small, tidal crests can stimulate this resonant oscillation, and the configuration of coasts around a basin can alter its rhythm.

For these and other reasons, some coastlines experience **semidiurnal** (twice daily) **tides**: two high tides and two low tides of nearly equal level each lunar day. Others have **diurnal** (daily) **tides**: one high and one low. The tidal pattern is called **mixed** if successive high tides or low tides are of significantly different heights through the cycle.[1] **Figure 10.13** gives an example of each tidal pattern. At Cape Cod two tidal crests arrive per lunar day, a semidiurnal pattern (**Figure 10.13a**). The natural tendency of water in an enclosed ocean basin to rock at a specific frequency modifies the pattern in the Gulf of Mexico, so Mobile sees one crest per lunar day, a diurnal pattern (**Figure 10.13b**). The Pacific coast of the United States has a mixed tidal pattern: often a *higher* high tide, followed by a *lower* low tide, a *lower* high tide, and a *higher* low tide each lunar day. (That's not as confusing as it first sounds—see **Figure 10.13c.**)

The Pacific Ocean has a unique pattern of diurnal, semidiurnal, and mixed tides. As **Figure 10.14** shows, it has the most complex of all tidal patterns. The east coast of Australia, all of New Zealand, and much of the west coasts of Central and South America have a semidiurnal tidal pattern. The Aleutians have diurnal tides. The Pacific coasts of North America and some of South America have mixed tides. *Why the differences?*

Remember the surface of Lake Geneva in our discussion of seiches? The water level at the center of the lake remains at the same height while water at the ends rises and falls (see again Figure 9.20). The long axis of the lake stretches east and west. Because of the Coriolis effect, water moving east at the center of the lake is deflected slightly to the right (to the south). If the lake were larger and the flow of water greater (and thus the Coriolis effect stronger), water would hug the southern shore as it traveled eastward. When the water began to rock the other way—back to the west—the Coriolis effect would move the water to the right toward (and along) the northern shore. Note that the overall movement of water would be counterclockwise.

Water moving in a tide wave tends to stay to the right of an ocean basin for the same reason. As water moves north in a Northern Hemisphere ocean, it moves toward the eastern boundary of the basin; as it moves south, it moves toward the

[1] Mixed tides are also called *mixed semidiurnal.*

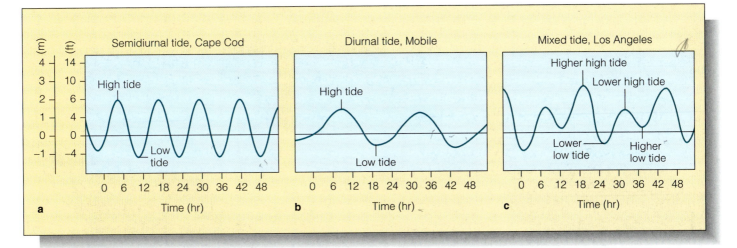

Figure 10.13 Tide curves for the three common types of tides. (**a**) A semidiurnal tide pattern at Cape Cod, Massachusetts. (**b**) A diurnal tide pattern at Mobile, Alabama. (**c**) A mixed tide pattern at Los Angeles, California.

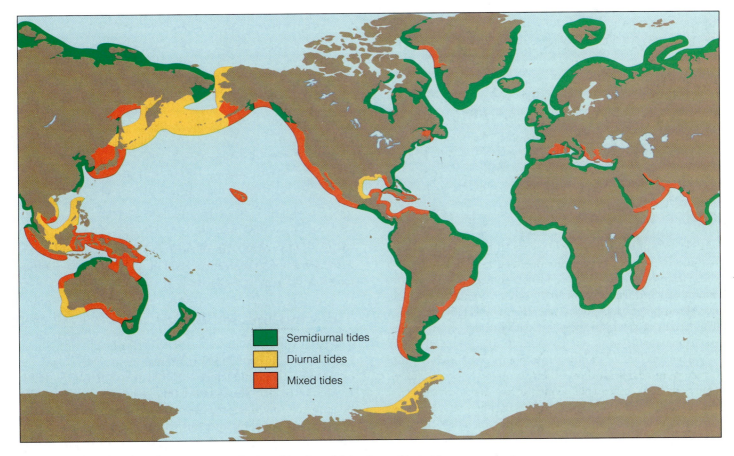

Figure 10.14 The worldwide geographic distribution of the three tidal patterns. Most of the world's ocean coasts have semidiurnal tides.

western boundary. A wave crest moving counterclockwise will develop around a node called an **amphidromic point** (*amphi* = around + *dromas* = running) if this motion continues to be stimulated by tidal forces. This rotary motion, known as *amphidromic circulation,* is shown in **Figure 10.15.**

Thus, an amphidromic point is a no-tide point in the ocean, around which the tidal crest rotates through one tidal cycle. Because of the shape and placement of land masses around ocean basins, the tidal crests and troughs cancel each other at these points. The crests sweep around amphidromic points like

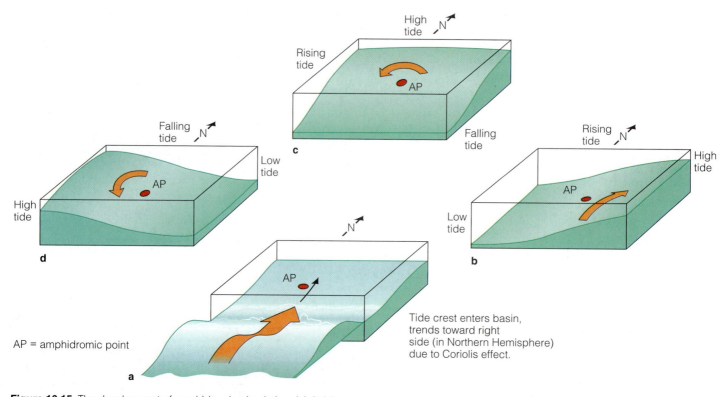

Figure 10.15 The development of amphidromic circulation. (**a**) A tide wave crest enters an ocean basin in the Northern Hemisphere. The wave trends to the right because of the Coriolis effect (**b**), causing a high tide on the basin's eastern shore. Unable to continue turning to the right because of the interference of the shore, the crest moves northward, following the shoreline (**c**) and causing a high tide on the basin's northern shore. The wave continues its progress around the basin in a counterclockwise direction (**d**), forming a high tide on the western shore and completing the circuit. The point around which the crest moves is an amphidromic point (AP).

AP = amphidromic point

Tide crest enters basin, trends toward right side (in Northern Hemisphere) due to Coriolis effect.

wheel spokes from a rotating hub, radiating crests toward distant shores. Tide waves are influenced by the Coriolis effect because a large volume of water moves with the wave; they move counterclockwise around an amphidromic point in the Northern Hemisphere, and clockwise in the Southern Hemisphere. The height of the tides increases with distance from an amphidromic point.

About a dozen amphidromic points exist in the world ocean; **Figure 10.16** shows their location. Notice the complexity of the Pacific, which contains five. It's no wonder that the arrival of tide wave crests at the Pacific's edge produces such a complex mixture of tide patterns, depending on shoreline location.

Tidal Datum

The reference level to which tidal height is compared is called the **tidal datum.** The tidal datum is the zero point (0.0) seen in tide graphs such as Figures 10.12 and 10.13. This reference plane is not always set at **mean sea level,** the height of the ocean surface averaged over a few years' time. On coasts with mixed tides, the zero tide level is the average level of the lower of the two daily low tides (*mean lower low water,* or MLLW). On coasts with diurnal and semidiurnal tides, the zero tide level is the average level of all low tides (*mean low water,* MLW).

Tides in Confined Basins

The **tidal range** (high-water to low-water height difference) varies with basin configuration. In small areas such as lakes, the tidal range is small. In larger enclosed areas such as the Baltic or Mediterranean Seas, the tidal range is also moderate. The tidal range is not the same over a whole ocean basin; it varies from the coasts to the centers of oceans. The largest tidal ranges occur at the edges of the largest ocean basins, especially in bays or inlets that concentrate tidal energy because of their shape.

If a basin is wide and symmetrical, like the Gulf of St. Lawrence in eastern Canada (**Figure 10.17**), a miniature amphidromic system develops that resembles the large systems of the open ocean. If the basin is narrow and restricted, the tidal crest cannot rotate around an amphidromic point and simply moves into and out of the bay (**Figure 10.18**). Extreme tides occur in places where arriving tidal crests stimulate natural oscillation periods of around 12 or 24 hours. In rare cases, water in the bay naturally resonates (seiches) at the

Figure 10.16 Amphidromic points in the world ocean. Tidal ranges generally increase with increasing distance from amphidromic points. Lines radiating from the points indicate tide waves moving around these points, counterclockwise in the Northern Hemisphere and clockwise in the Southern Hemisphere. The numbers represent the positions of a hypothetical tidal crest in passing hours.

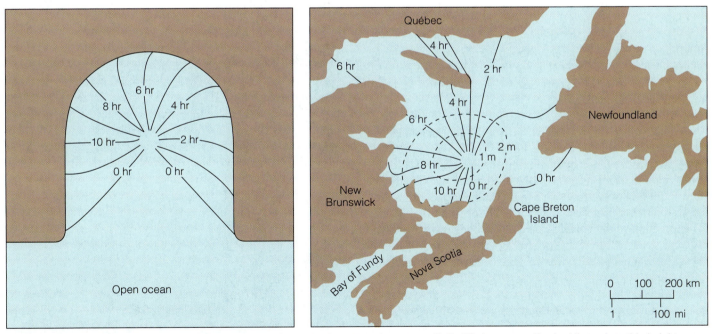

a Broad basin

b Amphidromic system: Gulf of St. Lawrence. Dashed lines show tide height

Figure 10.17 Tides in broad, confined basins. (**a**) An imaginary amphidromic system in a broad, shallow basin. The numbers indicate the hourly positions of tidal crests as a cycle progresses. (**b**) The amphidromic system for the Gulf of St. Lawrence between New Brunswick and Newfoundland, southeastern Canada. Compare the situation in this limited basin with the tides surrounding the amphidromic point in the open ocean south of Greenland in Figure 10.16.

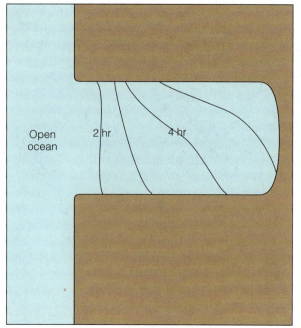

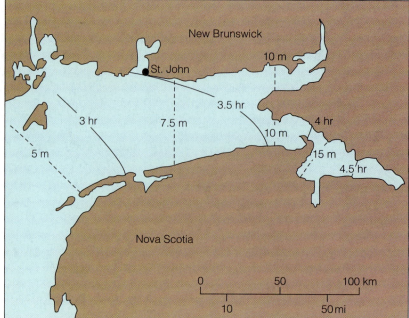

a Narrow basin

b Tide wave crests: Bay of Fundy

Figure 10.18 Tides in narrow, restricted basins. (**a**) True amphidromic systems do not develop in narrow basins because space for rotation is not available. (**b**) Tides in the Bay of Fundy, Nova Scotia, are extreme because water in the bay naturally resonates (seiches) at the same frequency as the lunar tide. (See also Figure 10.19.)

same frequency as the lunar tide (12 hours, 25 minutes). This rhythmic sloshing results in extreme tides. In the eastern reaches of the Bay of Fundy near Moncton, New Brunswick (Canada), the tidal range is especially wide: up to 15 meters (50 feet) from highs to lows (see **Figure 10.19**). The northern reaches of the Sea of Cortez east of Baja California have a tidal range of about 9 meters (30 feet). Tide waves sweeping toward the narrow southern end of the North Sea can build to great heights along the southeast coast of England and the north coast of France.

A **tidal bore** (*bara* = wave) will form in some inlets (and their associated rivers) exposed to great tidal fluctuation. Here, at last, is a true **tidal wave**—a steep wave moving upstream generated by the action of the tidal crest in the enclosed area of a river mouth (see **Figure 10.20**). The confining river mouth forces the tidal wave to move toward land at a speed that exceeds the theoretical shallow-water wave speed for that depth. The forced wave then breaks, forming a spilling wave front that

a

Figure 10.19 Tides in the eastern Bay of Fundy on the Atlantic coast of Canada. The tidal range is nearly 15 meters (50 feet). At the peak of the flood, water rises 1 meter (3.3 feet) in 23 minutes. These photos were taken near St. John, New Brunswick (see Figure 10.18b). Other places in the world have large tides: the Port of Bristol in England; the Sea of Okhotsk, northeast of Japan; Turnagain Arm in Alaska; the Bay of St. Malo in France; the Feuilles River in Ungava Bay, Quebec. All have tidal ranges of about 10 meters (33 feet). However, not as much water is moved at these locations as in the Bay of Fundy, where the tidal contribution equals the total daily discharge of all the world's rivers—about 100 cubic kilometers (24 cubic miles) of water.

b

Figure 10.20 A tidal bore on the Severn River in southwestern England. A tidal crest sweeping up the funnel-shaped Bristol Channel from the Atlantic Ocean encounters the increasingly shallow and narrow river passage, and it rushes forward to cause the bore. Like all tidal bores, the Severn bore reaches its greatest height during the days surrounding spring tides. The waves that form the bore in this example are nearly 1 meter (3.3 feet) high and are traveling at about 5 meters per second (11 miles per hour). Note that the smooth waves are followed by turbulent water also moving rapidly upriver. The tidal bore on the Severn River is large enough for surfing, and accomplished board surfers can ride upstream for many miles.

moves upriver. Though most are less than 1 meter (3 feet) high, bores on China's Qiantang River may be up to 8 meters (26 feet) high and move 11 meters per second (25 miles per hour). Their potential danger is lessened by their predictability. Accurately predicting the arrival of tidal bores is essential to safe navigation. In addition to the Bay of Fundy, tidal bores are common in the Amazon, the Ganges Delta, and England's Severn River. Some rivers in southern China may have three or four simultaneous bores at different places along the river's length.

Tidal Currents   10-11

The rise or fall in sea level as a tidal crest approaches and passes will cause a **tidal current** of water to flow into or out of bays and harbors. Water rushing into an enclosed area because of the rise in sea level as a tidal crest approaches is called a **flood current.** Water rushing out because of the fall in sea level as the tide trough approaches is called an **ebb current.** (The terms *ebb tide* and *flood tide* have no technical meaning.) Tidal currents reach maximum velocity midway between high tide and low tide. **Slack water,** a time of no currents, occurs at high and low tide when the current changes direction.

Anyone who has stood at the narrow mouth of a large bay or harbor cannot help but be impressed with the speed and volume of the tidal current that occurs between tidal extremes. Midway between high and low spring tides, the ebb current rushing from San Francisco Bay strikes the base of the south tower of the Golden Gate Bridge with such force that a bow wave is formed, giving the convincing illusion that the bridge

itself is moving rapidly. Tidal currents at the Golden Gate can reach 3 meters per second (about 7 miles per hour) because of the volume of enclosed water and the narrowness of the channel through which it must escape. Navigators must know the times of tidal currents to safely negotiate any harbor entrance or other narrow strait.

Tidal currents become more complex in the open sea. One's position relative to an amphidromic point, the shape of the basin, and the magnitude of gravitational forces and inertia must all be considered to calculate the speed and direction of tidal currents over a deep bottom. The velocity of tidal currents is less in the open sea because the water is not confined, as it is in a harbor. Open-sea tidal currents have been measured at a few centimeters per second, and their velocity tends to decrease with depth.

Tidal Friction   10-12

The daily rise and fall of the tides consumes a very large amount of energy, and this energy is ultimately dissipated as heat. Most of this energy comes directly from the rotation of Earth itself, and tidal friction is gradually slowing Earth's rotation by a few hundredths of a second per century. Even such a small change has long-term planetary effects, however. Geologists studying the daily growth rings of fossil corals and clams estimate that days have grown longer, so the number of days in a year has lessened as planetary rotation has slowed. Evidence suggests that 350 million years ago, a year contained between 400 and 410 days, with each day being about 22 hours long; and 280 million years ago there were about 390 22½-hour days in a year.

Tidal friction affects other bodies. Tidal forces have locked the rotation of the moon to that of Earth. As a result, the same side of the moon is always facing Earth, and a day on the moon is a month long.

Predicting Tides   10-13

There are at least 140 tide-generating and tide-altering forces and factors in addition to the ones discussed above. About seven of the most important of these must be considered and past records must be studied if we wish to predict tides mathematically. In fact, if a previously unknown continent were discovered on Earth, the coastal tide times and ranges on its shores could not be accurately predicted, because it is the study of past records that allows tide tables to be projected into the future. Experience permits prediction of tidal height to an accuracy of within about 3 centimeters (1.2 inches) for years in advance.

Even so, extraneous factors can affect the estimates. For example, arrival of a storm surge will greatly affect the height or timing of a tide—as will gentle, atmospherically induced seiching of the basin or excitement of large-scale resonances by a tsunami. Even a strong, steady wind onshore or offshore will affect tidal height and the arrival time of the crest. Weather-related alterations are sometimes called **meteorological tides** after their origin.

strong rise in electrical power generation), the length of a day would increase by an additional second in roughly a million and a half years. Clearly, the "cashing in" of Earth's rotational kinetic energy to supply energy needs is not of great concern. A much more serious hazard would arise in the disruption of ocean currents, biological cycles, and aesthetic values.

CHAPTER SUMMARY

Tides have the longest wavelengths of the ocean's waves. They are caused by a combination of the gravitational force of the moon and the sun, the motion of Earth, and the tendency of water in enclosed ocean basins to rock at a specific frequency. Unlike the other waves, these huge shallow-water waves are never free of the forces that cause them and so act in unusual but generally predictable ways. Basin resonances and other factors combine to cause different tidal patterns on different coasts. The rise and fall of the tides can be used to generate electrical power and are important in many physical and biological coastal processes.

TERMS AND CONCEPTS TO REMEMBER

amphidromic point	mixed tide
aphelion	neap tide
apogee	perigee
barycenter	perihelion
diurnal tide	semidiurnal tide
dynamic theory of tides	slack water
ebb current	solar tide
equilibrium theory of tides	spring tide
flood current	tidal bore
high tide	tidal current
low tide	tidal datum
lunar tide	tidal range
mean sea level	tidal wave
meteorological tide	tide

STUDY QUESTIONS

1. What causes the rise and fall of the tides? What celestial bodies are most important in determining tides? Are there such things as "tidal waves"?

2. What is a high tide? A low tide? A spring tide? A neap tide? A tidal bore?

3. What are the most important factors influencing the heights and times of tides? What tidal patterns are observed? Are there tides in the open ocean? If so, how do they behave?

4. How does the latitude of a coastal city affect the tides there—or does it?

5. From what you learned about tides in this chapter, where would you locate a plant that generated electricity from tidal power? What would be some advantages and disadvantages of using tides as an energy source?

FOR FURTHER STUDY

Bowditch, N. 1970. *American Practical Navigator.* Washington, DC: U.S. Navy Hydrographic Office. A one-volume nautical reference of unexcelled brevity and precision. The chapter on tides is clear and direct.

Cartwright, D. E. 1998. *Tides: A Scientific History.* Cambridge, UK: Cambridge University Press. A history of how we know what we know. Contemporary quotations and illustrations, clear writing.

Comins, N. F. 1993. *What If the Moon Didn't Exist: Voyages to Earths That Might Have Been.* New York: Harper-Collins. The influence of the moon on many things, tides included.

Defant, A. 1958. *Ebb and Flow: The Tides of Earth, Air and Water.* Ann Arbor: University of Michigan Press. A classic.

Dietrich, G., et al. 1980. *General Oceanography.* New York: Wiley. Careful mathematical treatment of tidal forces.

Greenberg, D. A. 1987. "Modeling Tidal Power." *Scientific American,* November, 128–38. What effect would a dam to extract power from tidal crests have in upper New England?

LeProvost, C., A. F. Bennett, and D. E. Cartwright. 1995. "Ocean Tides for and from *TOPEX/Poseidon.*" *Science* 267 (no. 5198): 639–42. Satellite data on tides are compared with data from 78 ground stations. Data from the satellite are shown to be nearly as accurate as those from the stations, a "ground truth" allowing researchers to trust satellite readings of sea surface elevation in the open sea.

Lynch, D. K. 1982. "Tidal Bores." *Scientific American,* October, 146–56. A bore is the hydraulic analogue of a sonic boom: a moving wall of water that carries the tide up some rivers that empty into the sea.

United States Geological Survey. 1988. "Catalogue of Worldwide Tidal Bore Occurrences and Characteristics." USGS Circular 1022. Washington, DC: U.S. Government Printing Office.

For additional readings, go to InfoTrac College Edition, your online research library at:

http://infotrac.thomsonlearning.com/

11

DYNAMIC, CHANGEABLE PLACES

Coasts are dynamic, changeable places, especially when viewed on a human scale. Henry David Thoreau wrote that to the people of Massachusetts, the sea is their garden, and the dog that growls at their door is the Atlantic Ocean. Few stretches of offshore water are more dangerous and turbulent than the winter Atlantic off New England, and this particular dog has an expensive appetite for the same shores favored by thousands of tourists and longtime seasonal residents. The resort town of Scituate, south and east of Boston, illustrates the ephemeral nature of some coasts. In February 1978, a particulary strong nor'easter combined with exceptionally high tides wreaked havoc on the town's sandy shore. Waves washed over the coastal zone, heavily built with vacation homes and businesses. The private property loss from the 36-hour storm was great, but in this one storm, the cost to public structures like seawalls, groins, and piers amounted to millions of dollars *more* than the value of the private property. In the 360 years since English settlers began recording the history of this stretch of shore, it has been inundated by storms and severely damaged at least 84 times. Since 1938 the Scituate seawall has been rebuilt 18 times at public expense.

A high-energy beach between winter storms. This stretch of secondary coast in northern California is exposed to violent surf driven by strong westerly winds.

The Pacific coast is also subject to coastal damage. An interesting example involved Ellen Scripps, heiress to a publishing fortune and benefactor of, among other good works, the land and first buildings of what is now the Scripps Institution of Oceanography. After World War I, Scripps developed land along the San Diego shore into Sunset Cliffs Park, a property landscaped with gazebos, walkways, and gardens, which she donated to the citizens of California. Unfortunately, the sandstone that formed the dramatic shore cliffs and adjacent park was so soft that a determined artist could scratch his initials into the rock using only his fingernails. The cliffs eroded at a rapid rate, and 55 years after the completion of the beautiful park, almost nothing remained of it. The coast in the area continues to erode, threatening a highway and a number of expensive homes.

Coastal change affects a great many Americans. More than 165 million of us will live in coastal areas by 2015. The value of coastal real estate in this country exceeded $3.1 trillion at the beginning of the year 2000.　　　　　　　　　　　　　11-1

- -

KEY CONCEPTS

1. Coasts are temporary structures, often subject to rapid change.

2. The *location* of a coast depends primarily on global tectonic activity and the volume of water in the ocean. The *shape* of a coast is a product of many processes: uplift and subsidence, the wearing-down of land by erosion, and the redistribution of material by sediment transport and deposition.

3. Coasts are classified as primary coasts (on which terrestrial influences dominate) or secondary coasts (on which marine influences dominate).

4. Secondary coasts often support beaches, accumulations of loose particles. Generally, the finer the particles on the beach, the flatter its slope. Beaches change shape and volume as a function of wave energy and the balance of sediment input and removal.

5. Estuaries are among the most complex and biologically productive coasts.

6. Human interference in coastal processes rarely increases the long-term stability of a coast.

CHAPTER AT A GLANCE

- -

Coastal areas join land and sea. The place where ocean meets land is usually called the **shore,** while the term **coast** refers to the larger zone affected by the processes occurring at this boundary. A sandy beach might form the shore in an area, but the coast (or coastal zone) includes the marshes, sand dunes, and cliffs just inland of the beach as well as the sand bars and troughs immediately offshore. The world ocean is bounded by about 440,000 kilometers (273,000 miles) of shore.

Because of its proximity to both ocean and land, a coast is subject to natural events and processes common to both realms. A coast is an active place. Here is the battleground on which wind waves break and expend their energy. Tides sweep water on and off the rim of land, rivers drop most of their sediments at the coasts, and ocean storms pound the continents. The *location* of a coast depends primarily on global tectonic activity and the volume of water in the ocean. The *shape* of a coast is a product of many processes: uplift and subsidence, the wearing-down of land by erosion, and the redistribution of material by sediment transport and deposition.

CLASSIFYING COASTS   11-2

As we saw in Chapter 3, no area of geology has been left undisturbed by the revelations of plate tectonics. In the 1960s geologists began to classify coasts according to tectonic position. *Active* coasts, near the leading edge of moving continental plates, were found to be fundamentally different from the more *passive* coasts near trailing edges. The shapes, composition, and ages of coasts are better understood by taking plate movements into account. But the slow forces of plate movement are frequently obscured by the more rapid action of waves, by the erosion of land, and by the transport of sediments.

Another important consideration in the classification of coasts is long-term change in sea level. Five factors can cause sea level to change. Three of these factors are responsible for **eustatic change**—variations in sea level that can be measured all over the world ocean:

- The amount of water in the world ocean can vary. Sea level is lower during periods of global glaciation (ice ages) because there is less water in the ocean. It is higher during warm periods, when the glaciers are smaller. Periods of abundant volcanic outgassing can also add water to the ocean and raise sea level.
- The volume of the ocean's "container" can vary. High rates of seafloor spreading are associated with the expansion in volume of the mid-ocean ridges. This displaces the ocean's water, which climbs higher on the edges of the continents. Sediments shed by the continents during periods of rapid erosion can also decrease the volume of ocean basins and raise sea level.
- The water itself may occupy more or less volume as its temperature varies. During times of global warming,

seawater expands and occupies more volume, raising sea level.

Of course, the continents rarely stay still as sea level rises and falls. Local changes are bound to occur, and two other factors produce variations in *local* sea level:

- Tectonic motions and isostatic adjustment can change the height and shape of the coast. Coasts can experience uplift as lithospheric plates converge or can be weighted down by masses of ice during a period of widespread glaciation. The continents slowly rise when the ice melts. (See Figure 3.7 for a review of the principle of isostatic equilibrium.)
- Wind and currents, seiches, storm surges, an El Niño or La Niña event, and other effects of water in motion can force water against the shore or draw it away.

Sea level has been at its current elevation (give or take 0.5 meter, 1.5 feet) for only about 2,500 years. Over the past 2 million years, worldwide sea level has varied from about 6 meters (20 feet) above to about 125 meters (410 feet) below its present position. The most recent low point occurred about 18,000 years ago at the height of the most recent glaciation. Indeed, sea level has been at the modern "high" only rarely in the past 2 million years—the predominant state for Earth is a much lower eustatic sea level position.

Changes in sea level produce major differences in the position and nature of coastlines, especially in areas where the edge of the continent slopes gradually or where the coast is rising or sinking. **Figure 11.1** shows an estimate of previous shore positions along the southeastern coast of the United States in the geologically recent past, and a prediction for the distant future should the present warming trend cause more of the polar ice to melt.

What coastal classification scheme best combines these elements? One of the most useful divides coasts into a *primary* and *secondary* classification. If a coast is essentially in almost the same condition as it was in when sea level stabilized after the last ice age, it is called a **primary coast.** Primary coasts are young coasts in which terrestrial influences (that is, processes that occur at the boundary between land and air) dominate. If the coast has been significantly changed by wave action and other marine processes since sea level stabilized, it is termed a **secondary coast.** Secondary coasts are usually older than primary coasts—they have been exposed to marine action for a longer time. Secondary coasts retain little (if any) evidence of the nonmarine processes that produced them. Some coasts show both characteristics: A single long shoreline can consist of both primary and secondary coasts.

PRIMARY COASTS   11-3

Primary coasts are often rough and irregular. The ocean has not had time to modify the terrestrial features provided by changes in sea level, the scouring of glaciers, deposition of sediment at

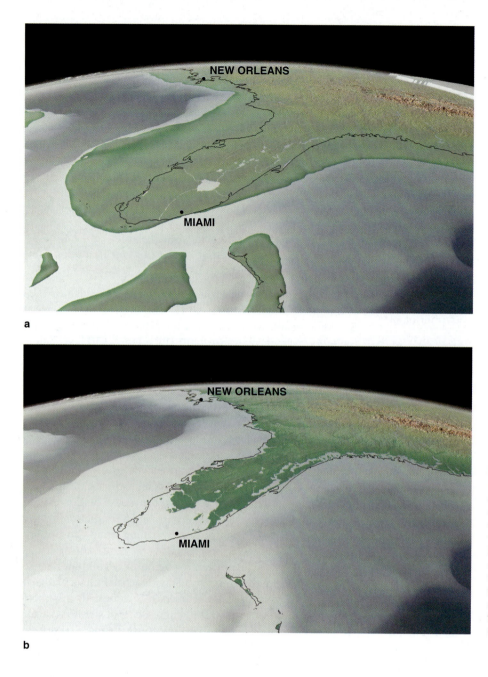

a

b

Figure 11.1 (**a**) The southeastern coast of the United States looked much different 18,000 years ago, during the last ice age. Because of lower sea level, the position of the gently sloping southeastern coast has been as much as 200 kilometers (125 miles) seaward from the present shoreline, leaving much of the continental shelf exposed. (**b**) In the distant future, if the ocean were to expand and the polar ice caps were to melt because of global warming, sea level could rise perhaps 60 meters (200 feet), driving the coast inland as much as 250 kilometers (160 miles).

the mouths of rivers, volcanic eruptions, or the movement of earth along faults.

Land Erosion Coasts

11-4

When sea level was lower during the last glaciations, rivers cut across the land and eroded sediment to form coastal river valleys. When higher sea level returned, the valleys were flooded, or *drowned*, with seawater. Chesapeake Bay (**Figure 11.2**) and the Hudson River valley are examples of drowned river valleys.

Glaciers sometimes form in river valleys when rivers cut through the edges of continents at high latitudes. Deep, narrow bays known as **fjords** are often formed by tectonic forces and later modified by glaciers eroding valleys into deep, U-shaped troughs. Fjords (**Figure 11.3**) are found in British Columbia,

Alaska, Greenland, Norway, New Zealand, and other cold, mountainous places.

Coasts Built Out by Land Processes

11-5

In a few places, sediments washing off the land have built out the coasts extensively. The shoreline in such places is much different from its configuration at the end of the last ice age. The most important of these coastal features are **deltas** (which are sometimes triangular in shape and thus named after the Greek letter delta, Δ).

Deltas do not form at the mouth of every sediment-laden river. A broad continental shelf must be present to provide a platform on which sediment can accumulate. In addition (as

Figure 11.2 A false-color photograph of Chesapeake Bay taken from space. The complex bay is an example of a drowned river valley.

Figure 11.3 Milford Sound, a deep fjord on the western coast of New Zealand's South Island.

Figure 11.4 River deltas form at places where sediment-laden rivers enter enclosed or semienclosed seas, where wave energy is limited. The bird's-foot shape of the Mississippi Delta is seen clearly in this photograph. Lobed and bird's-foot deltas form where deposition overwhelms the processes of coastal erosion and sediment transportation. The sediment-laden water looks brown or tan in this photograph taken from low orbit.

Figure 11.5 The Nile Delta, which has been smoothed into a more rounded shape by post-depositional coastal erosion and sediment transportation. (The tail of the space shuttle from which this photograph was taken can be seen in the upper left of the photo.)

befits a primary coast), marine processes must not dominate: Tidal range should be low, and waves and currents generally mild. There are no large deltas along the Atlantic coast of the United States because sediments arriving at the coast are deposited in the sunken river mouths or dispersed by tides and currents. Also, there are no large deltas along the western margins of North and South America because these coasts are converging margins, where an oceanic plate is being subducted and the continental shelf is very narrow; sediment that would form a delta is swept down the continental slope or dispersed along the coast by waves. Deltas are most common on the low-energy shores of enclosed seas (where the tidal range is not extreme) and along the tectonically stable trailing edges of some continents. The largest deltas are those in the Gulf of Mexico (the Mississippi, **Figure 11.4**), the Mediterranean Sea (the Nile, **Figure 11.5,** and the Rhône, Po, and Ebro), the Ganges–Brahmaputra river system in the Bay of Bengal, and the huge deltas formed by the rivers of China that empty into the South China Sea.

Deltas are not the only type of coast built out by the land. The glaciers that covered the poleward parts of the continents during the ice age deposited great quantities of sediments and rocks near their outer margins. When the glaciers retreated, they left **moraines**—hills and ridges of sediments—some of which still stand above sea level. Part of Long Island, New York, is a glacial moraine, as are the oval-shaped hills of Boston and sections of the Puget Sound coast at Seattle. Perhaps the most famous glacial moraine in the United States is the area around Cape Cod, Massachusetts. The cores of the islands of Martha's Vineyard and Nantucket represent the farthest advance of the glacial debris, and the spine of the cape itself is a remnant of material dropped by the glacier as it retreated northward when the climate warmed (see **Figure 11.6**).

Volcanic Coasts

As we saw in Chapter 4, most islands that rise from the deep ocean are of volcanic origin. If the volcanism has been recent, the coasts of a volcanic island will consist of lobed lava flows extending seaward, common features in the Hawaiian

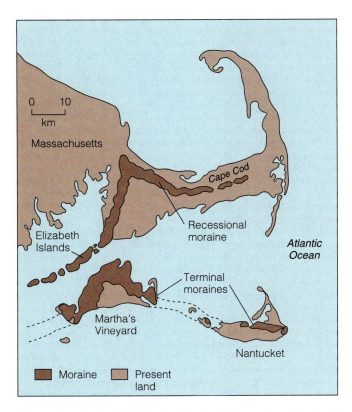

Figure 11.6 Most of the islands of Martha's Vineyard and Nantucket were formed from debris deposited by an advancing glacier. The spine of Cape Cod is built of material dropped as the glaciers retreated.

Figure 11.7 Concave shapes may also occur on a volcanic coast. The explosion and collapse of a volcano close to shore caused this indentation on the coast of Isla Encantada, Gulf of California, Mexico.

Islands. Concave shorelines (as in **Figure 11.7**) result when a small coastal volcano explodes or collapses and seawater fills the crater.

Coasts Shaped by Earth Movements

11-7

Coasts can coincide with places where Earth's crust is being warped or faulted. When the seabed on the seaward side of a coastal fault moves *downward,* a steep escarpment that continues to a greater depth than a wave-cut cliff can result. When the landward side of the fault moves *upward,* the part of the coast previously submerged during high tides can be left high and dry. The 1964 Alaska earthquake (described in Chapter 3) caused parts of Prince William Sound shore to rise more than 3.5 meters (12 feet). As can be seen in **Figure 11.8,** in some places a band of the old sea-cut bench nearly half a kilometer (¼ mile) wide was exposed, even at high tide.

Fault coasts sometimes occur along transform faults. The Pacific coast of North America provides two striking examples: the Gulf of California (located along the San Andreas Fault between Baja California and the mainland of Mexico) and the area around Point Reyes, north of San Francisco. Baja California was once a part of the North American continent, but movement at the boundary between the Pacific and North American Plates has torn the finger of land on the west side of the fault from the land mass to the east. The young, narrow,

Figure 11.8 This former seafloor at Prince William Sound, Alaska, was raised 3.5 meters (12 feet) above sea level by tectonic uplift during the great earthquake of 27 March 1964. The exposed surface, which slopes gently from the base of the sea cliffs to the water, is about 400 meters (¼ mile) wide. The light-colored coating on the rocks consists mainly of the dried remains of small marine organisms. The photo was taken at about 0.0 tide, 30 May 1964.

Figure 11.9 A characteristically straight fault coast at Tomales Bay, California. Point Reyes is visible to the left (west); the city of San Francisco is out of view to the south. The San Andreas Fault trace disappears below sea level in the bay (arrow); the straight sides of the bay closely parallel the submerged fault.

straight gulf has formed as seawater intruded. As can be seen in **Figure 11.9,** a photo of Tomales Bay north of San Francisco, fault coasts can be startlingly straight.

SECONDARY COASTS 11-8

Secondary coasts are those coasts that have been significantly changed by wave action and other marine processes after sea level has stabilized. Land erosion and marine erosion both work to change a primary coast to a secondary coast.

Secondary coasts are shaped and attacked from the land by stream erosion, the abrasion of wind-driven particles, the alternate freezing and thawing of water in rock cracks, the probing of plant roots, glacial activity, rainfall, dissolution by acids from soil, and slumping. Attack from the sea is by waves and currents. The continual onslaught of waves does most of the work, with currents distributing the results of the waves' labor.

Wave-Caused Erosion 11-9

Large storm surf routinely generates tremendous pressures. The crashing waves push air and water into tiny rock crevices. The repeated buildup and release of pressure within these crevices can weaken and fracture the rock. But it is not the hy-draulic pressure of moving water alone that abrades the coasts. Tiny pieces of sand, bits of gravel, or stones hurled by waves toward the shore are even more effective at eroding the shore. **Dissolution,** the dissolving of minerals in the rocks by water, contributes to the erosion of easily soluble coastal rocks such as limestone. Even the digging and scraping activities of marine organisms have an effect.

The rate at which a shore erodes depends on the hardness and resistance of the rock, the violence of the wave shock to which it is exposed, and the local range of tides. Hard rock resists wear. Coasts made of granite or basalt may retreat an insignificant amount over a human lifetime; the granite coast of Maine erodes only a few centimeters per decade. Coasts of soft sandstone or other weak (or soluble) materials, however, may disappear at a rate of a few meters per year.

Marine erosion is usually most rapid on **high-energy coasts,** areas frequently battered by large waves. High-energy coasts are most common adjacent to stormy ocean areas of great fetch and along the eastern edges of continents exposed to tropical storms. The coasts of Maine and British Columbia and the southern tips of South America and South Africa are typical high-energy coasts. **Low-energy coasts** are only infrequently attacked by large waves. Because of their generally protected location in the Gulf of Mexico, the U.S. Gulf states share a low-energy coast—at least between hurricanes!

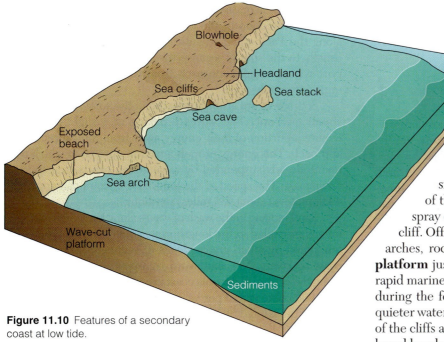

Figure 11.10 Features of a secondary coast at low tide.

Waves can only affect the coast where they strike, so erosion is concentrated near average sea level. A shore with little tidal variation can erode quickly because the wave action is concentrated near one level for longer times. Low-energy coasts protected by offshore islands usually erode slowly, as do areas below the low-tide line. Some erosion does occur below the surface because of the orbital motion of water in waves, but even the largest waves have little erosive effect at depths greater than about 15 meters (50 feet) below average sea level. Cliffs above shore are subject to pounding either directly from waves or by rocks hurled by waves.

Erosive forces can produce a wave-cut shore showing some or all of the features illustrated in **Figure 11.10.** Note

the complex, small-scale irregularities of this rocky coastline. **Sea cliffs** slope abruptly from land into the ocean, their steepness usually resulting from the collapse of undercut notches. The position of the sea cliffs marks the shoreward limit of marine erosion on a coast. The parade of waves cuts **sea caves** into the cliffs at local zones of weakness in the rocks. Most sea caves are accessible only at low tide. A blowhole can form if erosion follows a zone of weakness upward to the top of the cliff. When the tide is at just the right height, spray can blast from the fissure as waves crash into the cliff. Offshore features of rocky coasts can include natural arches, rock stacks, and a smooth, nearly level **wave-cut platform** just offshore, which marks the submerged limit of rapid marine erosion. Much of the debris removed from cliffs during the formation of these structures is deposited in the quieter water farther offshore, but some can rest at the bottom of the cliffs as exposed beaches. Indeed, as we shall soon see, broad beaches are often features of secondary coasts.

The first effect of marine erosion on a newly exposed coast is to intensify the irregularity of the coastline. This happens because coastal rocks are usually not uniform in composition over long horizontal distances. Some hard rocks will resist erosion well, while softer rocks on the same coast may disappear almost overnight. (This explains the uneven character of stacks, arches, and sea cliffs.)

Eventually, however, coastal erosion tends to produce a smooth shoreline. Because of wave refraction (see Figure 9.18), wave energy is focused onto headlands and away from bays (**Figure 11.11**). Sediment eroded from the headlands tends to collect as beaches in the relatively calm bays. As erosion continues, the deposits may eventually protect the base of

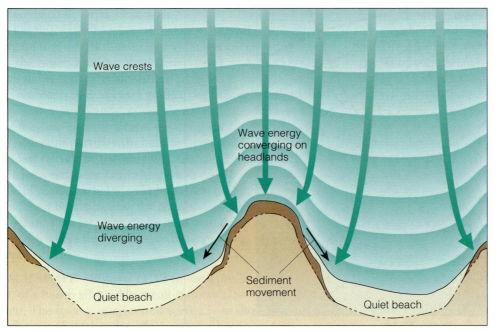

Figure 11.11 Wave energy converges on headlands and diverges in the adjoining bays. The accumulation of sediment derived from the headland in the tranquil bays eventually smooths the contours of the shore.

a

b

Figure 11.14 The movement of sand on Carlsbad State Beach, California. (**a**) The beach in the summer of 1978. (**b**) The same beach in the spring of 1983 after a winter of severe storms. The sand has moved offshore, and coastal structures have been damaged.

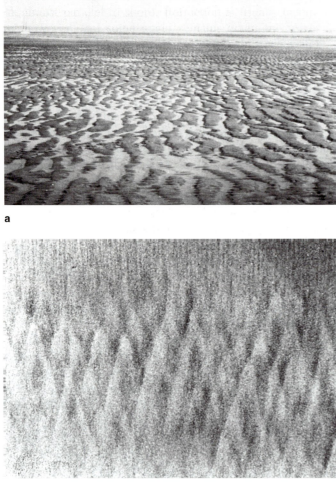

a

c

b

d

Figure 11.15 Minor features of beach sand. (**a**) Ripple marks. (**b**) The streamlike pattern of rill marks. (**c**) Backwash marks. Each V-shaped feature is caused by a small irregularity in the sand; each is about 15 centimeters (6 inches) long. (**d**) Layers of sediments are clearly seen in this freshly eroded beach face. The alternating layers were deposited by small waves during many ebb and flood tide cycles. Larger storm-driven waves have exposed the layers to view. The keys provide scale.

interrupt the backwash along the low tide terrace and cause uneven deposition or erosion of very fine sediment of a contrasting color. Another result of uneven deposition can be seen in beach layering. A vertical slice into a calm beach will often reveal layer upon layer of sediments. The sediments are often separated by their density and may be of different colors and textures (**Figure 11.15d**).

On a somewhat larger scale, the beach shoreline is sometimes scalloped by *cusps* (*cuspis* = point) (**Figure 11.16**). These repeating points, interrupted by miniature bays, are evenly spaced every several meters. Cusps tend to form at neap tide periods and are destroyed during the greater range of spring tides. Cusp size appears to be correlated with wave energy. Their origin is not well understood, but they are probably caused by interference between approaching waves.

COASTAL CELLS   As we have seen, most new sand on a coast is brought in by rivers. The sand is moved parallel to the beach by longshore drift, and it is moved onshore and offshore at right angles to the beach as the seasons change. If a beach is stable in size, neither growing nor shrinking, the amount of new sand entering must be balanced by the amount of old sand being removed. Sand that drifts below the reach of wave action is lost from the coast and may migrate farther out on the continental shelf. Some sand is driven by longshore currents into the nearshore heads of submarine canyons. Sand moving away from shore in these canyons sometimes forms impressive sandfalls (see Figure 4.16 for an example) and is lost from the beaches above. The bulk of this material is transported by gravity down the axis of the canyon and ultimately deposited on a submarine fan at the base of the slope.

The natural sector of a coastline in which sand *input* and sand *outflow* are balanced can be thought of as a **coastal cell.** The main features of such a cell are illustrated in **Figure 11.17a.** Coastal cells are usually bounded by submarine canyons that conduct sediments to the deep sea. Their size varies greatly. They are often very large along the relatively smooth, tectonically passive trailing edges of continents; coastal cells along the southeastern coast of the United States, for example, are hundreds of kilometers long. On the active leading edge of the continent, they are smaller. Four cells exist in the 360 kilometers (225 miles) between southern California's Point Conception and the Mexican border. Each terminates in a submarine canyon at the downcoast end (see **Figure 11.17b**).

Other Large-Scale Features of Secondary Coasts

 11-12

Aside from beaches, secondary coasts exhibit some other large-scale features resulting from the deposition of sediments. Some of these features are illustrated in **Figure 11.18.**

SAND SPITS AND BAY MOUTH BARS   **Sand spits** are among the most common of these. A sand spit forms where the longshore current slows as it clears a headland and approaches a quiet bay. The slower current in the mouth of the

Figure 11.16 Beach cusps at Point Reyes National Seashore, California, formed in a gently sloping fine-sand beach.

bay is unable to carry as much sediment, so sand and gravel are deposited in a line downcurrent of the headland. As can be seen in Figure 11.18, sand spits often have a curl at the tip. This is caused by the current-generating waves being refracted around the tip of the spit.

A **bay mouth bar** forms when a sand spit closes off a bay by attaching to a headland adjacent to the bay. The bay mouth bar protects the bay from waves and turbulence and encourages the accumulation of sediments there. An inlet—a passage to the ocean—may be cut through a bay mouth bar by tidal action, by water flowing from a river emptying into the bay, or by heavy rains. A bay mouth bar is shown in **Figure 11.19.**

BARRIER ISLANDS AND SEA ISLANDS   Secondary coasts can also develop narrow, exposed sandbars that are parallel to but separated from land. These are known as **barrier islands** (see **Figure 11.20**). About 13% of the world's coasts are fringed with barrier islands.

Barrier islands can form when sediments accumulate on submerged rises paralleling the shoreline. Some islands off the Mississippi–Alabama coast developed in this way. Larger barrier islands are thought to form in a different way, however. Near the end of the last major rise in sea level, about 6,000 years ago, coastal plains near the edge of the continental shelf were

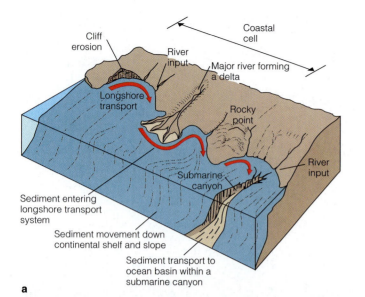

a

Cliff
erosion

River
input

Coastal
cell

Major river forming
a delta

Longshore
transport

Rocky
point

River
input

Submarine
canyon

Sediment entering
longshore transport
system

Sediment movement down
continental shelf and slope

Sediment transport to
ocean basin within a
submarine canyon

Figure 11.17 Coastal sediment transport cells. (**a**) The general features of coastal cells, in which sand is introduced by rivers, transported southward by the longshore drift, and trapped within the nearshore heads of submarine canyons. (**b**) A series of coastal cells in southern California. The white arrows show sand flowing toward the submarine canyons.

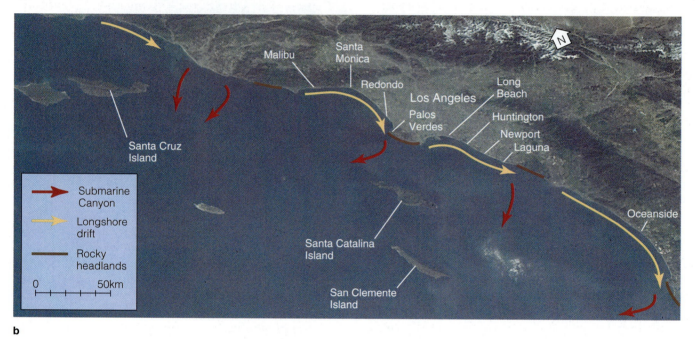

b

Malibu

Santa
Monica

Redondo

Long
Beach

Los Angeles

Huntington

Palos
Verdes

Newport
Laguna

Santa Cruz
Island

Oceanside

Submarine
Canyon

Longshore
drift

Rocky
headlands

0   50km

Santa Catalina
Island

San Clemente
Island

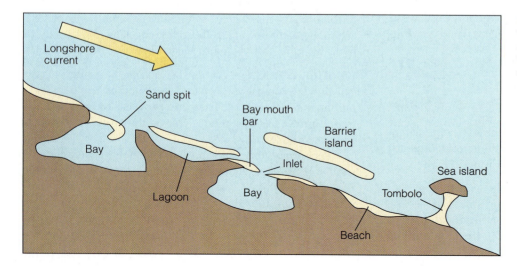

Figure 11.18 A composite diagram of the large-scale features of an imaginary secondary coast. Not all these features would be found in such close proximity on a real coast.

Longshore
current

Sand spit

Bay mouth
bar

Barrier
island

Bay

Inlet

Sea island

Lagoon

Bay

Tombolo

Beach

Figure 11.19 A bay mouth bar. The inlet is now closed, but increased river flow (from inland rainfall) or large waves combined with very high and low tides could break the bar.

fronted by lines of sand dunes. Rising sea level caused the ocean to break through the dunes and form a **lagoon**—a long, shallow body of seawater isolated from the ocean. The high lines of coastal dunes became islands. As sea level continued to rise, wave action caused the islands and lagoons to migrate landward. Most of the barrier islands off the southeast coast of the United States originated in this way. They are still migrating slowly landward as sea level continues to rise.

There are 295 barrier islands along the Atlantic and Gulf coasts of the United States, with a combined length of 2,591 kilometers (1,610 miles). Every year severe storms generate waves intense enough to erode barrier island beaches.

The largest of these storms can generate waves that overwash the low islands. Runoff from rivers swollen by rains, coupled with water driven by wind waves and storm surge, can rapidly flood a lagoon and cut new inlets through barrier islands.

Despite these dangers, about 70 of the barrier islands have been commercially developed, and millions of people live on them. The most famous barrier islands include Atlantic City, New Jersey; Ocean City, Maryland; Miami Beach and Palm Beach, Florida; and Galveston, Texas. Roughly once every hundred years a winter storm has catastrophic effects on populated areas of Atlantic barrier islands, and the southeastern Atlantic and Gulf coasts must contend with occasional

Figure 11.20 Barrier islands off the Texas Gulf coast. (This photo, taken from space, is on a much larger scale than Figure 11.19.)

a

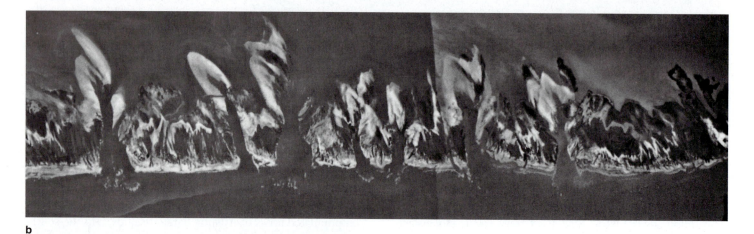

b

c

Figure 11.21 Barrier island modification, actual and potential. (**a**) The beach extending along the Matagorda Peninsula (Texas) barrier in September 1960. (**b**) The same area six days after the passage of Hurricane Carla in September 1961. The beach and island have been breached, and washover deltas are clearly seen. (**c**) Ocean City, Maryland, a developed barrier island. Host to 8 million visitors a year, this city (and others similarly situated) has no effective protection against flooding and damage from severe storms.

large hurricanes. The continuing subsidence of these passive coasts (combined with changes caused by commercial development and the ongoing rise in sea level) will undoubtedly cost lives and destroy property. **Figure 11.21** suggests the extent of the threat.

Unlike barrier islands, **sea islands** contain a firm central core that was part of the mainland when sea level was lower. The rising ocean separated these high points from land, and sedimentary processes surrounded them with beaches. Hilton Head, South Carolina, and Cumberland Island, Georgia, are sea islands. If the island is close to shore, a bridge of sediments called a **tombolo** may accumulate to connect the island to the mainland. Tombolos can also connect offshore rocky outcrops or volcanoes to the mainland. A sea island and a tombolo are shown in Figure 11.18.

COASTS FORMED BY BIOLOGICAL ACTIVITY

11-13

Coasts can be extensively modified by the activities of animals and plants. The most dramatic modifications occur in the tropics, where coral polyps form reefs around volcanic islands or along the margin of a continent. The greatest of all reefs is the Australian Great Barrier Reef (**Figure 11.22**), which begins

Figure 11.22 A small section of the Great Barrier Reef, Queensland, Australia. This coast has been extensively modified by biological activity.

in the Torres Strait separating New Guinea and Australia and runs down the northeastern coast of Australia for 2,500 kilometers (1,500 miles). The reef is not a single object, but a composite of more than 3,000 individual coral reefs covering 350,000 square kilometers (135,000 square miles)—collectively the largest structure made by living organisms on Earth.

The Florida Keys—a series of low islands extending south of the tip of Florida—are an excellent example of a coral reef coast in the continental United States. How can a coral coast extend above sea level? The Keys are relatively high because they were formed during a time between glaciations when sea level stood about 20 feet (6 meters) higher than it does today. Some coral reef islands of the Pacific and Indian Oceans are much lower—a great storm can submerge and fracture them. Those that extend above sea level do so because chunks of reef margin are thrown toward the center of these islands by storm waves and winds. The accumulated blocks are cemented together by the limestone (calcium carbonate) they contain.

Other coasts have been formed by mangroves, trees that can grow in salt water. The coast of southwestern Florida has been extended and shaped by the activity of mangroves, whose root systems trap and hold sediments around the plant (**Figure 11.23**). The root complex forms an impenetrable barrier and safe haven for organisms around the base of the trees.

Figure 11.23 A mangrove coast in Florida. Mangrove trees trap sediments, building and stabilizing the coast.

ESTUARIES

An **estuary** (*œstus* = tide) is a body of water partially surrounded by land, where fresh water from a river mixes with ocean water. Estuaries are areas of remarkable biological productivity and diversity. The coasts of the United States contain about 15,150 square kilometers (5,850 square miles) of estuarine waters. Chesapeake Bay, San Francisco Bay, and Puget Sound are all estuaries.

Classification of Estuaries

Estuaries are classified into four types depending on their origins:

- Drowned river mouths
- Fjords
- Bar-built
- Tectonic

Each type of estuary is shown in **Figure 11.24.**

Estuaries formed at drowned river mouths are common throughout the world, particularly along the Atlantic Coast of the United States. Remember, sea level has risen about 125 meters (410 feet) since the end of the last major period of glaciation some 18,000 years ago, and this has resulted in the incursion of seawater into river mouths. The mouths of the York, James, and Susquehanna rivers, and Chesapeake Bay, are examples of this type of estuary.

As Figure 11.3 suggests, fjords are steep, glacially eroded U-shaped troughs. They are often about 300 to 400 meters (1,000 to 1,300 feet) deep but typically terminate in a shallow lip, or sill, of terminal glacial deposits. In fjords with shallow sills, little vertical mixing occurs below the sill depth, and the bottom waters can become stagnant (look ahead to Figure 11.25d). In fjords with deeper sills, the bottom waters mix slowly with adjacent oceanic waters. While fjords are common in Norway, Greenland, New Zealand, Alaska, and western Canada, they are not common in the lower 48 states. The Strait of Juan de Fuca in Washington is a good example.

Bar-built estuaries form when a barrier island or a barrier spit is built parallel to the coast above sea level. Since these estuaries are shallow and usually have only a small inlet connecting them to the ocean, tidal action is limited. Waters in bar-built estuaries are mainly mixed by the wind. Albemarle and Pamlico Sounds in North Carolina, and Chincoteague Bay in Maryland, are bar-built estuaries.

Estuaries produced by tectonic processes are coastal indentations formed by faulting and local subsidence. Fresh water and seawater both flow into the depression, and an estuary results. San Francisco Bay is, in part, a tectonic estuary.

Characteristics of Estuaries

Three factors determine the characteristics of estuaries: the shape of the estuary, the volume of river flow at the head of the estuary, and the range of tides at the estuary's mouth. The mingling of waters of different densities, the rise and fall of the tide, and the variations in river flow—along with the actions of wind, ice, and the Coriolis effect—guarantee that patterns of water circulation in an estuary will be complex.

Estuaries are categorized by their circulation patterns. The simplest circulation patterns are found in **salt wedge estuaries,** which form where a rapidly flowing large river enters the ocean in an area where tidal range is low or moderate. The exiting fresh water holds back a wedge of intruding seawater (**Figure 11.25a**). Note that density differences cause fresh water to flow over salt water. The seawater wedge retreats seaward at times of low tide or strong river flow, and it returns landward as the tide rises or when river flow diminishes. Some seawater from the wedge joins the seaward-flowing fresh water at the steeply sloped upper boundary of the wedge, and new seawater from the ocean replaces it. Nutrients and sediments from the ocean can enter the estuary in this way. Examples of salt wedge estuaries are the mouths of the Hudson and Mississippi Rivers.

A different pattern occurs where the river flows more slowly and tidal range is moderate to high. As their name implies, **well-mixed estuaries** contain differing mixtures of fresh and salt water through most of their length. Tidal turbulence stirs the waters together as river runoff pushes the mixtures to sea. A well-mixed estuary is illustrated in **Figure 11.25b.** The mouth of the Columbia River is an example.

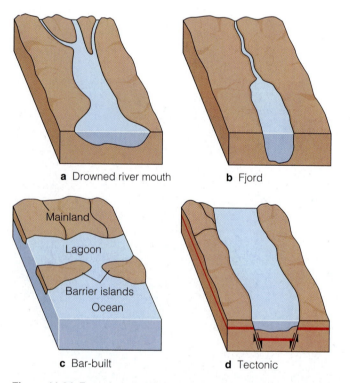

a Drowned river mouth

b Fjord

Mainland

Lagoon

Barrier islands

Ocean

c Bar-built

d Tectonic

Figure 11.24 Estuaries classified by their origins. (**a**) Drowned river mouths: Chesapeake Bay and the mouths of the James, York, and Susquehanna Rivers. (**b**) Fjords: Milford (see Figure 11.3), the Strait of Juan de Fuca north of Washington State. (**c**) Bar-built: Albemarle and Pamlico Sounds in North Carolina. (**d**) Tectonic: San Francisco Bay, Tomales Bay (see Figure 11.9).

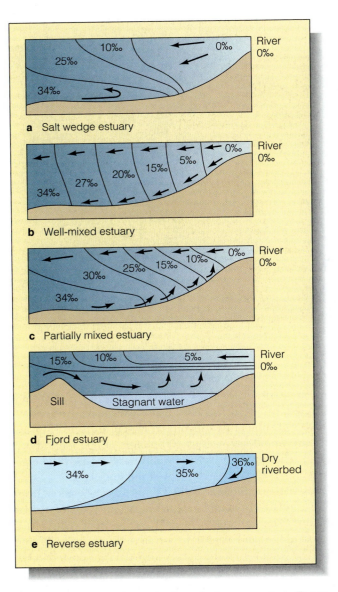

Figure 11.25 Types of estuaries in vertical cross sections. The salinity values show the amount of mixing between fresh water (0‰) and seawater (34‰) in the various types. (**a**) Salt wedge estuary. (**b**) Well-mixed estuary. (**c**) Partially mixed estuary. (**d**) Fjord estuary. (**e**) Reverse estuary, in which evaporation plays a major role.

Deeper estuaries exposed to similar tidal conditions but greater river flow become **partially mixed estuaries.** Partially mixed estuaries share some of the properties of salt wedge and well-mixed estuaries. Note in **Figure 11.25c** the influx of seawater beneath a surface layer of fresh water flowing seaward; mixing occurs along the junction. Energy for mixing comes from both tidal turbulence and river flow. England's Thames River, San Francisco Bay, and Chesapeake Bay are examples.

In well-mixed and partially mixed estuaries in the Northern Hemisphere, the incoming seawater will press against the right side of the estuary because of the Coriolis effect. Outflowing river water will also trend to the right of its direction of travel. This rightward drift can be seen in the contour of lines representing surface salinity in Chesapeake Bay (**Figure 11.26**).

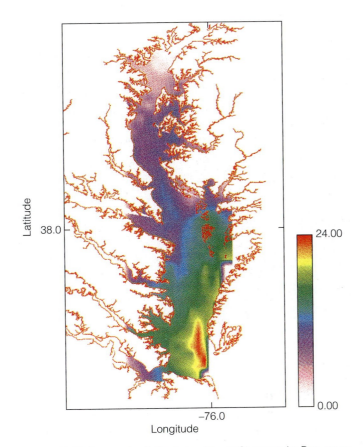

Figure 11.26 Range of salinity during May in Chesapeake Bay, an example of a partially mixed estuary. The colors indicate salinity in parts per thousand. The typical distribution of surface salinity in the estuary ranges from 28‰ at the mouth to 1‰ near the upper reaches. The Coriolis effect forces the inflowing salt water against the right (eastern) bank. (Notice how the salinity contours trend toward the right bank.) Compare this diagram to the photograph in Figure 11.2.

Fjord estuaries form where glaciers have gouged steep U-shaped valleys below sea level. Typically, fjord estuaries have small surface areas, high river input, and little tidal mixing. River water tends to flow seaward at the surface and have little contact with the seawater below (**Figure 11.25d**). In fjord estuaries with steep sills, a layer of stagnant water—cold water containing little oxygen and few nutrients—can form above the floor.

Reverse estuaries can form along arid coasts when rivers cease to flow. The evaporation of seawater in the uppermost reaches of these estuaries will cause water to flow from the ocean into the estuary, producing a gradient of *increasing* salinity from the ocean to the estuary's upper reaches (see **Figure 11.25e**). Reverse estuaries are common on the Pacific coast of Mexico's Baja Peninsula and along the U.S. Gulf coast.

The Value of Estuaries   11-17

Some of the oldest continuous civilizations have flourished in estuarine environments. The lower regions of the Tigris and Euphrates Rivers, the Po River Delta region of Italy, the Nile Delta, the mouths of the Ganges, and the lower Hwang Ho

Figure 11.27 Wave-cut terraces on San Clemente Island.

Valley have supported dense human habitation for thousands of years. Estuaries continue to be irresistible to developers. In areas of high population density, estuaries are routinely dredged to provide harbors, marinas, and recreational areas, and filled to make space for homes and agricultural land.

Estuaries often support a tremendous number of living organisms. The easy availability of nutrients and sunlight, protection from wave shock, and the presence of many habitats permit the growth of many species and individuals. Estuaries are frequently nurseries for marine animals; several species of perch, anchovy, and Pacific herring take advantage of the abundant food in estuaries during their first weeks of life. Unfortunately for their inhabitants, the high demand for development is incompatible with a healthy estuarine ecosystem. More than half the nation's estuaries and other wetlands have been lost. Of the original 870,000 square kilometers (215 million acres) of wetlands that once existed in the lower 48 states, only about 360,000 square kilometers (90 million acres) remain.

CHARACTERISTICS OF U.S. COASTS 11-18

Plate tectonic forces have had immense influence on the margins of continents, and the edges of the United States are no exception. The results of plate movement on the Pacific coast differ greatly from those on the Atlantic and Gulf coasts, primarily because the Pacific coast is near an active plate margin while the Atlantic and Gulf coasts are not.

The Pacific Coast 11-19

The Pacific coast is an actively rising margin on which volcanoes, earthquakes, and other indications of recent tectonic

activity are easily observed. Pacific coast beaches are typically interrupted by jagged rocky headlands, volcanic intrusions, or the effects of submarine canyons. Wave-cut terraces (**Figure 11.27**) are found as much as 400 meters (1,300 feet) above sea level, evidence that tectonic uplift has exceeded the general rise in sea level through the past million years.

Most of the sediments on the Pacific coast originated from erosion of relatively young granitic or volcanic rocks of nearby mountains. The particles of quartz and feldspar that constitute most of the sand were transported to the shore by flowing rivers. The volume of sedimentary material transported to Pacific coast beaches from inland areas greatly exceeds the amount originating at the coastal cliffs. Deltas tend not to form at Pacific coast river mouths because the continental shelf is narrow, river flow is generally low (except for the Columbia River), and beaches are usually high in wave energy. The predominant direction of longshore drift is to the south because northern storms provide most of the wave energy.

The Atlantic Coast 11-20

The Atlantic coast is a passive margin, tectonically calm and subsiding because of its trailing position on the North American Plate. Subsidence along the coast has been considerable—3,000 meters (10,000 feet) over the last 150 million years. A deep layer of sediment has built up offshore, material that helped produce today's barrier islands. Relatively recent subsidence has been more important in shaping the present coast, however. Except for the coast of Maine (which is still in isostatic rebound after the recent departure of the glaciers), coastal sinking and rising sea level have combined to submerge some parts of the Atlantic coast at a rate of about 0.3 meter (1 foot) per century. This process has formed the huge flooded valleys of Chesapeake and Delaware Bays, the landward-migrating barrier islands, and the shrinking lowlands of Florida and Georgia.

Rocks to the north (in Maine, for example) are among the hardest and most resistant to erosion of any on the continent, so beaches are uncommon in Maine. But from New Jersey southward the rocks are more easily fragmented and weathered, and beaches are much more common. As on the Pacific coast, sediments are transported coastward by rivers from eroding inland mountains, but the transported material is trapped in estuaries and therefore plays a less important role on beaches. Eastern beaches are typically formed of sediments from shores eroding nearby, or from the shoreward movement of offshore deposits laid down when the sea level was lower. The amount of sand in an area thus depends in part on the resistance or susceptibility of nearby shores to erosion. Sand moves generally south on these beaches just as it does on the Pacific coast, but the volume of moving sand is less here.

As we have seen, glaciers have also contributed to shaping the northern part of the Atlantic coast: large portions of Long Island and all of Cape Cod are remnants of debris deposited by glaciers.

The Gulf Coast 11-21

The Gulf coast experiences a smaller tidal range and—hurricanes excepted—a smaller average wave size than either the Pacific or Atlantic coast. Reduced longshore drift and an absence of interrupting submarine canyons allow the great volume of accumulated sediments from the Mississippi and other rivers to form large deltas, barrier islands, and a long raised "super berm" that prevents the ocean from inundating much of this sinking coast.

These are fortunate conditions because the rate of subsidence in the Gulf coast is greater than that for most of the Atlantic coast. Subsidence here is not the result of tectonic activity but rather due to sediment compaction, de-watering, and the removal of oil and natural gas. Sediment starvation and dredging have exacerbated the situation around some large cities. At Galveston, Texas, for example, sea level appears nearly 64 centimeters (25 inches) higher than it was a century ago, and parts of New Orleans are now 2 meters (6.6 feet) below sea level. As we have seen, the results of hurricanes at such places can be tragic. The protective natural berm can easily be breached, and floodwaters can surge far inland.

HUMAN INTERFERENCE IN COASTAL PROCESSES 11-22

Beaches exist in a tenuous balance between accumulation and destruction. Human activity can tip the balance one way or the other. For example, consider the rocky **breakwater** shown in **Figure 11.28.** The breakwater interrupts the progress of waves to the beach, weakening the longshore current and allowing sand to accumulate there. Without dredging, the beach

a
b

Figure 11.28 Growth of a beach protected by a breakwater: Santa Monica, California. (**a**) The shoreline as it appeared in 1931. (**b**) The same shoreline in 1949 after the breakwater was built. The boat anchorage formed by the breakwater is filling with sand deposited by disruption of the longshore current.

will eventually reach the breakwater and fill the small boat anchorage the breakwater was built to provide. This is a minor example of human alteration of a beach, yet it serves to introduce the growing problem of human influences on coastal processes.

We often divert or dam rivers, build harbors, and develop property with surprisingly little understanding of the impact our actions will have on the adjacent coast. Our role then becomes that of powerless observers. Residents of eroding coasts can only accept the inevitable loss of their property to the attack of natural forces, but residents of coasts in which deposition exceeds erosion are sometimes presented with alternatives. The choices are almost never simple. For example, should rivers be dammed to control devastating floods? If the dams are built, they will trap sediments on their way from mountains to coast. Beaches within the coastal cell fed by the dammed river will shrink because the sand on which they depend (to replenish losses at the shore) is blocked. Alarmed coastal residents will then take steps to hang onto whatever

sand remains. They may try to trap "their" beaches by erecting **groins**: short extensions of rock or other material placed at right angles to longshore drift, to stop the longshore transport of sediments. This temporary expedient usually accelerates erosion downcoast (**Figure 11.29a**). Diminished beaches then expose shore cliffs to accelerated erosion. Wind-wave energy that would have harmlessly churned sand grains now speeds the destruction of natural and artificial structures. Seawalls don't help, either (**Figure 11.29b**). They increase beach erosion by deflecting wave energy onto the sand. Churning by this increased energy eventually undermines the seawall, causing it to collapse. The importation of sand trapped behind dams (or from other sources) is also only a temporary—and very expensive—expedient (**Figure 11.29c**). Scenes like that shown in **Figure 11.30** will be more common.

What are the implications of these unlooked-for sand movements? Douglas Inman, director of the Center for Coastal Studies at Scripps Institution of Oceanography, believes that *at*

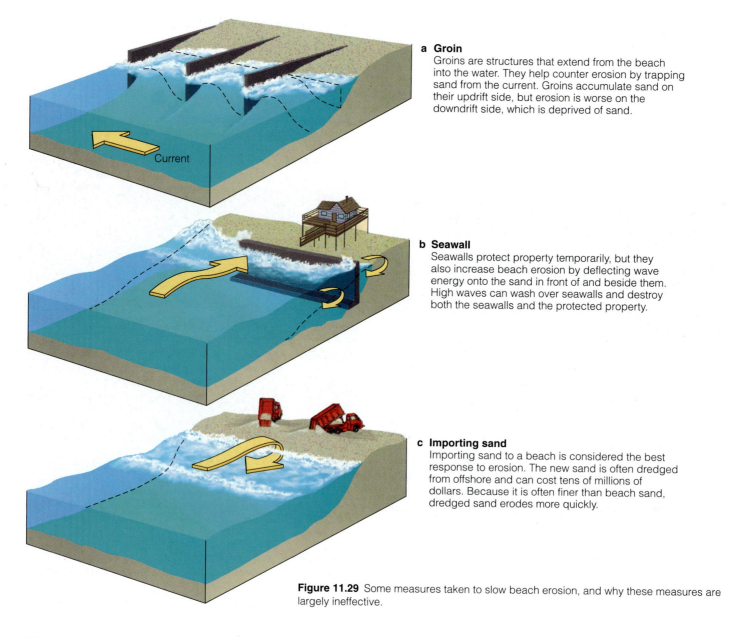

a Groin
Groins are structures that extend from the beach into the water. They help counter erosion by trapping sand from the current. Groins accumulate sand on their updrift side, but erosion is worse on the downdrift side, which is deprived of sand.

b Seawall
Seawalls protect property temporarily, but they also increase beach erosion by deflecting wave energy onto the sand in front of and beside them. High waves can wash over seawalls and destroy both the seawalls and the protected property.

c Importing sand
Importing sand to a beach is considered the best response to erosion. The new sand is often dredged from offshore and can cost tens of millions of dollars. Because it is often finer than beach sand, dredged sand erodes more quickly.

Figure 11.29 Some measures taken to slow beach erosion, and why these measures are largely ineffective.

Figure 11.30 A resident of Rodanthe, North Carolina, rides his bicycle past a house undercut by Hurricane Dennis in September, 1999. The extent of beach erosion is dramatically clear.

least 20% of the beach-bounded coastline of the United States is in danger of serious or catastrophic alteration. On the West Coast, a 30-year period of relatively mild weather may be ending. During this time, people felt it was safe to build close to the shore. Increased dam building and breakwater, jetty, and groin construction have made southern California's beaches more vulnerable. In the 1997–98 El Niño, coastal California alone suffered losses exceeding $750 million. The barrier islands of the Atlantic and Gulf coasts are at least as vulnerable.

Shores that look permanent through the short perspective of a human lifetime are in fact among the most temporary of all marine structures. Let's enjoy them in their present stages.

1. I've noticed currents moving seaward through the surf on high-energy beaches. What causes these so-called rip tides?

These small currents are caused by the movement offshore of a large amount of water at one time. Because they're not caused by gravitational forces, *rip tide* isn't a good term for them; they're properly called **rip currents.**

Rip currents form when a group of incoming waves piles an excess of water on the landward side of the surf zone faster than the longshore current can carry it. The water breaks through the wave line in a few places and flows rapidly through the surf back to sea. Rip currents are often made visible by the muddy color of their suspended sediments contrasting with the cleaner water just offshore (**Figure 11.31**). A strong swim-

Figure 11.31 A rip current. Rip currents are created when waves pile water inside the surf zone. The excess water then returns seaward as a swift rip current, ending in turbulence outside the surf. The rip current can slice through offshore sandbars, and the narrow opening can focus the current into an even faster-moving mass.

mer can use the rip current to his or her advantage, hitching a ride seaward through the churning surf to where the body surfing is best. To escape this narrow ocean-going band, however, a swimmer is advised to swim slowly parallel to shore and then return to shallow water. The higher the surf, the greater the probability of rip currents.

Rip currents are sometimes called *undertows,* a word as deceptive and inaccurate as *rip tide.* There aren't any small-scale, nearshore features that suck swimmers beneath the surface; even the legendary whirlpools would have trouble accomplishing that task (see, for example, Chapter 10's opener).

2. My foot tends to sink whenever I stand on the beach and let water from a wave run over it. The sand moves away from the edges of my foot, and I sink in. Why?

This is a good example of water's ability to carry more sediment as its speed of flow increases. Your foot interrupts the flow of water up or down the beach after a wave breaks, and the water must speed up to get around your foot. Fast-running water moves sand more effectively than slow-running water; so the sand immediately next to your foot is removed. This process, termed *scouring*, becomes a problem when structures are placed in shallow water.

3. What are those little white pellets I find on the beach along the high-tide line? Surfers call them "nurdles."

Those ubiquitous, insidious particles are the raw material for molded plastic goods. They are transported from producers to fabricators in containers loaded onto container ships. The pellets escape if a container is mishandled, breaks open during a storm, or is lost overboard. Virtually indestructible and able to float, these small "nurdles" float with the winds and currents until they encounter a shore. One researcher has calculated that just 25 containers would carry enough plastic pellets to spread 100,000 "nurdles" per mile along all the seashores of the world!

4. I hope someday to live near the ocean. What should I look for in buying property there?

Firm ground! Coastal Maine would be an ideal bet. The dense metamorphic rock of much of the Maine shore is stable (within the human time frame) and hard—ideal footings for a house. Make sure the site is far enough inland to avoid high surf, storm surge, and tides. If the winters in Maine don't appeal to you, coastal Florida might make a good choice—if your children aren't hoping to inherit the property. If you insist on building on a barrier island, make sure your home is on the mainland side of the southern end! Parts of the Pacific coast are all right, but local variability on that active margin makes some knowledge of the geological history of the area very valuable. For example, some parts of the San Diego shoreline are eroding about 3 meters (10 feet) per year—hardly a solid investment.

CHAPTER SUMMARY

Our personal experience with the ocean usually begins at the coast. These temporary, often beautiful junctions of land and sea are subject to rapid rearrangement by waves and tides, by gradual changes in sea level, and by biological processes. The *location* of a coast depends primarily on global tectonic activity and the ocean's water volume, while the *shape* of a coast is a product of many processes: uplift and subsidence, the wearing-down of land by erosion, and the redistribution of material by sediment transport and deposition. Coasts are classified as primary coasts (on which terrestrial influences dominate) or secondary coasts (on which marine influences dominate). Secondary coasts often support beaches, accumulations of loose particles. Generally, the finer the particles on the beach, the flatter its slope. Beaches change shape and volume as a function of wave energy and the balance of sediment input and removal. Estuaries are among the most complex and biologically productive coasts. Human interference with coastal processes has generally accelerated the erosion of coasts near inhabited areas.

TERMS AND CONCEPTS TO REMEMBER

backshore	longshore bar
backwash	longshore current
barrier island	longshore drift
bay mouth bar	longshore trough
beach	low-energy coast
beach scarp	moraine
berm	partially mixed estuary
berm crest	primary coast
breakwater	reverse estuary
coast	rip current
coastal cell	salt wedge estuary
delta	sand spit
dissolution	sea cave
erosion	sea cliff
estuary	sea island
eustatic change	secondary coast
fjord	shore
foreshore	swash
groin	tombolo
high-energy coast	wave-cut platform
lagoon	well-mixed estuary

STUDY QUESTIONS

1. How is a primary coast different from a secondary coast?

2. What features would you expect to see along a primary coast? A secondary coast? What determines how long the features will last?

3. What two processes contribute to longshore drift? What powers longshore drift? What is the predominant direction of drift on U.S. coasts? Why?

4. What are some of the features of a sandy beach? Are they temporary or permanent? Is there a relationship between wave energy on a coast and the size (or slope, or grain size) of beaches found there?

5. How are deltas classified? Why are there deltas at the mouths of the Mississippi and Nile Rivers, but not at the mouth of the Columbia River?

6. What is a coastal cell? Where does sand in a coastal cell come from? Where does it go?

7. How are estuaries classified? Upon what does the classification depend? Why are estuaries important?

8. Compare and contrast the U.S. Pacific, Atlantic, and Gulf coasts.

9. How do human activities interfere with coastal processes? What steps can be taken to minimize loss of life and property along U.S. coasts?

FOR FURTHER STUDY

Bascom, W. 1980. *Waves and Beaches.* Rev. ed. New York: Anchor/Doubleday. A valuable reference, well and clearly written, nicely illustrated.

Beardsley, T. 2000. "Dissecting a Hurricane." *Scientific American,* March, 81–85. How Hurricane Dennis was studied from an aircraft in August, 1999.

Boicourt, W. C. 1993. "Estuaries: Where River Meets the Sea." *Oceanus* 36 (no. 2): 29–37. An overview of the main estuaries of the continental United States, the processes that shape them, and the human impact on them.

Clark, J. R. 1996. *Coastal Zone Management.* Boca Raton, FL: CRC Press. A thorough look at the subject.

Davis, R. A. 1994. *The Evolving Coast.* New York: Scientific American Library. A clearly written and very well illustrated overview of coastal processes and problems.

Dean, C. 1999. *Against the Tide: The Battle for America's Beaches.* New York: Columbia University Press. Specific cases and controversies surrounding the curation of coastlines.

Dolan, R., and H. Lins. 1987. "Beaches and Barrier Islands." *Scientific American,* July, 68–77. Excellent summary of the evolution of these ephemeral structures.

Flanagan, R. 1993. "Beaches on the Brink." *Earth,* November, 24–33.

Oceanus 36, nos. 1 and 2 (Spring and Summer 1993) are both dedicated to coastal science and policy.

Pilkey, O. H. 1983. *Coastal Design: A Guide for Builders, Planners, and Home Owners.* New York: Van Nostrand-Reinhold. Pilkey is a leading force for responsible coastal development in the Southeast. His guidelines hold for any coast.

Schneider, D. 1997. "The Rising Seas." *Scientific American,* March, 112–7. There is concern that global warming will lead to a meltdown of polar ice, flooding coastlines everywhere. Is the concern justified? It depends.

Shepard, F. P., and H. R. Wanless. 1971. *Our Changing Coastlines.* New York: McGraw-Hill.

U. S. Geological Survey. 1990. "Coasts in Crisis." Circular #1075. A clear and often alarming look at the U. S. situation.

For additional readings, go to InfoTrac College Edition, your online research library at:

http://infotrac.thomsonlearning.com/

Life in the Ocean 12

THE FOREST BENEATH THE WAVES

The ocean often moves placidly in a kelp forest—the long seaweeds interfere with the circular movement of water molecules as the waves pass. Because kelp secretes a lubricant that smooths a diver's way, visitors can easily glide between the closely entwined seaweeds, gently nudging the strands from their paths. Sunlight filters between the fronds, sending illuminating shafts flickering into the deeper water below. Schools of fishes often move through the kelp, and countless animals—many too small to be seen—nestle around their dark bases. The feeling is like that of being in a tall grove of redwoods, and a diver's relative weightlessness allows nearly all parts of the forest to be explored. It is usually quiet here, peaceful and calm.

A biologist sees even more beauty in the scene and notes the rapid growth; the seaweeds are longer on this visit than a few weeks ago, and their stalks are thicker. Some worms have begun to make spiral tracings on the kelp surfaces. The animals on the seabed are different now and arrayed in new patterns. A family of otters has moved to the area, attracted by the many nutritious sea urchins scattered across the gravel-covered bottom. Schools of anchovetta flash overhead, swimming between the kelp blades in long follow-the-leader schools, efficiently sieving the ocean for food. There are many more plantlike organisms living here than just the big seaweeds swaying in the currents: a living haze crowds the water immedi-

Sunlight penetrates a kelp forest, one of the ocean's most productive habitats.

ately ahead of the diver's faceplate. A flashlight beam reveals swarms of dust-sized organisms in every direction. Countless millions of organisms drift unobserved, too small to reflect the light. Life abounds.

The ocean here fairly hums with productivity. Food is being produced rapidly and in great quantity. Big seaweeds and microscopic algae are harnessing light from the sun to assemble carbohydrates as their forebears have done since the ocean was young. The temperate coastal ocean brims with life, nearly all of it dependent on the subtle light-driven biochemistry proceeding in this lovely, gracefully moving place. 🔲 **12-1**

. . . .

KEY CONCEPTS

1. Life on Earth is notable for both unity and diversity: *diversity* because there are at least 5 million different species (kinds) of living things on Earth; *unity* because each species shares the same underlying basic life processes.

2. Primary producers—autotrophs—are organisms that synthesize food from inorganic substances by photosynthesis and chemosynthesis.

3. A variety of physical factors affects the density, variety, and success of marine life. These factors include water's transparency, temperature, dissolved nutrients, salinity, dissolved gases, hydrostatic pressure, and acid–base balance.

4. Feeding relationships in a community resemble complex webs.

5. The marine environments populated by marine life can be classified by physical characteristics.

6. Marine organisms are naturally classified by their physical characteristics and by the degree to which they resemble other organisms.

7. Organisms are distributed through the marine environment in specific communities—groups of interacting producers, consumers, and recyclers that share a common living space.

CHAPTER AT A GLANCE

Energy and Marine Life
Primary Productivity
Feeding (Trophic) Relationships

Physical Factors Affecting Marine Life
Light
Temperature
Dissolved Nutrients
Salinity
Dissolved Gases
Acid–Base Balance
Hydrostatic Pressure
Limiting Factors

Classification of the Marine Environment
Classification by Light
Classification by Location

Classification of Oceanic Life
Systems of Classification
Scientific Names

Marine Communities
Organisms Within Communities
Competition
Box 12.1: Mass Extinctions
Change in Marine Communities

. . . .

To the weekend sailor, the ocean may seem interesting mainly because of its winds and currents and waves, but the seemingly empty seawater next to a small sailboat may support millions of invisible plantlike organisms. The waters beneath the hull can conceal wonderful worms and colorful crustaceans, the gently swaying forests just described, whole communities of microscopic creatures drifting with the currents, schools of fishes, maybe even whales. On the dark distant bottom, animals locate one another with glowing lures or jostle for food near volcanic vents. The great wide sea is the ideal habitat for life.

The ocean can support such a bewildering array of life forms because of water's unique physical properties—its density, its dissolving power, its ability to absorb large quantities of heat yet rise very little in temperature. The oceanic environment is a relatively easy place for cells to live. All life on our planet is water-based and shares the same basic underlying life processes; life almost certainly began in the ocean. Life and Earth have changed together, generation by generation, over about 4 billion years.

There are at least 5 million different species (kinds) of living things on Earth. Yet despite their astonishing diversity in form and lifestyle, all species share the same underlying mechanisms for capturing and storing energy, manufacturing proteins, and transmitting information between generations. Biologists know there is nothing special about the atoms or energy of life, no way to distinguish the physical components of life from their nonliving counterparts. What *does* distinguish life from nonlife is the ability of living things to capture, store, and transmit energy—and the ability to reproduce.

ENERGY AND MARINE LIFE

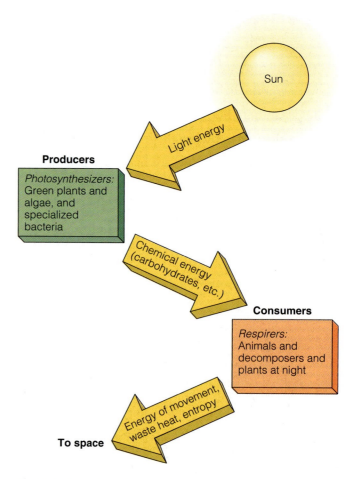

12-2

The energy marine organisms need to function comes directly or indirectly from the sun. The sun produces enormous quantities of energy—some in the form of visible light, a tiny portion of which strikes Earth. Only about 1 part in 2,000 of the light that reaches Earth's surface is captured by organisms, but that "small" input of energy powers nearly all the growth and activity of living things. Light energy from the sun is trapped by chlorophyll in organisms called *producers* (certain bacteria, algae, and green plants) and changed into chemical energy. The chemical energy is used to build simple carbohydrates and other organic molecules—**food**—which is then used by the producer or eaten by animals (or other organisms) called *consumers*. Because light energy is used to synthesize molecules rich in stored energy, the process is called **photosynthesis** (*photos* = light, *syn* + *tithenai* = to place together). Here is a general formula for photosynthesis:

		sunlight			
$6\,CO_2$	+	$6\,H_2O$	$\rightarrow$	$C_6H_{12}O_6$	+ $6\,O_2$
6 molecules of carbon dioxide	+	6 molecules of water	(yields by photosynthesis)	a molecule of glucose (a carbohydrate)	+ 6 molecules of oxygen

Figure 12.1 The flow of energy through living systems. At each step, energy is degraded (transformed into a less useful form).

The energy is released when food such as glucose is used for growth, repair, movement, reproduction, and the other functions of organisms. The breakdown of food eventually produces waste heat, which flows away from Earth into the coldness of space. This one-way flow of energy is shown in **Figure 12.1.**

Photosynthesis is the dominant method of binding energy into carbohydrates, but there is another. **Chemosynthesis,** employed by a few relatively simple forms of life, is the production of usable energy directly from energy-rich inorganic molecules available in the environment rather than from the sun. As we will see in Chapter 14, some unusual forms of marine life depend on chemosynthesis. Overall, chemosynthetic production of food in the ocean is very small in comparison to photosynthetic production.

Primary Productivity

12-3

The synthesis of organic materials from inorganic substances by photosynthesis or chemosynthesis is called **primary productivity (Figure 12.2).** Primary productivity is typically expressed in *grams of carbon bound into organic material per square meter of ocean surface area per year* ($gC/m^2/yr$). The organic material produced is usually glucose, a carbohydrate. The source of carbon for glucose is dissolved CO_2.

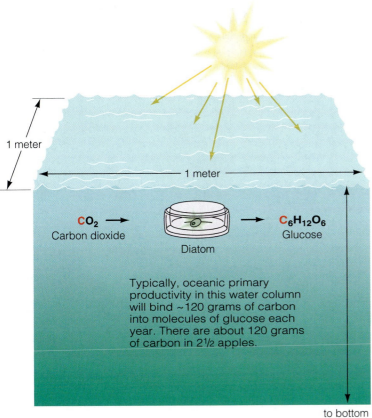

1 meter

1 meter

CO_2 →
Carbon dioxide

Diatom

→ $C_6H_{12}O_6$
Glucose

Typically, oceanic primary productivity in this water column will bind ~120 grams of carbon into molecules of glucose each year. There are about 120 grams of carbon in 2½ apples.

to bottom of ocean

a

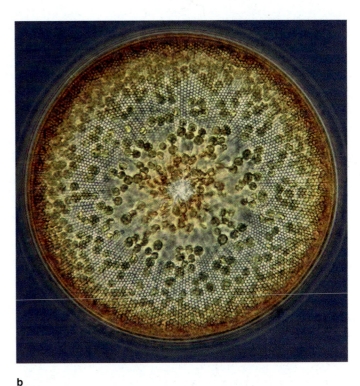

b

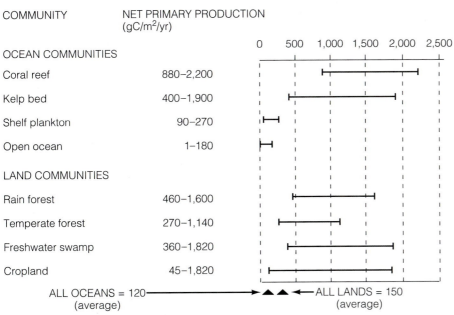

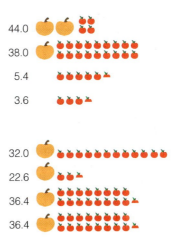

COMMUNITY	NET PRIMARY PRODUCTION (gC/m²/yr)		EQUIVALENT AMOUNT OF CARBON IN PUMPKINS AND APPLES/m²/yr (maximum values) (1 pumpkin = 20 apples)
		0 500 1,000 1,500 2,000 2,500	
OCEAN COMMUNITIES			
Coral reef	880–2,200		44.0
Kelp bed	400–1,900		38.0
Shelf plankton	90–270		5.4
Open ocean	1–180		3.6
LAND COMMUNITIES			
Rain forest	460–1,600		32.0
Temperate forest	270–1,140		22.6
Freshwater swamp	360–1,820		36.4
Cropland	45–1,820		36.4

ALL OCEANS = 120 (average) ▲▲ ← ALL LANDS = 150 (average)

c

Figure 12.2 (**a**) Oceanic productivity—the incorporation of carbon atoms into carbohydrates—is measured in grams of carbon bound into carbohydrates per square meter of ocean surface area per year (= gC/m²/year). (**b**) The diatom *Coscinodiscus,* an important marine primary producer. In a very bright light this single-celled plantlike organism would barely be visible to the unaided eye. (**c**) Net annual primary productivity in some marine and terrestrial communities. For visualization, an apple contains about 50 grams of carbon (50gC). High productivities exceeding the equivalent of 20 apples/m²/year are represented here by pumpkins, with each pumpkin equal to 20 apples. (*Source*: Milne, 1995)

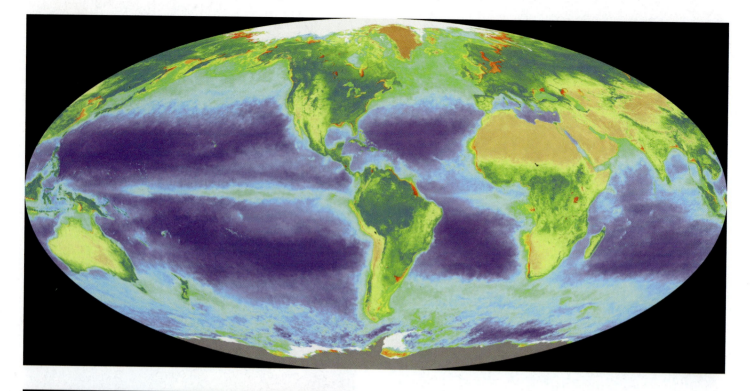

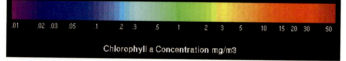

.01 .02 .03 .05 .1 .2 .3 .5 1 2 3 5 10 15 20 30 50

Chlorophyll a Concentration mg/m3

Figure 12.3 Oceanic productivity can be observed from space. NASA's *SeaWiFS* satellite, launched in 1997, can detect the amount of chlorophyll in ocean surface water. Chlorophyll content allows an estimate of productivity. Red and yellow areas indicate high primary productivity; blue areas indicate low. This image was derived from measurements made from September 1997 through August 1998.

Phytoplankton—minute, drifting photosynthetic organisms you will meet in Chapter 13—produce between 90% and 96% of oceanic carbohydrates. Seaweeds—larger marine plants that we discuss in Chapter 14—contribute from 2% to 5% of the ocean's primary productivity. Chemosynthetic organisms probably account for between 2% and 5% of the total. Though estimates vary widely, recent studies suggest that total ocean productivity ranges from 75 to 150 grams of carbon bound into carbohydrates per square meter of ocean surface per year (75 to 150 gC/m^2/yr). (For comparison, a well-tended alfalfa field produces about 1,600 gC/m^2/yr.)

How does marine net productivity compare with terrestrial net productivity? Recent research suggests the global net productivity in *marine* ecosystems is 35 to 50 billion metric tons of carbon bound into carbohydrates per year; global net *terrestrial* productivity is roughly similar at 50 to 70 billion metric tons per year.[1] However, the total producer **biomass**

(the mass of living tissue) in the ocean is only 1 to 2 billion metric tons, compared with 600 to 1,000 billion metric tons of living biomass on land! Clearly, marine producers are *much* more efficient in their production of food than their land-based counterparts.

The total weight of a **primary producer** is assumed to be about ten times the mass of the carbon it has bound into carbohydrates. Thus, a primary productivity of 100 gC/m^2/yr represents the yearly growth of about 1,000 grams of primary producers for each square meter of ocean surface (see **Figure 12.3**). Between 35 and 50 billion metric tons of carbon is believed to be bound into carbohydrates in the ocean each year, so between 350 and 500 billion metric tons of marine plants and plantlike organisms are produced annually. Each year this vast bulk is consumed by the metabolic activity of the producers themselves and by the consumers that graze on them. The component atoms are then reassembled by photosynthesis into carbohydrates in a continuous solar-powered cycle.

Feeding (Trophic) Relationships   12-4

Photosynthetic and chemosynthetic organisms can be called either primary producers or **autotrophs** (*auto* = self + *trophe* = nourishment), because they make their own food. The bodies of autotrophs are rich sources of chemical energy for any organisms capable of consuming them. **Heterotrophs** (*hetero* = other, different) are organisms that must consume other organisms because they are unable to synthesize their own food molecules. Some heterotrophs consume autotrophs, and some consume other heterotrophs.

We can label organisms by their positions in a "who eats whom" feeding hierarchy called a **trophic pyramid** (*trophos*

---

[1] 1 billion metric tons = 1.1 billion tons.

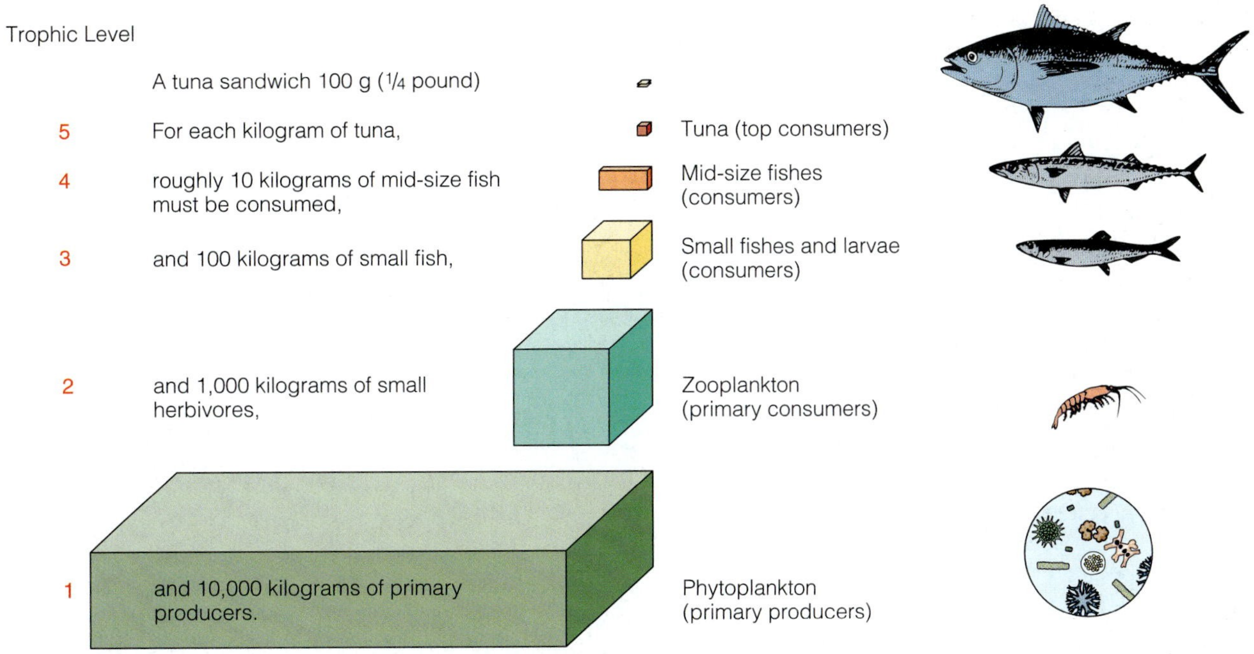

Trophic Level

	A tuna sandwich 100 g (¼ pound)	
5	For each kilogram of tuna,	Tuna (top consumers)
4	roughly 10 kilograms of mid-size fish must be consumed,	Mid-size fishes (consumers)
3	and 100 kilograms of small fish,	Small fishes and larvae (consumers)
2	and 1,000 kilograms of small herbivores,	Zooplankton (primary consumers)
1	and 10,000 kilograms of primary producers.	Phytoplankton (primary producers)

Figure 12.4 A generalized trophic pyramid, showing the long and energetic history of a tuna sandwich. How many kilograms of primary producers are necessary to maintain 1 kilogram of tuna, a top carnivore? Using the trophic pyramid model shown here, you can see that 1 kilogram of tuna (enough to make ten ¼-pound tuna sandwiches) at the fifth trophic level (the fifth feeding step of the pyramid) is supported by 10 kilograms of mid-sized fish at the fourth, which in turn is supported by 100 kilograms of small fish at the third, who have fed on 1,000 kilograms of zooplankton (primary consumers) at the second, which have eaten 10,000 kilograms of phytoplankton (small autotrophs, primary producers) at the first. (These figures have been rounded off to illustrate the general principle. The actual measurements are difficult to make and quite variable.)

= one who feeds). The primary producers shown at the bottom of the pyramid in **Figure 12.4** are mostly chlorophyll-containing photosynthesizers. The animal heterotrophs that eat them are called **primary consumers** (or herbivores), the animals that eat them are called secondary consumers, and so on to the **top consumer** (or top carnivore).

Note that the mass of consumers becomes smaller as energy flows toward the top of the pyramid. There are many small primary producers at the base, and a very few large top consumers at the apex. Only about 10% of the energy from the organisms consumed is stored in the consumers as flesh, so each level is about one-tenth the mass of the level directly below. The rest of the energy is lost as waste heat as organisms live and work to maintain themselves.

Pyramids such as the one in Figure 12.4 can lead to the misconception that one kind of fish eats only one other kind of fish, and so on. Real communities are more accurately described as food webs, an example of which is included as **Figure 12.5**. A **food web** is a group of organisms linked by complex feeding relationships in which the flow of energy can be followed from primary producers through consumers. Organisms in a food web almost always have some choices of food species.

So organisms interact with each other, feed on one another, and transfer energy as food from producing autotrophs through a web of consuming heterotrophs. Nearly all are ultimately dependent on sunlight and photosynthesis. What is the physical role of the ocean in this?

PHYSICAL FACTORS AFFECTING MARINE LIFE   12-5

Marine organisms depend on the ocean's chemical composition and physical characteristics for life support. Any aspect of the physical environment that affects living organisms is called a **physical factor.** Living in the ocean often has advantages over living on land—physical conditions in the sea are usually milder and less variable than physical conditions on land. The most important physical factors for marine organisms are light, temperature, dissolved nutrients, salinity, dissolved gases, acid–base balance, and hydrostatic pressure.

Light   12-6

On land, most photosynthesis proceeds at or just above ground level. But seawater, unlike soil, is relatively transparent, which allows photosynthesis to proceed for some distance below the ocean surface. Incoming sunlight must run a gauntlet of

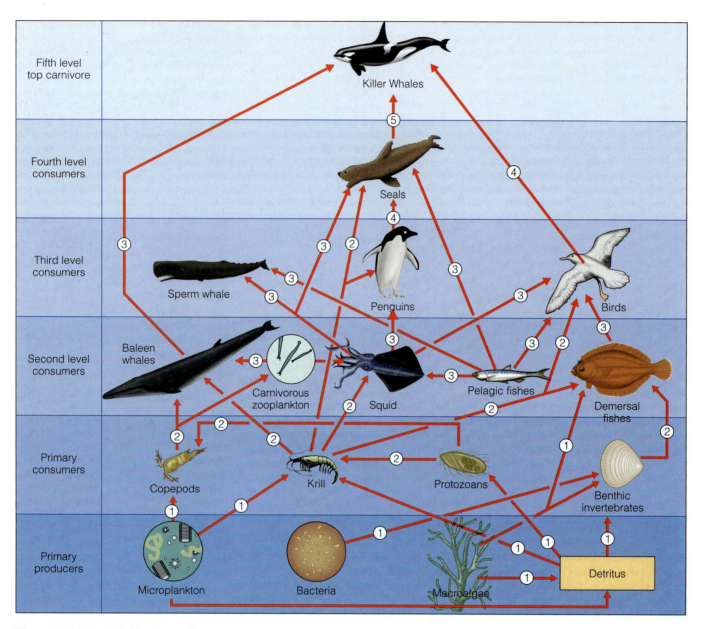

Figure 12.5 A simplified food web, illustrating the major trophic relationships leading to an adult killer whale. The arrows show the direction of energy flow; the numbers on each area represent the trophic level at which the organism is feeding. Note that feeding relationships are not as simple as one might assume from Figure 12.4.

difficulties, however, before it can be absorbed by the chlorophyll in marine autotrophs.

Most sunlight approaching at a low angle (near sunrise or sunset, or in the polar regions) reflects off the water surface and doesn't enter the ocean. Once through the surface, light is selectively absorbed—water is more transparent to some colors of light than others. In clear water, blue light penetrates to the greatest depth, while red light is absorbed near the surface. **Figure 12.6** shows the depths attained by light of various wavelengths (colors) in clear ocean water. Light energy absorbed by water turns to heat.

The depth to which light penetrates is also limited by the number and characteristics of particles in the water. These particles, which may include suspended sediments, dustlike bits of once-living tissue, or the organisms themselves, scatter and absorb light. High concentrations of particles quickly absorb most blue and ultraviolet light. This absorption, combined with the reflection of green light by chlorophyll within the producers, changes the color of productive coastal waters to green.

How far down does light penetrate? Near the coasts, conditions become uncomfortably dark for divers at depths of around 30 to 40 meters (100 to 130 feet). The human eye is most sensitive to the blue wavelengths, and in very clear tropical waters with a smooth surface and a high solar angle, human observers in submersibles have seen dark blue light at about 200 meters (660 feet). There is evidence to suggest that some

Color	Wavelength (nm)	% Absorbed in 1 m of Water	Depth by Which 99% Is Absorbed (m)
Infrared	800	82.0	3
Red	725	71.0	4
Orange	600	16.7	25
Yellow	575	8.7	51
Green	525	4.0	113
Blue	475	1.8	254
Violet	400	4.2	107
Ultraviolet	310	14.0	31

a

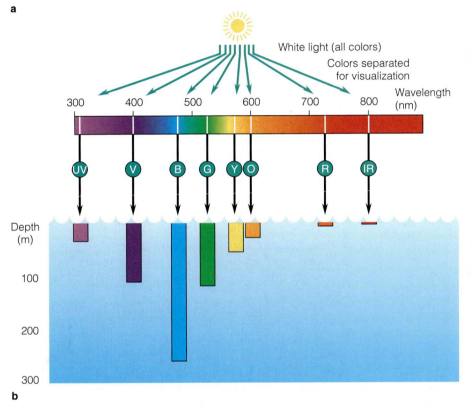

b

c

d

Figure 12.6 Absorption of light of different wavelengths (colors) by seawater. (**a**) The table shows the percentage of light absorbed in the uppermost meter of the ocean, and the depths at which only 1% of the light of each wavelength remains. (**b**) The bars show the depths of penetration of 1% of the light of each wavelength (as in the last column of the table). (**c**) A fish photographed in normal oceanic light—the color blue predominates. (**d**) The same fish photographed with a strobe light. The flash contains all colors, and the distance from the strobe to the fish and back to the camera is not far enough to absorb all the red light. The fish shows bright warm colors otherwise invisible.

deep-water fish use even this dim light for body orientation, feeding, and predator avoidance. Photometers much more sensitive than the human eye have detected light at even greater depths—the present record is 590 meters (1,935 feet) in the tropical Pacific!

Photosynthesis proceeds slowly at low light levels. Most of the biological productivity of the ocean occurs in an area near the surface called the **euphotic zone** (*eu* = good + *photos* = light), shown in **Figure 12.7.** This is where marine autotrophs trap more energy than they use. Though it is difficult to generalize for the ocean as a whole, the euphotic zone typically

extends to a depth of approximately 70 meters (230 feet) in mid-latitudes, averaged over the whole year. The upper productive layer of ocean is a very thin skin indeed—the water within this zone amounts to less than 1% of world ocean volume—and yet nearly all marine life depends on this fine illuminated band.

Below the euphotic zone lies the **disphotic zone** (*dys* = difficult, bad + *photos* = light), also seen in Figure 12.7. Though light is present in this zone, it is not bright enough to allow photosynthesis to generate as much carbohydrate as would be used by an autotroph through a day. Below the

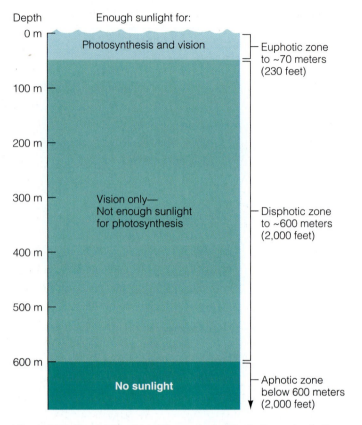

Depth | Enough sunlight for:

0 m — Photosynthesis and vision — Euphotic zone to ~70 meters (230 feet)

100 m

200 m

300 m — Vision only— Not enough sunlight for photosynthesis — Disphotic zone to ~600 meters (2,000 feet)

400 m

500 m

600 m

No sunlight — Aphotic zone below 600 meters (2,000 feet)

Figure 12.7 The relationship of the euphotic, disphotic, and aphotic zones to one another. (The euphotic zone statistic is for mid-latitude waters averaged through a year.)

disphotic zone—below around 600 meters (2,000 feet)—lies the dark **aphotic zone** (*a* = without + *photos* = light), the vast bulk of the ocean where sunlight never reaches. Another view of these layers is shown in Figure 12.9.

Temperature

12-7

The rate at which chemical reactions occur in a living organism is largely dependent on the molecular vibration we call heat. Since agitation brings reactants together, warmer temperatures increase the rate at which chemical reactions occur. Thus, an organism's **metabolic rate,** the rate at which energy-releasing reactions proceed within an organism, increases with temperature. The metabolic rate approximately doubles with a 10°C (18°F) temperature rise. The interior temperature of an organism is directly related to the rate at which it moves, reacts, and lives.

The great majority of marine organisms are "cold blooded," or **ectothermic** (*ektos* = outside + *therme* = heat), having an internal temperature that stays very close to that of their surroundings. A few complex animals—mammals and birds and some of the larger, faster fishes—are "warm blooded," or **endothermic** (*endon* = within), meaning that they have a stable, high internal temperature.

In general, the warmer the environment of an ectotherm within its tolerance range, the more rapidly its metabolic processes will proceed. Tropical fish in a heated aquarium will therefore eat more food and require more oxygen than goldfish of the same size living in an unheated but otherwise identical aquarium. The tropical fish will generally grow more rapidly, have a faster heartbeat, reproduce more rapidly, swim more swiftly, and live shorter lives. But you can't just crank the heater up another notch for even faster fish—eventually the little fellows will cook. The upper limit of temperature that an ectotherm can tolerate is often not much higher than its optimum temperature. The lower limit is usually more forgiving because molecules are merely slowed.

Do endotherms have narrow temperature requirements? Yes and no. Endotherms can tolerate a tremendous range of *external* temperature compared to ectotherms; think of a whale migrating from polar waters to the tropics, or an Emperor penguin incubating an egg at −51°C (−60°F). Their *internal* temperatures, however, vary only slightly. In our own case, consider the temperatures of places inhabited by humans in contrast to the narrow internal temperature range physicians consider normal. Sophisticated thermal regulation mechanisms make it possible for endotherms to live in a variety of habitats, but they pay a price. Their high metabolic rates make proportionally high demands on food supply and gas transport, but the benefit of having a biochemistry finely tuned to a single efficient temperature is worth the regulatory difficulties involved.

Ocean temperature varies with depth and latitude. The average temperature of the world ocean is only a few degrees above freezing, with warmer water found only in the lighted surface zones of the temperate and tropical ocean, and in rare, deep, warm chemosynthetic communities. Though temperature ranges of the ocean are considerable (**Figure 12.8**), they are much narrower than comparable ranges on land.

Dissolved Nutrients

12-8

A **nutrient** is a compound required for the production of organic matter. Some nutrients help form the structural parts of organisms, some make up the chemicals that directly manipulate energy, and some have other functions. A few of these necessary nutrients are always present in seawater, but most are not readily available.

The main inorganic nutrients required in primary productivity include nitrogen (as nitrate, NO_3^-) and phosphorus (as phosphate, PO_4^{3-}). As any gardener knows, plants require fertilizer—mainly nitrates and phosphates—for success. Ocean gardeners would have more trouble raising crops than their terrestrial counterparts, though, because the most fertile ocean water contains only about 1/10,000 the available nitrogen of topsoil. Phosphorus is even more scarce in the ocean, but fortunately less of it is required by living things, which have about 1 atom of phosphorus for every 16 atoms of nitrogen.

Nitrogen and phosphorus are often depleted by autotrophs during times of high productivity and rapid reproduction. Also in short supply during rapid growth are dissolved sil-

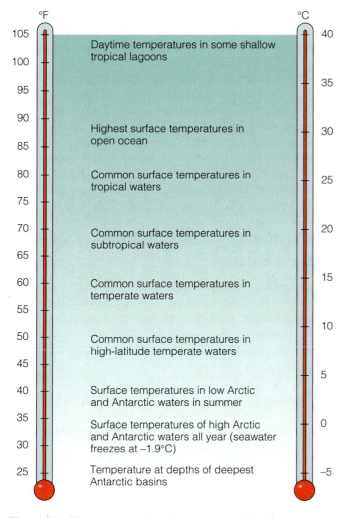

°F / °C

- 105 / 40 — Daytime temperatures in some shallow tropical lagoons
- 100 / 35
- 95
- 90 / 30 — Highest surface temperatures in open ocean
- 85
- 80 / 25 — Common surface temperatures in tropical waters
- 75
- 70 / 20 — Common surface temperatures in subtropical waters
- 65
- 60 / 15 — Common surface temperatures in temperate waters
- 55
- 50 / 10 — Common surface temperatures in high-latitude temperate waters
- 45 / 5
- 40 — Surface temperatures in low Arctic and Antarctic waters in summer
- 35 / 0 — Surface temperatures of high Arctic and Antarctic waters all year (seawater freezes at −1.9°C)
- 30
- 25 / −5 — Temperature at depths of deepest Antarctic basins

Figure 12.8 Temperatures of marine waters capable of supporting life. Some isolated areas of the ocean, notably within and beneath hydrothermal vents, may support specialized living organisms at temperatures of up to 113°C (238°F).

icates and calcium compounds (used for shells and other hard parts) and trace elements such as iron, copper, and magnesium (used in enzymes, vitamins, and other large molecules). Marine plants have no choice but to recycle these nutrients.

Salinity
12-9

Cell membranes are greatly affected by the salinity of surrounding water. The salinity of seawater (see Chapter 6) can vary in places because of rainfall, evaporation, runoff of water and salts from land, and other factors. Surface salinity varies most, with lows of 6‰ or less along the coast of the outer Baltic Sea in early summer, to year-round highs exceeding 40‰ in the Red Sea. Salinity is less variable with increasing depth, with the ocean typically becoming slightly saltier with depth.

Changing salinity can physically damage membranes, and concentrated salts can alter protein structure. Salinity can affect the specific gravity, the density, and therefore the buoyancy of an organism. Salinity is also important because it can cause water to enter or leave a cell through the membrane, changing the cell's overall water balance. Seawater is nearly identical in salinity to the interior of all but the most advanced forms of marine life; this means that maintaining salt balance, and therefore water balance, is easy for most marine species.

Dissolved Gases
12-10

Nearly all marine organisms require dissolved gases—in particular, carbon dioxide and oxygen—to stay alive. Oxygen does not dissolve easily in water, and as a result there is about a hundred times more gaseous oxygen in the atmosphere than in the ocean. But CO_2, essential to primary productivity, is much more soluble and reactive in seawater than is oxygen (as **Table 12.1** shows). Although as much as a thousand times more carbon dioxide than oxygen can dissolve in water, normal values at the ocean surface average around 50 milliliters per liter for CO_2 and around 6 milliliters per liter for oxygen. At present the ocean holds about 60 times as much carbon dioxide as does the atmosphere. Because of this abundance, marine plants almost never run out of CO_2.

Deep water tends to contain more carbon dioxide than surface water. Why should this be? Table 12.1 also shows the relationship between water temperature and its ability to dissolve gases. Note that colder water contains more gas at saturation. You may recall that the deepest and densest seawater masses are formed at the surface in the cold polar regions, and, as we have seen, more CO_2 can dissolve in that low-temperature environment. The dense water sinks, taking its large load of CO_2 to the bottom, and the pressure at depth helps to keep it in solution. CO_2 also builds up in deep water because only heterotrophs (animals) live and metabolize there, and because CO_2 is produced as decomposers consume falling organic matter. No photosynthetic primary producers are present in the dark depths to use this excess CO_2, since there is not sufficient sunlight to permit photosynthesis to occur.

Rapid photosynthesis at the surface lowers CO_2 concentrations and increases the quantity of dissolved oxygen. Oxygen is least plentiful just below the limit of photosynthesis

Table 12.1	Solubility of Gases in Seawater as a Function of Temperature (Salinity = 33‰)		
Temperature	Solubility (ml/l at atmospheric pressure)[a]		
	N_2	O_2	CO_2
0°C (32°F)	14.47	8.14	8,700.0
10°C (50°F)	11.59	6.42	8,030.0
20°C (68°F)	9.65	5.26	7,350.0
30°C (86°F)	8.26	4.41	6,660.0

[a]Figures are given at *saturation,* the maximum amount of gas held in solution before bubbling begins.
Source: F. G. Walton-Smith, *CRC Handbook of Marine Science* (Cleveland, OH: CRC Press, 1974).

because of respiration by many small animals at middle depths. (These relationships were shown in Figure 6.7.

Low oxygen levels can sometimes be a problem at the ocean surface. Plants produce more oxygen than they use, but they produce it only during daylight hours. The continuing respiration of plants at night will sometimes remove much of the oxygen from the surrounding water. In extreme cases this oxygen depletion may lead to the death of the plants and animals in the area, a phenomenon most noticeable in enclosed coastal waters during spring and fall plankton blooms.

The greatest variability in levels of dissolved gas is found at the surface near shore. Less dramatic changes occur in the open sea.

Acid–Base Balance 12-11

The complex chemistry of Earth's life forms depends on precisely shaped enzymes, large protein molecules that speed up the rate of chemical reactions. Like heat, strong acids or bases distort the shapes of these vital proteins and they lose their ability to function normally.

The acidity or alkalinity of a solution is expressed in terms of a *pH scale*, a logarithmic measure of the concentration of hydrogen ions in a solution. Recall (from Figure 6.8) that 7 on the pH scale is neutral, with smaller numbers indicating greater acidity and larger numbers indicating greater alkalinity.

Seawater is slightly alkaline; its average pH is about 8. The dissolved substances in seawater act to *buffer* pH changes, preventing broad swings of pH when acids or bases are introduced. The normal pH range of seawater is much less variable than that of soil—terrestrial organisms are sometimes limited by the presence of harsh alkali soils that damage cell components.

Though seawater remains slightly alkaline, it is subject to some variation. When dissolved in water, some CO_2 becomes carbonic acid. In areas of rapid plant growth, pH will rise because CO_2 is used by the plants for photosynthesis. And because temperatures are generally warmer at the surface, less CO_2 can dissolve in the first place. Thus, surface pH in warm productive water is usually around 8.5.

At middle depths and in deep water, more CO_2 may be present. Its source is the respiration of animals and bacteria. With cold temperatures, high pressure, and no photosynthetic plants to remove it, this CO_2 will lower the pH of the water, making it more acidic with depth. Thus, deep, cold seawater below 4,500 meters (15,000 feet) has a pH of around 7.5. This lower pH can dissolve calcium-containing marine sediments. A drop to pH 7 can occur at the deep-ocean floor when bottom bacteria consume oxygen and produce hydrogen sulfide.

Hydrostatic Pressure 12-12

Marine organisms are often subject to great pressure from the constant weight of water above them, but this so-called **hydrostatic pressure** presents very little difficulty to them. In fact, the situation in the ocean is parallel to that on land. Land animals live in air pressurized by the weight of the atmosphere above them (1 kilogram per square centimeter, or 14.7 pounds per square inch, at sea level) without experiencing any problems. Indeed, atmospheric pressure is necessary for breathing, flight, and some other physical necessities of life.

Pressures inside and outside an organism are virtually the same, both in the ocean and at the bottom of the atmosphere. Thus marine organisms do not need heavy shells to keep from being crushed by hydrostatic pressure. Great pressure does have some chemical effects: Gases become more soluble at high pressure, some enzymes are inactivated, and metabolic rates for a given temperature tend to be slightly higher. These effects are felt only at great depth, though. Unless marine organisms have gas-filled spaces in their bodies, a moderate change in pressure has little effect.

Limiting Factors 12-13

Often too much or too little of a single physical factor can adversely affect the function of an organism. We call that factor a **limiting factor,** a physical or biological necessity whose presence in inappropriate amounts limits the normal action of the organism. Imagine, for example, an ocean area in which everything (warmth, nutrients, CO_2) is perfect for photosynthesis—everything, that is, *except light*. In that circumstance no photosynthesis will occur; light is the limiting factor. If light were present but nitrates were absent, nitrate nutrients would be the limiting factor. Sometimes *too much* of something can be limiting—heat, for instance.

CLASSIFICATION OF THE MARINE ENVIRONMENT 12-14

Scientists have found it useful to divide the marine environment into **zones,** areas with homogeneous physical features. Divisions can be made on the basis of light, temperature, salinity, depth, latitude, water density, or almost any of the other physical dimensions we have discussed. Some classifications are more useful than others, however, and we will survey those classifications in this section and in **Figure 12.9.**

Classification by Light 12-15

The sunlit layer of water at the ocean's surface is called the **photic zone** (*photos* = light). At noon in the clear tropics the photic zone may extend to a depth of around 600 meters (2,000 feet), but in mid-latitude water small organisms (and other small scattering particles) are more abundant, so light reaches down to only about 150 meters (500 feet). As noted above (and in Figure 12.7), the upper part of the photic zone—where there is sufficient light for carbohydrate production by photosynthesis to exceed loss of carbohydrates through respiration—is called the *euphotic zone*. The lower part of the photic zone is the *disphotic zone,* a region where animals can see, but there is insufficient light for productive photosynthe-

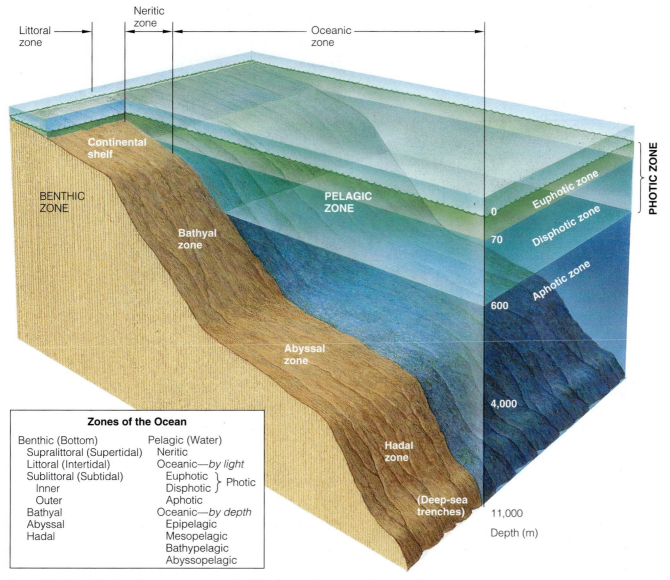

Littoral zone

Neritic zone

Oceanic zone

PHOTIC ZONE

Continental shelf

BENTHIC ZONE

PELAGIC ZONE

Euphotic zone

Bathyal zone

Disphotic zone

Aphotic zone

0

70

600

Abyssal zone

4,000

Hadal zone

(Deep-sea trenches)

11,000

Depth (m)

Zones of the Ocean

Benthic (Bottom)	Pelagic (Water)
Supralittoral (Supertidal)	Neritic
Littoral (Intertidal)	Oceanic—*by light*
Sublittoral (Subtidal)	Euphotic ⎫ Photic
Inner	Disphotic ⎭
Outer	Aphotic
Bathyal	Oceanic—*by depth*
Abyssal	Epipelagic
Hadal	Mesopelagic
	Bathypelagic
	Abyssopelagic

Figure 12.9 Classification of marine environments. This diagram is designed to show divisions and is somewhat exaggerated in indicating the proper proportions.

sis. Below the dysphotic zone is the deepest, largest region of the open ocean extending to the seabed, the *aphotic zone,* the dark zone that extends to the bottom.

Classification by Location

12-16

The primary division is between water and ocean bottom. Open water is called the **pelagic zone** (*pelagius* = of the sea) and is divided into two subsections: the **neritic zone** (*neritos* = shallow), over the continental shelf, and the deep-water **oceanic zone,** beyond the continental shelf. The oceanic zone is further divided by depth into zones. The *epipelagic zone* (*epi* = atop) corresponds to the lighted photic zone. In the aphotic depths are layered the *mesopelagic* (*mesos* = in the middle), *bathypelagic* (*bathos* = depth), and *abyssopelagic*

(*a* = without + *byssos* = bottom) zones. Abyssopelagic water is the water in the deep trenches.

Divisions of the bottom are labeled **benthic** (*benthos* = bottom) and begin with the intertidal **littoral zone** (*litoral* = of shore), the band of coast alternately covered and uncovered by tidal action. (The **supralittoral zone,** the splash zone *above* the high intertidal, is not technically part of the ocean bottom.) Past the littoral is the **sublittoral zone** (*sub* = below), which is further divided into inner and outer segments: The *inner sublittoral* is the ocean bottom near shore, and the *outer sublittoral* is the ocean floor out to the edge of the continental shelf.[2] The **bathyal zone** covers the seabed on the slopes and

---

[2] The words *supertidal, intertidal,* and *subtidal* are often used instead of the *littoral* terms.

down to great depths, where the **abyssal zone** begins. The **hadal zone** (*Hades* = underworld) is the deepest seabed of all, the trench walls and floors.

CLASSIFICATION OF OCEANIC LIFE

Just as oceanographers found it necessary to develop a standard classification and naming system for position in the oceanic realm, biologists centuries earlier had realized the value of being able to classify living things into categories and give them universally understood names.

Systems of Classification

The study of biological classification is called **taxonomy** (*taxo* = to put in order + *onoma* = name). Classification schemes have been around for as long as people have looked at living things. Putting animals in one category and plants in another is an ancient distinction, for example.

The Greek philosopher Aristotle proposed a system of classifying animals based on their exterior similarities, but his results were not very useful. Using his system we would place airline pilots, gliding squirrels, flying fish, and grasshoppers into the same group because each can fly! Such a system is an **artificial system of classification.** (Another artificial system of classification would be grouping books by jacket color, or page size, or typeface.) By contrast, the **natural system of classification** for living organisms biologists use today relies on structural and biochemical similarities among organisms. We place all insects together regardless of their flying ability, just as we place all books by Melville together, all compositions of Beethoven together, and all sea stars together, because each group has a common underlying natural origin. The groups are arranged *systematically*—that is, in some order that makes structural and evolutionary sense.

One of the first persons to classify groups of organisms into natural categories was the eighteenth-century Swedish naturalist Carl von Linné, or as he called himself, **Carolus Linnaeus** (**Figure 12.10**). In his zeal to classify every aspect of the natural world, Linnaeus invented three supreme categories, or **kingdoms**: animal, vegetable, and mineral. Today's biologists leave the mineral kingdom to the geologists and have expanded Linnaeus's two living kingdoms to five.

Linnaeus's great contribution was a system of classification based on **hierarchy,** a grouping of objects by degrees of complexity, grade, or class. In this boxes-within-boxes approach, sets of small categories are nested within larger categories. Linnaeus devised names for the categories, starting with kingdom (the largest category) and passing down through phylum, class, order, family, and genus, to species (the smallest category). In 1758 he published a catalog of all animals then known, his monumental *Systema Naturæ* (*The System of Nature*). **Figure 12.11** shows the classification of a familiar sea

Figure 12.10 Carolus Linnaeus—the father of modern taxonomy—in Laplander costume. (He went on a scientific expedition to Lapland in 1732.)

gull using the Linnaean method. Note the nested arrangement of category-within-category, each category becoming more specific with every downward step.

Scientific Names

Linnaeus also perfected the technique of naming animals. The *genus* and *species* names—the names of the last two nested categories—constitute an organism's **scientific name,** or binomen. *Octopus bimaculatus* is the scientific name of a common West Coast octopus: *Octopus* is the generic name, *bimaculatus* the specific name. A closely related species, *Octopus dofleini*, is a larger animal that ranges to Alaska. *Octopus bimaculatus* and *Octopus dofleini* are not interfertile (they're not the same species), but, as their shared generic name suggests, they *are* closely related.

The advantage of a scientific name over a common name is immediately apparent to anyone trying to identify a shell found on the beach. The same shell may have many different common names in many different languages, but it will have *only one scientific name.* When you discover that name in a good key to shells you can use it to find references that will tell you what is known about the animal, its lifestyle, its range, and its evolutionary history.

MARINE COMMUNITIES

Organisms are distributed through the marine environment in specific groups of interacting producers, consumers, and recyclers that share a common living space. These groups are

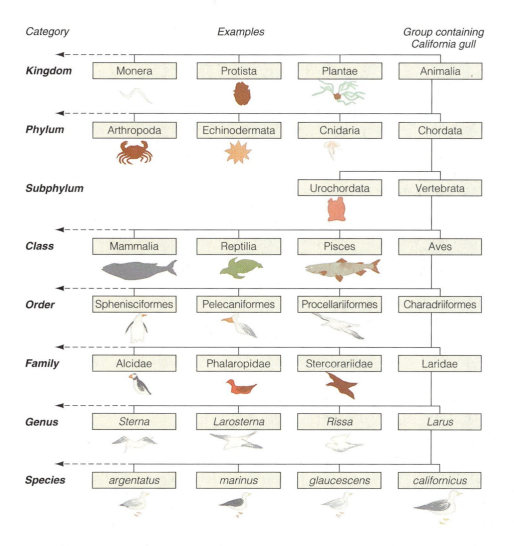

Category	Examples			Group containing California gull
Kingdom	Monera	Protista	Plantae	Animalia
Phylum	Arthropoda	Echinodermata	Cnidaria	Chordata
Subphylum			Urochordata	Vertebrata
Class	Mammalia	Reptilia	Pisces	Aves
Order	Sphenisciformes	Pelecaniformes	Procellariiformes	Charadriiformes
Family	Alcidae	Phalaropidae	Stercorariidae	Laridae
Genus	*Sterna*	*Larosterna*	*Rissa*	*Larus*
Species	*argentatus*	*marinus*	*glaucescens*	*californicus*

Figure 12.11 The modern system of biological classification using the California gull (*Larus californicus*) as an example. Note the "boxes-within-boxes" approach—a hierarchy.

called *communities* (**Figure 12.12**). A **community** is comprised of the many populations of organisms that interact with one another at a particular location. A **population** is a group of organisms of the same species occupying a specific area. The location of a community, and the populations that comprise it, depend on the physical and biological characteristics of that living space. In the next two chapters we will survey the organisms in the ocean's two great realms: the pelagic and benthic environments. Pelagic organisms (*pelagios* = of the sea) live suspended in the water; benthic organisms (*benthos* = seabed) live on or in the ocean bottom.

The largest marine community—and the most sparsely populated—is the pelagic community lying within the uniform mass of permanently dark water between the sunlit surface and the deep bottom. Few animals live there because so little food is available, but those organisms that survive are

Figure 12.12 Students inspect a rocky intertidal community. In spite of wave shock, periodic exposure to drying wind and sun, and a broad temperature range, intertidal communities can be among the ocean's most diverse and productive. Nutrients are usually available here, and there are a large number of niches and habitats to be occupied.

among the strangest in the ocean. Opportunities for feeding in the deep open-ocean community are few and far between, so some animals are able to consume prey larger than themselves should the occasion arise. Because so few animals are present, mating is also a rare event—in a few species males and females become permanently bonded during their first encounter, the male burrowing into the female's body for a life-long free ride.

In contrast, the smallest obvious marine communities may be those benthic communities established against solitary rocks on an otherwise flat, featureless seabed. Drifting larvae will colonize the place; the established community can seem an oasis of life and activity in an otherwise static sedimentary desert. Seaweeds will grow, worms will burrow, snails will scrape algae from the hard surfaces, and small fishes will nestle in crevices. Hundreds of small plants and animals can live their lives within a meter of each other, interacting in a compact solitary community with no similar environment available for thousands of meters. The larvae of the next generation drift away with little chance of finding a suitable place to carry on their lives. Microscopic communities also exist—interacting populations can exist on a single grain of sand or on one decomposing fish scale.

Organisms Within Communities

 12-21

There are many different places to live and many different "jobs" for organisms within even a simple community. A **habitat** is an organism's "address" within its community, its physical *location*. Each habitat has a degree of environmental uniformity. An organism's **niche** (*nidus* = nest) is its "occupation" within that habitat, its relationship to food and enemies, an expression of what the organism is *doing*. For example, the small fishes living among the coral heads in a coral reef community share the same habitat, but each species has a slightly different niche. Each population in the community has a different "job" for which its shape, size, color, behavior, feeding habits, and other characteristics particularly suit it.

Competition

 12-22

The availability of resources such as food, light, and space in a community determines the number and composition of the populations of organisms within that community. Competition for the necessities of life may occur within the community between members of the *same* population or between members of *different* populations. Physical or biological factors may give one population the advantage for a time, then shift subtly to favor another.

When *members of the same population* (all members of the same species) compete with each other, some individuals will be larger, stronger, or more adept at gathering food, avoiding enemies, or mating. These animals tend to prosper, forcing their less successful relatives to emigrate, fight, or die in the course of competition. This kind of competition continually

adjusts the characteristics of individuals in a population to their environment.

When *members of different populations* compete, one population may be so successful in its "job" that it eliminates competing populations. In a stable community, two populations cannot occupy the same niche for long. Eventually the more effective competitor overwhelms the less effective one. For example, the little barnacle *Chthamalus* lives on the uppermost rocks in many intertidal communities while the larger barnacle *Balanus* lives on the lower rocks (see **Figure 12.13**). Planktonic larvae of both species can attach themselves to rocks anywhere in the intertidal zone and begin to grow. In the lower zone the more rapidly growing *Balanus* pushes the weaker *Chthamalus* off the rocks, while at higher positions *Balanus* cannot survive because it is less resistant to drying and exposure than the tough little *Chthamalus*. At the top and bottom of their distribution the two species do not compete for food or space. The competition at the intersection of their ranges prevents each species from occupying as much of the habitat as might otherwise be possible.

As we have seen, physical and biological factors affect the number and positions of organisms in a community. The number of individuals per unit area (or volume) is known as the **population density**—rare individuals have a much lower population density than dominant ones. In general, *more* different species exist in benign habitats where physical factors stay near optimal values (like a coral reef or a rain forest), and *fewer* species exist in rigorous habitats where physical factors range to extremes (like a beach or a desert). That is, "easy" habitats typically have high **species diversity** (they contain more species in more niches within a given area) and harsh habitats usually have a lower species diversity. Relatively few species can cope with the stressful environment of the polar ocean, for example, but many species have adapted to the relatively benevolent environment of a tropical reef.

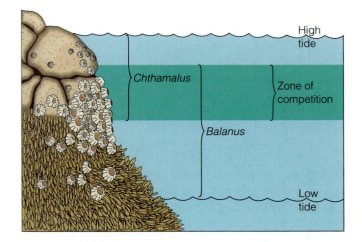

Figure 12.13 Competition between two species of barnacles prevents either from occupying as much of the intertidal zone as possible. The central zone of overlap indicates the area where *Chthamalus* and *Balanus* compete for food and space.

The environment for life changes as time passes, and the changes are not always gradual. The biological history of Earth has been interrupted—catastrophically—at least six times in the last 450 million years. In these events, known as **mass extinctions,** a great many species died off simultaneously (in geological terms). Scientists are not certain of the causes of the mass extinctions, but leading candidates for a couple of these events include the collision of Earth with an asteroid or comet.

Flocks of asteroids (Figure **a**) orbit the sun along with Earth and the other planets. Some of these asteroids have orbits that cross our own, and meetings are inevitable. Earth and its neighbors are pocked with impact craters (Figure **b**) as evidence of these meetings. The consequences to Earth of a collision with even a small asteroid are all but unimaginable. An asteroid only 10 kilometers (6 miles) in diameter would strike with an energy equivalent to the explosion of half a trillion tons of TNT (Figure **c**). More than 100 million metric tons of Earth's crust would be thrown into the atmosphere, obscuring the sun for decades and causing acid rain that would pollute the planet's surface. Concussive shock waves would shatter structures, crush large organisms, and trigger earthquakes for a radius of hundreds of kilometers. If the impact occurred in the Atlantic Ocean, say 1,600 kilometers (1,000 miles) east of Bermuda, the resulting tsunami would wash away the resort islands and swamp most of Florida. Boston would be struck by a 100-meter (300-foot) wall of water. A hole more than 25 kilometers (16 miles) wide and perhaps 10 kilometers (6 miles) deep would mark the point of impact. Clouds of fine particles, accelerated to escape velocity, would travel around the sun in orbits that would intersect Earth's; a steady rain of fine debris might fall for tens of thousands of years.

These kinds of cataclysmic events have been disquietingly common in Earth's past. Where are the craters? As you may recall from Chapter 3's discussion of plate tectonics, much of the ocean floor has been recycled by the movement of lithospheric plates, so any undersea impact craters more than about 100 million years old have disappeared. The distortion and erosion of continents has obscured the outlines of ancient craters on land, although they are more readily visible from space than from the surface—especially if we know what to look for (as in Figure b). Evidence for one massive impact has been bolstered by the discovery of a thin, worldwide layer of

a What one asteroid looks like, courtesy of the *Galileo* spacecraft en-route to study Jupiter. This asteroid would extend from Washington, D.C., halfway to Baltimore. Asteroids are rocky, metallic bodies with diameters ranging from a few meters to 1,000 kilometers (600 miles). Most were swept into the planets during their formation, but about 6,000 large asteroids are still orbiting the sun in a belt between Mars and Jupiter. Unfortunately, the orbits of many dozens of others take them across Earth's orbit.

b What an impact crater looks like. Aerial view of Manicouagan Crater, Quebec, where an asteroid struck Earth some 210 million years ago. The crater is about the size of Rhode Island.

c An artist's conception of a large cometary nucleus, 10 kilometers (6 miles) across, striking Earth 65 million years ago. The cataclysmic explosion is thought to have propelled shock waves and huge clouds of seabed and crust all over Earth, producing a time of cold and dark that contributed to the extinction of many species, including the dinosaurs.

iridium-rich continental rock dated at the boundary between the Cretaceous and Tertiary periods (see Appendix II). Iridium is rare on Earth but common in asteroids. The thin iridium-rich layer may have formed from the dust settling after a collision some 65 million years ago.

As can be seen in Table **a,** vast numbers of marine organisms have perished in mass extinctions. (Extinctions of land families and genera are thought to have been roughly comparable.) At the end of the Cretaceous period, about 65 million years ago, almost one of five families, half the genera, and three-quarters of the species disappeared. Many scientists believe an asteroid impact triggered the mass extinction. Clearly, these tumultuous interruptions do not represent biological business as usual. In a few instances, rocky visitors from space appear to have massively disrupted the environment for life on this planet. The animals, plants, bacteria, and single-celled organisms we see on Earth are descendants of the survivors.   12-23

Table a The Six Great Mass Extinctions			
		Percent Marine Extinctions for:[a]	
Geological Period in Which Extinction Occurred	**Millions of Years Ago**	**Families**	**Genera**
Late Ordovician	435	27	57
Late Devonian	365	19	50
Late Permian	245	57	83
Late Triassic[b]	220	23	48
Late Cretaceous[b]	65	17	50
Late Eocene	35	2	16

Source: C. Sagan and A. Druyan. *Comet*. New York: Random House, 1985.
[a]Rough estimate of percent of all families and genera of marine animals with hard parts (so we have fossil evidence of their existence) rendered extinct; numbers are given to the nearest 5 million years (see column 2).
[b]Mass extinctions for which asteroid or comet strikes may be responsible.

Like the organisms that comprise them, communities change through time. The slow changes associated with seafloor spreading, climate cycles, atmospheric composition, or newly evolved species have shaped this generally slow evolution. As on land, the species, community composition, and location of a marine community are changed by the environmental factors to which members of the community are exposed. Communities themselves can gradually modify the physical aspects of their environment: a coral reef is an extreme example. The massive accumulation of coral and sediments within the reef can alter current patterns, influence ocean temperature, and change the proportions of dissolved gases.

But rapid changes can also occur in marine communities. A natural catastrophe—a volcano erupting, a landslide that blocks a river, or the collision of an asteroid with Earth, for example—can disrupt a community. Similarly, human activities such as altering an estuary by damming a river, dumping excess nutrients into a nearshore area, or stressing organisms with toxic wastes can cause rapid, disruptive changes. The composition of offshore communities changes abruptly near new sewage outfalls, for example.

A stable, long-established community is known as a **climax community.** This self-perpetuating aggregation of species tends not to change unless disrupted by severe external forces such as violent storms, significant changes in current patterns, epidemic diseases, or influx of great amounts of fresh water or pollutants. A disrupted climax community can be reestablished through the process of **succession,** the orderly changes of a community's species composition from temporary inhabitants to long-term inhabitants. Disruption makes the environment more hostile to the original species, but destruction of species in the original community leaves open habitats and niches. A few highly tolerant species will move into the area, eventually drawing in other species that depend on them. If the environment is permanently changed by the disruption, a climax community different from the previous one will be established.

In the next two chapters we turn our attention to the climax communities of the ocean's vast pelagic and benthic domains.

1. Total terrestrial primary productivity appears to be roughly the same as total marine primary productivity. But the total *biomass* of producers in the ocean is at least 300 times smaller (and maybe 1,000 times smaller) than the total *biomass* of producers on land! How can that be? How can terrestrial and marine productivity be so similar?

Because of the astonishing efficiency of small marine autotrophs (phytoplankton), turnover time of molecules is exceedingly fast—nutrients and carbohydrates are cycled with speed and efficiency. There may not be nearly as great a biomass of producers in the ocean, but they appear to be *very* busy indeed!

2. What would terrestrial productivity and food chains be like if 99% of the *land* environment were too dark for successful photosynthesis? In other words, what if land plants had to contend with an environment as dark as the ocean?

Productivity would plummet, of course. Milne (1995) estimates that if the land were lighted as the ocean is, all land animals would be dependent on the plant growth in a lighted area the size of the United States east of the Mississippi River. Life on land would be much less abundant than at present, and animals would be concentrated in or around the lighted region. Because land plants are less efficient than aquatic ones (in part because of the infrastructure needed for support, to pump juices around their bodies, and to hold leaves to the light and air), total land productivity would be reduced to less than 1% of present values.

3. If humans have a fluid much like seawater bathing our cells, why can't we drink seawater and survive?

Human cells function in an environment hypotonic to seawater; that is, blood plasma is less saline than seawater. Drinking seawater therefore causes water to leave the intestinal walls, flood the intestine, and leave the body. There is a net loss via intestine or kidneys even if the seawater is diluted with fresh water before drinking. Moral: Never drink seawater in a survival situation at sea, and never dilute fresh water with seawater (in any proportion) to stretch your supply.

4. What proportion of total world productivity is achieved by chemosynthesis?

An interesting and controversial question! Chemosynthesis occurs in the marine environment primarily at hydrothermal vents associated with the 65,000-kilometer (41,000-mile) network of mid-ocean ridges. Specialized communities of organisms have evolved in the zones of warm or hot mineral-rich water percolating through the seafloor in these places. Once thought to be rare and isolated, vent communities have now been discovered in broadly separate locations and may be quite common to mid-ocean ridges. The trophic pyramid in these deep communities is based on chemosynthetic bacteria. Some biological oceanographers have suggested that if vent communities are abundant on ridges, or if bacteria can grow at warmer temperatures than previously thought, chemosynthesis there may account for a much larger proportion of total oceanic productivity than was previously thought. And recent discoveries have shown chemosynthetic communities deep *within* the seabed itself!

Life on Earth is notable for both unity and diversity: *diversity* because there are at least 5 million different species (kinds) of living things on Earth; *unity* because each species shares the

Pelagic Communities **13**

"... SILVER-SHINING, SWIFT, STRONG, STREAMLINED ..."

Among the ranks of marine drifters and swimmers, members of the tuna family are the ocean's fastest and widest-ranging animals. The body of a tuna is dedicated to speed. Its fins retract into slots, its eyes form a smooth surface with the rest of the head, and it may consume as much as 25% of its weight in food each day. Indeed, a tuna uses so much energy that one of its greatest physiological problems is to avoid overheating! The biological equivalent of the legendary Flying Dutchman, these powerful fishes are fated to travel continuously. If they ever stopped they would suffocate, and their massive bodies would fall to the depths.

Tuna and their relatives swim enormous distances and exhibit astonishing bursts of speed. Studies have shown that albacore tuna regularly migrate from the coast of California to Japan and back, a one-way trip of 8,500 kilometers (5,300 miles), moving at an average speed of not less than 26 kilometers (16 miles) per day. Tagged bluefin tuna have traveled at least 7,770 kilometers (4,830 miles) across the North Atlantic in 119 days—that is, more than 65 kilometers (40 miles) each day. In reality, the bluefins must have traveled much farther, continually detouring from a straight-line course to hunt for food. The fastest tuna can maintain speeds of more than 75 kilometers (47 miles) per hour, and the sailfish, a close relative, can rocket to 110 kilometers (68 miles) per hour for a short time.

Yellowfin tuna and happy fisherman. Among the strongest and fastest of pelagic animals, yellowfin migrate across great distances. The record yellowfin was caught in 1977 off the coast of Mexico and weighed 153 kilograms (338 pounds).

These magnificent fishes are valued for their meat, especially in Japan where they are highly prized for sashimi, the raw fish component of sushi. In the Tokyo fish market, old and fat bluefin tuna have sold for as much as $70,000 a ton! In 1989 a school of huge bluefin appeared off the California Channel Islands—the largest weighed 458 kilograms (1,008 pounds). Many lucky fishermen paid off home mortgages and boat loans in a week of heroic fishing. But it is the living animal that provides the greatest inspiration: silver-shining, swift, strong, and streamlined, these silent nomads slip through the ocean more than a million miles in a lifetime. 13-1

KEY CONCEPTS

1. Pelagic organisms live suspended in the water. Planktonic organisms drift or swim weakly; nektonic organisms actively swim.

2. The term *plankton* is not a collective natural category; it is a description of a lifestyle. The plankton include many plantlike species and nearly every major group of animals.

3. The plantlike organisms that make up the phytoplankton are responsible for most of the ocean's primary productivity. Of these, diatoms are the most numerous and efficient. Most primary productivity occurs in nearshore areas of the temperate and south subpolar zones.

4. Important nektonic animals include cephalopods, shrimps, fishes, seabirds, and marine mammals. Each has a continuously evolving suite of adaptations that allows it to meet the challenges of the marine environment.

CHAPTER AT A GLANCE

Plankton
Collecting and Studying Plankton
Phytoplankton: The Autotrophs
Phytoplankton Productivity
Zooplankton: The Heterotrophs
Plankton and Food Webs

Nekton
Squids and Nautiluses
Shrimps and Their Relatives
Fishes
The Problems of Fishes
Marine Reptiles
Marine Birds
Marine Mammals

Pelagic organisms live suspended in seawater. They are immensely varied, but all have common problems of maintaining their vertical position, producing or obtaining food, and surviving long enough to reproduce. They can be divided into two broad groups based on their lifestyle: The **plankton** drift or swim weakly, going where the ocean goes, unable to move consistently against waves or current flow. The **nekton** are pelagic organisms that actively swim. **Figure 13.1** shows a few representative pelagic organisms.

PLANKTON 13-2

The pelagic organisms that constitute plankton (*planktos* = wandering) are as important as they are inconspicuous. The diversity of planktonic organisms is astonishing—there are giant drifting jellyfish with tentacles 8 meters (25 feet) long, small but voracious arrow worms, many single-celled creatures that glow brightly when disturbed, mollusks with slowly beating flaps that resemble butterfly wings, crustaceans that look like microscopic shrimp, miniature jet-propelled animals that live in jellylike houses and filter food from water, and shimmering crystal-shelled algae. The only feature common to all plankton is their inability to move consistently laterally through the ocean. However, many can and do move vertically in the water column.

Plankton include many different plantlike species and virtually every major group of animals. Thus, the term *plankton* is not a collective natural category like mollusks or algae, which would imply an ancestral relationship between the organisms; instead it describes a basic ecological connection. Members of the plankton community, informally referred to as *plankters,* can and do interact with one another: There is grazing, predation, parasitism, and competition among members of this dynamic group. The organisms within the ovals in Figure 13.1 are plankton.

Collecting and Studying Plankton 13-3

The first large-scale, systematic study of plankton was carried out by biologists aboard the research vessel *Meteor* during the German Atlantic Oceanographic Expedition of 1925–26. Many of the tools and techniques they pioneered are still in use today. **Plankton nets** (**Figure 13.2**) of the type perfected aboard *Meteor* are essential to plankton studies. These conical nets are customarily made of nylon or Dacron cloth woven in a fine interlocking pattern to assure consistent spacing between threads. The net is hauled slowly for a known distance behind a ship, or cast to a set depth, then reeled in. Trapped organisms are flushed to the net's pointed end and gently removed for analysis. Because very small plankton can slip through a plankton net, their capture and study requires concentration by centrifuge, or entrapment by a plankton filter through which water is drawn; the filter is later disassembled and the plankton studied in place.

Quantitative analysis of plankton requires the organisms and an estimate of the sampled volume of water. Measurements

35
mm

5
mm

a

1 dolphins, *Delphinus*
2 tropic birds, *Phaëthon*
3 paper nautilus, *Argonauta*
4 anchovies, *Engraulis*
5 mackerel, *Pneumatophorus*, and sardines, *Sardinops*
6 squid, *Onykia*
7 *Sargassum*
8 sargassum fish (*Histro*)
9 pilot fish (*Naucrates*)
10 white-tipped shark *Carcharhinus*
11 pompano, *Palometa*
12 ocean sunfish, *Mola mola*
13 squids, *Loligo*
14 rabbitfish, *Chimaera*
15 eel larva, *Leptocephalus*
16 deep sea fish
17 deep sea angler, *Melanocetus*
18 lantern fish, *Diaphus*
19 hatchetfish, *Polyipnus*
20 "widemouth," *Malacosteus*
21 euphausid shrimp, *Nematoscelis*
22 arrowworm, *Sagitta*
23 amphipod, *Hyperoche*
24 sole larva, *Solea*
25 sunfish larva, *Mola mola*

26 mullet larva, *Mullus*
27 sea butterfuly, *Clione*
28 copepods, *Calanus*
29 assorted fish eggs
30 stomatopod larva
31 hydromedusa, *Hybocodon*
32 hydromedusa, *Bougainvilli*
33 salp (pelagic tunicate), *Doliolum*
34 brittle star larva
35 copepod, *Calocalaus*
36 cladoceran, *Podon*
37 foraminifer, *Hastigerina*
38 luminescent dinoflagellates, *Noctiluca*
39 dinoflagellates, *Ceratium*
40 diatom, *Coscinodiscus*
41 diatoms, *Chaetoceras*
42 diatoms, *Cerautulus*
43 diatom, *Fragilaria*
44 diatom, *Melosira*
45 dinoflagellate, *Dinophysis*
46 diatoms, *Biddulphia regia*
47 diatoms, *B. arctica*
48 dinoflagellate, *Gonyaulax*
49 diatom, *Thalassiosira*
50 diatom, *Eucampia*
51 diatom, *B. vesiculosa*

b

Figure 13.1 (**a**) Pelagic communities: representative plankton and nekton of the subtropical Atlantic Ocean. (**b**) Key. Deep-sea fishes and the sargassum fish (8) are shown near their actual size. Surface-dwelling fishes are at least 20 times larger in life than shown in the figure. This stylized representation shows organisms to be much more crowded than they would be in real life.

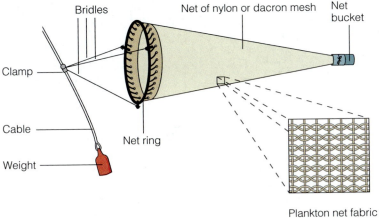

Plankton net fabric as seen under a microscope

a

Figure 13.2 Plankton nets. (**a**) The standard conical net is made of fine mesh and has a mouth up to 1 meter (3.3 feet) in diameter. The net is towed behind a ship for a set distance. The number of organisms present in the water can be estimated if the trapped organisms are counted and the volume of sampled water is known. (**b**) The net shown here has a somewhat coarser mesh because its target organisms, shrimplike crustaceans known as *krill*, are relatively large.

b

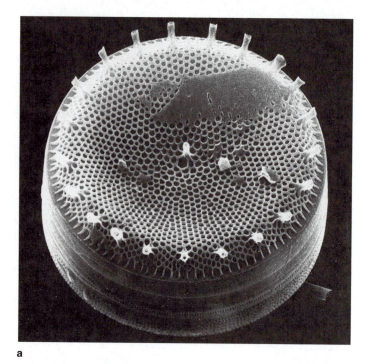

a

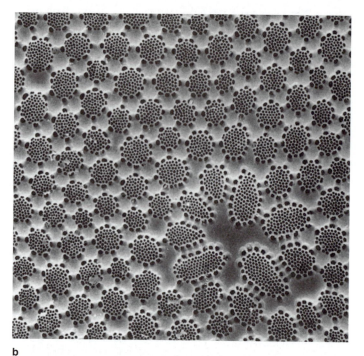

b

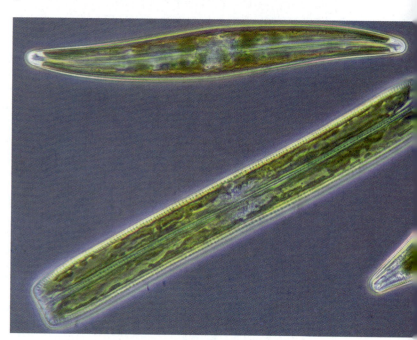

c

Figure 13.3 Diatoms. (**a**) The valve, or "shell," of the diatom *Coscinodiscus* as shown in a scanning electron micrograph of the beautiful silica frustule. (**b**) A closer view, showing the intricate perforations from which diatoms take their name. (**c**) A pennate (elongated) diatom. Each of the cells in this figure is about 15 micrometers (300 millionths of an inch) across.

of physical ocean conditions such as dissolved carbon dioxide and oxygen content, pH, temperature, and light intensity at the time and place of sampling are also important in interpreting the samples.

Phytoplankton: The Autotrophs   13-4

Autotrophic Plankton are generally called **phytoplankton** (*phyton* = plant). A huge, nearly invisible mass of phytoplankton drifts within the sunlit surface layer of the world ocean. Phytoplankton are critical to all life on Earth because of their great contribution to food webs and their generation of large amounts of atmospheric oxygen through photosynthesis. Planktonic autotrophs are thought to bind *at least* 35 billion metric tons of carbon into carbohydrates each year, at least 40% of the food made by photosynthesis on Earth! These easily overlooked, mostly single-celled, drifting photosynthesizers are much more important to marine productivity than the larger and more conspicuous seaweeds.

There are at least eight major types of phytoplankton, of which the most prominent are the diatoms and dinoflagellates. In the past decade, however, the small coccolithophores and silicoflagellates, and extremely small cyanobacteria and autotrophic picoplankton have been found to be more important in world productivity than most researchers previously imagined.

DIATOMS   The dominant photosynthetic organisms in the plankton—and in the world—are the **diatoms.** More than 5,600 species of diatoms are known to exist. The larger species

are barely visible to the unaided eye. Most are round, but some are elongated, branched, or triangular.

Typical diatoms are shown in **Figure 13.3.** The name means "cut through" (*dia* = through + *tomos* = to cut), a reference to the patterns of perforations through the diatom's rigid cell wall, or **frustule** (*frustulum* = a small piece). As much as 95% of the mass of the frustule consists of silica (SiO_2), giving this beautiful covering the optical, physical, and chemical characteristics of glass—an ideal protective window for a photosynthesizer. Magnification reveals that the frustule consists of two closely fitting halves, or **valves,** which fit together like a well-made gift box, the top valve adhering tightly over the lip of the bottom one. The pattern of perforations, slits, striations, dots, and lines on the surface of the valves is different for each diatom species. When diatoms die, their valves can drift to the seafloor to accumulate as layers of siliceous ooze (see Chapter 5.)

Inside the diatom's tailored valves lies an extraordinary photosynthetic machine. Fully 55% of the energy of sunlight absorbed by a diatom can be converted into the energy of carbohydrate chemical bonds, one of the most efficient energy conversion rates known. Oxygen not needed in the cell's respiration is released through the perforations in the frustule into the water. Some oxygen is absorbed by marine animals, some is incorporated into bottom sediments, and some diffuses into the atmosphere. Most of the oxygen we breathe has moved recently through the many glistening pores of diatoms.

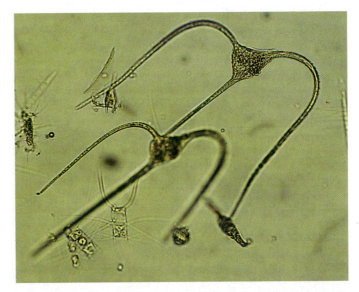

Diatoms store energy as fatty acids and oils, compounds that are lighter than their equivalent volume of water and thus assist in flotation. As you might guess, flotation is a potential problem for diatoms because the weight of their heavy silica frustule seems at odds with their requirement for staying near the sunlit ocean surface. Oil floats, glass sinks, and a balanced amount of both produces neutral buoyancy. Not all diatoms need to float, however. Many nonplanktonic species lie on shallow bottoms where light and nutrients are able to support photosynthesis. These benthic species are nearly always elongated (or *pennate*) in shape.

DINOFLAGELLATES Most **dinoflagellates (Figure 13.4a)** are single-celled autotrophs. A few species live within the tissues of other organisms, but the great majority of dinoflagellates live free in the water. Most have two whiplike projections called **flagella** in channels grooved in their protective outer covering of cellulose. One flagellum drives the organism forward while the other causes it to rotate in the water (hence the name: *dino* = whirling + *flagellum* = whip). Their flagella allow dinoflagellates to adjust orientation and vertical position to make the best photosynthetic use of available light.

Some species of dinoflagellates can become so numerous that the water turns a rusty red as light reflects from the accessory pigments within each cell. These species are responsible for the phenomenon of *red tide* (**Figure 13.4b**). During times of such rapid growth (usually in springtime), concentration of these microscopic organisms may briefly reach 6 million per liter (23 million per gallon)! At night, the huge numbers of dinoflagellates in a red tide can cause breaking waves to glow a bright blue, a phenomenon known as bioluminescence (*bios* = life + *luminis* = light).

Figure 13.4 (**a**) Two representatives of the genus *Ceratium,* a common dinoflagellate. Two flagella beat in opposing grooves in the armor-plated body: one groove is visible at the equator of the upper cell, between the two "horns." These specimens are about 0.5 millimeter (0.02 inch) across. (**b**) During a red tide, the presence of millions of dinoflagellates turns seawater a brownish-red color.

Red tide can be dangerous because some dinoflagellate species synthesize potent toxins as byproducts of metabolism. Among the most effective poisons known, these toxins may affect nearby marine life or (if the dinoflagellates are dried on the beach and blown inland) even humans. Some of the toxins are similar in chemical structure to the muscle relaxant curare but are tens of times more powerful. Humans should avoid eating certain species of clams, mussels, and other filter feeders

a

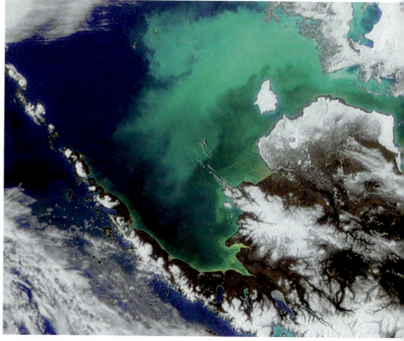

b

Figure 13.5 (**a**) Calcium carbonate plates in place on *Umbili-cosphaera*, a coccolithophore. The tiny plates are 6 micrometers (120 millionths of an inch) across. (**b**) A bloom of coccolithophores in April 1998 turns the ocean surface turquoise in the Bering Sea, north of Alaska's Aleutian Islands.

during summer months when toxin-producing dinoflagellates are abundant in the plankton. If shellfish from a particular area are unsafe, a state agency will issue an advisory that may remain in effect for six weeks or more until the danger is past.

COCCOLITHOPHORES AND OTHER PHYTOPLANKTON

Most other types of phytoplankton are extraordinarily small and so are called **nanoplankton** (*nanus* = dwarf). The **coccolithophores,** for example, are tiny single cells covered with discs of calcium carbonate (coccoliths) fixed to the outside of their cell walls (**Figure 13.5a**). Coccolithophores live near the ocean surface in brightly lit areas. The translucent covering of coccoliths may act as a sunshade to prevent absorption of too much light. In areas of high coccolithophore productivity, most notably in the Mediterranean and Bering Seas (**Figure 13.5b**), their numbers occasionally become so great that the water appears milky or chalky. Coccoliths can also build seabed deposits of ooze. The famous White Cliffs of Dover in England consist largely of fossil coccolith ooze deposits uplifted by geological forces.

Other very small phytoplankters, the *silicoflagellates,* possess filamentous internal supporting structures of silica, the same material from which diatoms construct their valves. One or two flagella are always present. Compared to diatoms, however, the overall structure and biochemistry of silicoflagellates seem primitive. Little is known about their worldwide distribution or abundance. Recent evidence suggests that silicoflagellates, coccolithophores, and other, even tinier photosynthesizers may contribute much more to overall plankton productivity than previously thought. The very smallest photosynthesizers—organisms less than 1.5 micrometers (30 millionths of an inch) in diameter—are only now being studied.

Preliminary estimates suggest very small autotrophs may be responsible for up to 70% of all the photosynthetic activity in the world ocean!

Phytoplankton Productivity   13-5

Where is phytoplankton productivity the greatest? This question is among the most important in biological oceanography. Since phytoplankton form the base of nearly all oceanic food webs, the biological characteristics of any ocean area will depend heavily on the presence and success of phytoplankton.

With some exceptions, the distribution of phytoplankton corresponds to the distribution of macronutrients. Because of coastal upwelling and land runoff, nutrient levels are highest near the continents. Plankton are most abundant there, and productivity is highest. The water above some continental shelves sustains productivity in excess of 1 gC/m²/*day*! (See Figure 12.2 for a review of primary productivity notation and estimates.) But what of the open ocean? Where is productivity greatest away from land?

In the tropics? The open tropical oceans have abundant sunlight and CO_2 but are generally deficient in surface nutrients because the strong thermocline discourages the vertical mixing necessary to bring nutrients from the lower depths. The tropical oceans away from land are therefore oceanic deserts nearly devoid of visible plankton. The typical clarity of tropical water underscores this point. In most of the tropics productivity rarely exceeds 30 gC/m²/yr, and seasonal fluctuation in productivity is low.

Tropical coral reefs—benthic habitats—are exceptions to this general rule. Reef areas, which account for less than 2%

of the tropical ocean surface, are productive places because autotrophic dinoflagellates live *within* the tissues of coral animals and don't drift as plankton. Nutrients are made available by coastal upwelling and by the coral's own metabolism. These nutrients are cycled tightly through the reef and not lost to sinking.

In the polar regions? At very high latitudes the low sun angle, reduced light penetration due to ice cover, and weeks or months of darkness in winter severely limit productivity. At the height of summer, however, 24-hour daylight, a lack of surface ice, and the presence of upwelled nutrients can lead to spectacular plankton blooms. The surface of some sheltered bays can look like tomato soup because dinoflagellates and other plankton are so abundant. This bloom cannot last, because nutrients are not quickly recycled and because the sun is above the critical angle for a few weeks at best. The short-lived summer peak does not compensate for the long, unproductive winter months.

Average productivity at very high northern latitudes tends to be lower than at high southern latitudes, averaging less than 25 gC/m²/yr. The Arctic Ocean is almost surrounded by land masses, which limit water circulation and, therefore, nutrient upwelling. The southern ocean, on the other hand, is enriched by water upwelling to replace sinking Antarctic Bottom Water. This rich mixture is stirred by the West Wind Drift. The Antarctic accounts for a much greater share of high-latitude production than the Arctic because more nutrients are available.

In the temperate and subpolar zones? With the tropics generally out of the running for reasons of nutrient deficiency and the north polar ocean suffering from slow nutrient turnover and low illumination, the overall productivity prize is left to the temperate and southern subpolar zones. Thanks to the dependable light and the moderate nutrient supply, annual production in the nearshore temperate and southern subpolar ocean areas is the greatest of any open ocean area. Typical productivity in the temperate zone is about 120 gC/m²/yr. In ideal conditions southern subpolar productivity can approach 250 gC/m²/yr!

Figure 13.6 shows the levels of productivity in tropical, temperate, and polar ocean areas. Note that *nearshore* productivity is nearly always higher than *open ocean* productivity, even in the relatively productive temperate and south subpolar zones.

Curiously, the open ocean area with the greatest annual productivity is an exception to the general picture developed in this section. The slender, cold finger of high productivity pointing west from South America along the equator is a result of wind-propelled upwelling due to Ekman transport on either side of the geographic equator. Look for this area in Figure 13.6.

Does productivity change with the seasons? **Figure 13.7** shows the relationship of phytoplankton biomass to season. The low, flat line representing annual tropical productivity contrasts with the high, thin peak representing the Arctic summer. The higher of the two peaks for the temperate zone

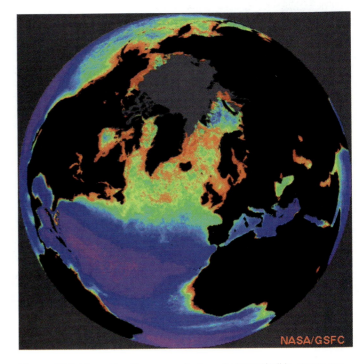

Figure 13.6 Measuring the concentration of chlorophyll in ocean water by satellite. This false color image from the Coastal Zone Color Scanner aboard the *Nimbus 7* satellite represents average conditions in late spring in the Northern Hemisphere. It shows the concentration of chlorophyll in the upper layer of the ocean, with higher amounts indicated by green, orange, and red colors. Note the high phytoplankton concentrations induced by increased nutrient availability along the coasts. The centers of the oceanic gyres contain relatively few phytoplankton, as shown by their purple hue.

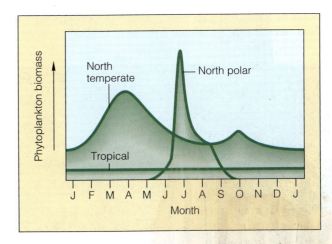

Figure 13.7 Variation in the biomass of phytoplankton by season and latitude. (Note that the total area under each curve represents productivity.)

indicates the plankton bloom of northern spring caused by increasing illumination, while the smaller peak representing the northern fall is caused by nutrients returning to the surface.

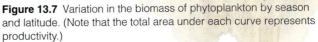

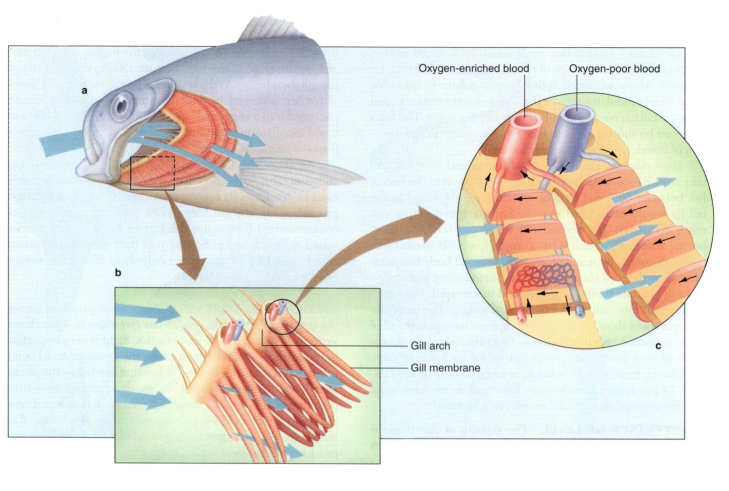

Oxygen-enriched blood   Oxygen-poor blood

a

b

Gill arch

Gill membrane

c

Figure 13.16 Cutaway of a mackerel, showing the position of the gills (**a**). Broad arrows in (**b**) and (**c**) indicate the flow of water over the gill membranes of a single gill arch. Small arrows in (**c**) indicate the direction of blood flow through the capillaries of the gill filament in a direction opposite to that of the incoming water. This mechanism is called *countercurrent flow.*

a

b

Figure 13.17 Passive cryptic coloration in a kelp fish. (**a**) The fish resembles a seaweed blade in shape, color, and pattern. (**b**) The fish nestles within the seaweed blades, nearly disappearing from view, moving with the blades as they sway with the water.

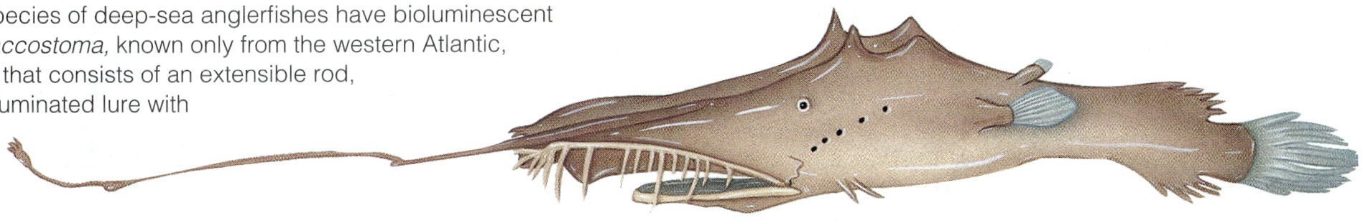

Figure 13.18 Some species of deep-sea anglerfishes have bioluminescent lures. *Lasiognathus saccostoma*, known only from the western Atlantic, has fishing equipment that consists of an extensible rod, filament, "float," and illuminated lure with little hooks. Its body is about 20 centimeters (8 inches) long.

a fish school, or is it a single large animal? Many predators might not stay around long enough to find out! Schools have the added benefits of reducing chance detection by a predator, providing ready mates at the appropriate time, and increasing feeding efficiency.

Fishes living in the twilight world at the bottom of the photic zone use bioluminescence in feeding, avoiding being eaten, and mate attraction. Some members of this sparsely populated community have built-in luminescent organs that cast dim blue light downward; this light masks their own shadows, so they have less chance of being detected and eaten. Other deep-swimming organisms attract their infrequent meals with a luminous lure (**Figure 13.18**). These animals also use patterns of glowing spots or lines to identify themselves to members of the same species, a necessary first step in mating. Some use flashes of light to dazzle or frighten potential predators.

Marine Reptiles 13-13

Each of the three main groups of reptiles has marine representatives: turtles, sea snakes and marine lizards (iguanas), and marine crocodiles. Like all reptiles, marine reptiles are ectothermic, breathe air with lungs, are covered with scales and a relatively impermeable skin, and are equipped with special **salt glands** to concentrate and excrete excess salts from body fluids. Except for one widely ranging species of turtle, all marine reptiles require the warmth of tropical or subtropical waters.

The best-known and most successful living marine reptiles are the eight species of sea turtles. Unlike land turtles, sea turtles have relatively small streamlined shells without enough interior space to retract head or limbs. The shell provides an effective passive defense, and adult sea turtles have no predators except humans. Their forelimbs are modified as flippers and provide propulsive power, their hindlimbs act as rudders. The two species of green sea turtles (genus *Chelonia*) (**Figure 13.19**) are the most abundant and widespread species. Green turtles range over great distances looking for the marine algae, turtle grass, and other plants on which they feed. The largest living turtle is the carnivorous Atlantic leatherback, a streamlined animal with a soft skin-covered "shell"; it reaches lengths in excess of 2 meters (6½ feet) and may weigh more than 600 kilograms (1,300 pounds).

Sea turtles are justly famous for their remarkable feats of navigation. Sea turtles return at two-, three-, or four-year intervals to lay eggs on the beaches at which they were hatched. Homing behavior can be a great advantage to any animal: If

the parent survived its earliest childhood at this location, it will probably be a suitable place for hatching the next generation. The navigation of green turtles to tiny Ascension Island, an emergent point of the Mid-Atlantic Ridge between Brazil and Africa, has been extensively studied. Researchers have found that the turtles use solar angle (to derive latitude), wave direction, smell, and visual cues first to find the island, then to discover the spot on the beach where they hatched perhaps 20 years before!

All marine turtles are in danger of extinction. Though protected and no longer used extensively for human food, their breeding beaches are being developed or invaded by noisy recreational pursuits, their eggs and shells are in great demand, they are drowned in fishing and shrimping nets, and their feeding areas are increasingly damaged by pollutants. Much of the tropical turtle grass has disappeared because of human interference. In addition, some species mistake floating plastic bags and other debris for the jellyfish on which they normally feed; the plastic clogs their digestive tract and they starve to death. The outlook for these creatures is not bright.

Marine Birds 13-14

Birds probably evolved from small, fast-running dinosaurs about 160 million years ago. Their reptilian heritage is clearly visible in their scaly legs and claws, and in the configuration of their internal organs and skeletons. The success of the 8,600 living species of birds is due in large part to the evolution of feathers (derivatives of reptilian scales), used to insulate the

Figure 13.19 A green sea turtle, *Chelonia mydas*.

body and to provide aerodynamic surfaces for flight. Birds (and mammals) are *endothermic*; they generate and regulate metabolic heat to maintain a constant internal temperature that is generally higher than their surroundings.

Flying birds have light, thin, hollow bones without fatty insulation; they have forsaken the heavy teeth and jaws of reptiles for a lightweight beak. Their highly efficient respiratory system can accept great quantities of oxygen, and their large four-chambered heart circulates blood under high pressure. All birds lay eggs on land, and most incubate them and provide care for the young. Some seabirds may stay at sea for years, but all must eventually return to land to breed.

Only about 270 kinds of birds, about 3% of known bird species, qualify as seabirds. Most seabirds live in the Southern Hemisphere. Like the marine reptiles, seabirds have special salt-excreting glands in their heads to eliminate the excess salt taken in with their food. Salty brine from these glands may sometimes be seen dripping from the tips of their beaks. Marine birds are voracious feeders, and the ocean will usually be teeming with life wherever they are found. True seabirds generally avoid land unless they are breeding, obtain virtually all their food from the sea, and seek isolated areas for reproduction.

Of the four groups of seabirds, the gulls and the pelicans may be most familiar to us because there are many Northern Hemisphere species and because they spend much of their time near shore. But the groups best adapted to the pelagic world are the tubenoses (albatrosses, petrels), and penguins.

THE TUBENOSES The 100 species of tubenoses of order Procellariiformes (*procella* = storm) are the world's most oceanic birds. Their prosaic common name does not convey any sense of their beauty and grace; it refers to the plumbing in their beak responsible for sensing air speed, detecting smells, and ducting saline water from the salt glands. Foraging for months across the ocean in conditions of strong winds and high waves, seeking no shelter during storms, being exposed to tropical heat and polar sleet, soaring continually through air with a gliding efficiency exceeding that of the most perfectly built human sailplane, the beautiful albatrosses, petrels, and shearwaters are masters of the sky.

The largest of the tubenoses are the magnificent wandering albatrosses of genus *Diomedea*, which reach a wingspan of 3.6 meters (12 feet) and a weight of 10 kilograms (22 pounds). The key to the albatross's success lies in its aerodynamically efficient wing (**Figure 13.20**), a very long, thin, narrow, cupped and pointed structure ideal for high-speed gliding and soaring. This beautiful wing enables albatrosses to cover great distances in search of food with very little expense of energy, flying continuously for weeks at a time. They use the uplift from wind deflected by ocean waves to stay aloft, and soar in long looping arcs. Albatrosses rarely flap their wings.

Satellite tracking data indicate that wandering albatrosses routinely cover 15,000 kilometers (9,300 miles) on foraging trips and reach speeds of 80 kilometers (50 miles) per hour. High speeds over long distances are their specialty—one bird was observed to travel 808 kilometers (502 miles) at an average speed of 56 kilometers (35 miles) per hour.

Figure 13.20 A wandering albatross (*Diomedea exulans*). The largest of these birds has a wingspan of 3.6 meters (12 feet). This is not a seagull-sized animal—one of these wings is as long as an average man is tall!

Albatrosses were once thought to locate food exclusively by sight, but recent research suggests their extraordinarily acute sense of smell also plays a role in feeding. They can evidently find schools of fish by the odor of fish oil wafted tens of kilometers downwind. Tubenoses catch fish (and squids) by dipping their bill into the water during flight or during brief stops on the surface. Albatrosses take shore leave only during breeding periods. Chicks are hatched and raised on remote islands to which albatrosses regularly migrate from virtually any point over the world ocean. They are astonishing animals—no one who has seen one sweeping over the sea surface soon forgets the experience.

THE PENGUINS Penguins have completely lost the ability to fly, but they use their reduced wings to swim for long distances and with great maneuverability. The name of their order, Sphenisciformes (*spheniskos* = little wedge), refers to the shortness and shape of their wings. Their flightlessness makes it practical to have fatty insulation, greasy peglike feathers, stubby appendages, and large size and weight; indeed, such heat-conserving adaptations are critical to survival of aquatic organisms in very cold climates. Their neutral buoyancy is an advantage as they forage for food underwater. Emperor penguins, the largest of the living penguin species, may dive to depths of 265 meters (875 feet) and stay submerged for ten minutes or more. Penguins feed on fish, large zooplankters, bottom-dwelling mollusks or crustaceans, and squid. A few of the 18 species spend two uninterrupted years at sea between breedings.

Penguins are native only to the Southern Hemisphere and range from the size of a large duck to a height of more than a meter (3.3 feet) and a weight exceeding 36 kilograms (80 pounds). They are thought to consume about 86% of all food taken by birds in the southern ocean—about 34 million metric tons (37 million tons) per year, mostly larger zooplanktonic crustaceans. The small Galápagos penguin lives a comparatively easy life fishing the cold, nutrient-rich Humboldt Current at the equator, but its Antarctic relatives lead what must surely be the most rigorous existence of any seabird. For example, Emperor penguins breed and incubate during the bitterly cold Antarctic winter (**Figure 13.21**). Un-

Figure 13.21 Emperor penguins and a maturing chick. One of the larger penguin species, Emperors subsist mainly on a diet of fish and squid.

like most other birds, Emperors do not establish a territory, instead huddling together in tight crowds to conserve heat. The mass of penguins moves slowly around the breeding area as warm penguins from the center of the mob circulate to the outside to be replaced at the core by their chilled peripheral friends.

Marine Mammals

13-15

The class **Mammalia** (*mamma* = breast), to which humans belong, is the most advanced vertebrate group. About 4,300 species of mammals are known. The three living groups of marine mammals are the porpoises, dolphins, and whales of order **Cetacea**; the seals, sea lions, walruses, and sea otters of order **Carnivora**; and the manatees and dugongs of order **Sirenia.**

Each of these orders arose independently from land ancestors. They exhibit the mammalian traits of being endothermic, breathing air, giving birth to living young that they suckle with milk from mammary glands, and having hair at some time in their lives. Unlike other mammals, however, these extraordinary creatures have become adapted to life in the ocean.

All marine mammals share four common features:

1. Their *streamlined body shape* with limbs adapted for swimming makes an aquatic lifestyle possible. Efficient locomotion depends on minimum drag and maximum ability to transfer propulsive energy from the muscles to the water. Thin, stiff flippers and tail flukes situated at the rear of the animal drive it forward, and similarly shaped forelimbs act as rudders for directional control. Drag is reduced by a slippery skin or hair covering.
2. They *generate internal body heat* from a high metabolic rate and *conserve* this heat with layers of insulating fat and, in some cases, fur. Their large size gives them a favorable surface-to-volume ratio; with less surface area per unit of volume they lose less heat through the skin. This is why there are no marine mammals smaller than a sea otter; a small mammal would lose body heat too rapidly. These adaptations are critical to animals living in cold water where food is readily available.
3. The *respiratory system is modified* to collect and retain large quantities of oxygen. The air duct "plumbing" of marine mammals is typically much different from that of land mammals, and the lungs can be more thoroughly emptied before drawing a fresh breath. The biochemistry of blood and muscle is optimized for the retention of oxygen during deep, prolonged dives. Some whales can stay submerged for 90 minutes!
4. A number of *osmotic adaptations* free marine mammals from any requirement for fresh water. Unlike other marine vertebrates, the marine mammals do not have salt-excreting glands or tissues. They swallow little water during feeding (or at any other time), and their skin is impervious to water. This minimal seawater intake, coupled with their kidneys' ability to excrete a concentrated and highly saline urine, permit them to meet their water needs with the metabolic water derived from the oxidation of food.

ORDER CETACEA The 79 living species of cetaceans (*ketos* = whale) are thought to have evolved from an early line of ungulates—hooved land mammals related to today's horses and sheep—whose descendants spent more and more time in productive shallow waters searching for food. Modern whales range in size from 1.8 meters (6 feet) to 33 meters (110 feet) in length and weigh up to 100,000 kilograms (110 tons). Their paddle-shaped forelimbs are used primarily for steering, and their hind limbs are reduced to vestigial bones that do not protrude from the body. They are propelled mainly by horizontal tail flukes, moved up and down by powerful muscles at the animal's posterior end. A thick layer of oily blubber provides insulation, buoyancy, and energy storage. One or two nostrils are located at the top of the head and have special valves to prevent intake of water when submerged. Whales have large, deeply convoluted brains and are thought to form complex family and social groupings.

Modern cetaceans are further divided into two suborders. **Figure 13.22** shows representative whales in each division. Whales in the suborder **Odontoceti** (*odontos* = tooth), the toothed whales, are active predators and possess teeth to

Humpback whale

Bowhead whale

Right whale

Minke whale

Blue whale

Fin whale

Feeding on krill

Sei whale

Gray whale

Mysticetes (baleen whales)

Figure 13.22 Some representatives of the order Cetacea.

Atlantic white-sided dolphin

Harbor porpoise

Common dolphin

Killer whale

Beluga whale

Bottle-nosed dolphin

False killer whale

Cuvier's beaked whale

Pilot whale

Narwhal

Pygmy sperm whale

Sperm whale

Squid

Baird's beaked whale

| 0 | | 5 | | 10 | | 15 | | 20 | | 25 | | 30 m |

| 0 | 10 | 20 | 30 | 40 | 50 | 60 | 70 | 80 | 90 | 100 ft |

Odontocetes (toothed whales)

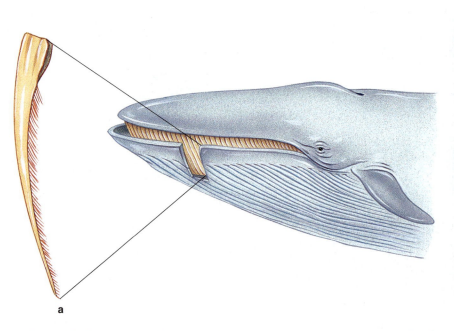

Figure 13.23 (**a**) A plate of baleen and its position in the jaw of a baleen whale. For clarity, the illustration shows part of the mouth cut away. (**b**) A student and a whale, up close and personal. This gray whale uses its stiff, coarse baleen plates to sieve crustaceans from shallow bottom mud.

subdue their prey. Toothed whales have a high brain-weight-to-body-weight ratio, and though much of their "extra" brain tissue is involved in formulating and receiving the sounds on which they depend for feeding and socializing, many researchers believe them to be quite intelligent. Smaller whales in this group include the killer whale and the familiar dolphins and porpoises of oceanarium shows. The largest toothed whale is the 18-meter (60-foot) sperm whale, which can dive to at least 1,140 meters (3,740 feet) in search of the large squids that provide much of its diet.

Toothed whales search for prey using **echolocation,** the biological equivalent of sonar; they generate sharp clicks and other sounds that bounce off prey species and return to be recognized. Odontocete whales are now thought to use sound offensively as well. Recent research indicates that some odontocetes can generate sounds loud enough to stun, debilitate, or even kill their prey. In one experiment dolphins produced clicks as loud as 229 decibels, equivalent to a blasting cap exploding close to the target organism. Sperm whales, it has been calculated, may generate sounds exceeding 260 decibels! (The decibel scale is logarithmic; compare this figure to the 130-decibel noise of a military jet engine at full power 20 feet away!) How this prodigious noise is generated is not yet known, nor do we know how such energy is radiated from the whale without damaging the organs that produce and focus it.

Whales in the suborder **Mysticeti** (*mystidos* = unknowable), the whalebone or baleen whales, have no teeth. Filter feeders rather than active predators, these whales subsist primarily on krill, a thumb-sized shrimplike crustacean zooplankter obtained in productive polar or subpolar waters. They do not dive deep, but commonly feed a few meters below the surface. Their mouths contain interleaving triangular plates of bristly hornlike **baleen** (**Figure 13.23**) used to filter

the zooplankton from great mouthfuls of water. The plankton are concentrated as water is expelled, swept from the baleen plates by the whale's tongue, compressed to wring out as much seawater as possible, and swallowed through a throat not much larger in diameter than a grapefruit. A great blue whale (**Figure 13.24**), largest of all animals, requires about 3 metric tons (13,200 pounds) of krill each day during the feeding season . The short, efficient food chain from phytoplankton to zooplankton to whale provides the vast quantity of food required for their survival.

Mysticeti is an excellent name for these odd and wonderful animals. We know comparatively little of their social structure, intelligence, sound-producing abilities, navigational skills, or physiology. We do know that humpback whales use complex songs in group communication, and that blue whales may use very low frequency sound to communicate over tremendous distances. Some species migrate annually from polar to tropical waters and back. Until recently our primary response to all whales has been to slaughter them in countless numbers for meat and oil with little thought for their extraordinary abilities and assets. More of this depressing history can be found in the discussion of marine resources in Chapter 15.

ORDER CARNIVORA The order Carnivora (*carnis* = flesh + *vorare* = to devour) includes land predators ranging from dogs and cats to bears and weasels, but the members of the carnivoran suborder **Pinnipedia**—the seals, sea lions, and walruses (see **Figure 13.25**)—are almost exclusively marine. Unlike the cetaceans, the gregarious pinnipeds (*pinna* = wing + *pedalis* = foot) leave the ocean for varying periods of time to mate and raise their young.

True seals have a smooth head with no external ear flaps, the external part of the ear having been sacrificed to further

Figure 13.24 A full-sized female blue whale photographed from the deck of a fishing boat. This whale is nearly 30 meters (100 feet) long.

streamline the body. They are covered with a short coarse hair without soft underfur. Seals are graceful swimmers that pursue small fish, their usual prey, with powerful side-to-side strokes of their hind limbs. These rear appendages are partially fused and always point back from the hind end of the body; thus, they are of very little use for locomotion on land. The elephant seal, named for its long snout and large size, holds the diving depth record for all air-breathing vertebrates: 1,560 meters (5,120 feet). Sea lions, familiar to many as the performers in "seal" shows, have hind limbs with a greater range of motion and thus are more mobile on land. They have a streamlined head with small external ears and a pelt with soft underfur; unlike seals, they use their front flippers for propulsion. Walruses are much larger than either seals or sea lions and may reach weights of 1,800 kilograms (2 tons).

Walruses use their tusks like sled runners to guide their sensitive "whiskers" just above the sediment surface looking for clam siphons at depths of up to 90 meters (300 feet). They dig up the clams with their mouths, crush the clamshells and remove the meat, then eject the inedible fragments. The large tusks are also useful for hauling their heavy bodies onto ice floes.

The suborder **Fissipedia** (*fissus* = split + *pedalis* = foot) has many members (including cats, dogs, raccoons, and bears) but only one truly marine representative, the sea otter (**Figure 13.26**), a relative newcomer to the marine environment. This, the smallest of marine mammals, is an active and pleasing creature. Human demand for the fur, the densest and warmest of any animal, caused its near-extermination. The modern population of the Pacific sea otter descends from a very few individuals accidentally overlooked by fur hunters of the late nineteenth and early twentieth centuries. Playful and intelligent, otters rarely exceed 120 centimeters (4 feet) in length; they eat voraciously, consuming up to 20% of their body weight in mollusks, crustaceans, and echinoderms each day. Sometimes they lie on their back in the water, balance a

a

b

Figure 13.25 Some representatives of the suborder Pinnipedia. (**a**) Three juvenile harbor seals anxiously await lunch in a marine mammal care facility. (**b**) A California sea lion, the "seal" of seal shows.

rock on their chest, and hammer the shell of the prey against it until it cracks. Morsels of food are extracted with small, nimble fingers, and rolling over in the water cleans away the debris. A morning spent watching sea otters in their coastal habitat is a morning well spent!

ORDER SIRENIA The bulky, lethargic, small-brained dugongs and manatees, collectively called *sirenians* (*siricis* = a mermaid) (**Figure 13.27**), are the only herbivorous marine mammals. Like the cetaceans, they appear to have evolved

from the same ancestors as modern ungulates. They make their living grazing on sea grasses, marine algae, and estuarine plants in coastal temperate and tropical waters of North America, Asia, and Africa. Some species live in fresh water. The largest sirenians reach 4.5 meters (15 feet) in length and weigh 680 kilograms (1,500 pounds). They were first compared to mermaids by early Greeks who noted the manatee's habit of resting in an upright position in the water and holding a suckling calf to her breast. Sirenians have been hunted extensively—only about 10,000 individuals are thought to

Figure 13.26 Female sea otter and nearly mature offspring.

Figure 13.27 A manatee, or sea cow.

exist worldwide. Even though protected now, many are killed or wounded each year in Florida by the propellers of powerboats.

QUESTIONS FROM STUDENTS

1. Phytoplanktonic organisms account for a significant percentage of world primary productivity. Why are phytoplankton so successful? Is there something about the planktonic lifestyle that lends itself to high efficiency?

Phytoplankton are successful because of their size. Because water supports them, small marine autotrophs don't require the elaborate support systems of large terrestrial plants. Their small size allows easy diffusion of required nutrients into their single cells and prompt transport of wastes out; vessels and sap are not needed. Nearly all of their tissue is photosynthetic.

With transparent siliceous valves to protect them, diatoms are especially successful. The relatively recent evolution of diatoms was a high point in the development of marine autotrophs, and the additional oxygen these organisms contributed to the atmosphere has greatly influenced the success of life on land. More plants have led to more animals. After a few hundred million years of evolution, the planktonic lifestyle is elegantly tuned and interlocked into the complex and productive system we now observe in the world as a whole.

2. How do birds like the wandering albatross navigate across the trackless ocean?

No one is certain. Experiments done at Cornell University with homing pigeons suggest that homing birds use a combination of magnetic and optical cues to return to their starting points. Even polarized light and the positions of certain stars might be involved. However it works, the behavior is not learned but is instinctive to the bird. Study continues.

3. What's the difference between a dolphin fish and a dolphin mammal?

The name confusion began in Aristotle's time and has not abated. Dolphin mammals are small toothed whales; dolphin fish are teleosts (bony fish). The dolphin fish appears on restaurant menus and in seafood markets as mahi-mahi, dorado, and other names. Dolphin mammals are slaughtered in dismaying numbers in association with tuna fishing by some non-U.S. fishing fleets, but none of their meat appears on U.S. plates.

4. How is a porpoise different from a dolphin?

Dolphin and *porpoise* are common names of two subtly different groups of odontocetes. *Porpoise,* as a term, refers to the smaller members of the group, which have spade-shaped teeth, a triangular dorsal fin, and a smooth front end tapering to a point. *Dolphins* are usually larger and have an extended bottlelike jaw filled with sharp, round teeth. The small jumping whales in most oceanarium shows are dolphins, but killer whales are a species of large porpoise. To make matters even more complicated, the common dolphin seen in ocean-themed amusement parks, *Tursiops truncatus,* is often referred to as a porpoise, even by show announcers. This confusion between common names points out how useful scientific names can be: the real name of the animal, *Tursiops,* is clear and unambiguous.

5. If some seals and whales dive to great depths, why don't they get the bends?

The bends is a painful and occasionally fatal condition brought on by nitrogen gas—the most plentiful component of air—leaving solution and forming bubbles in the blood. The condition is analogous to what happens in a newly opened soda bottle. Human divers take a source of air with them to depth, breathe the air under pressure, and dissolve excess nitrogen gas in their blood. If they have been down long enough, or deep enough, the release of pressure upon surfacing will be like taking the cap off the soda bottle. Whales don't use scuba tanks—they have no source of supplemental air at depth. They "tank up" at the surface by oxygenating their blood and tissues, but they have no excess gas to bubble from the blood at the end of their dives.

CHAPTER SUMMARY

The organisms of the pelagic world drift or swim in the ocean. (Animals and plants associated with the bottom are known as benthic organisms.)

The organisms that drift in the ocean are known collectively as plankton. The plantlike organisms that comprise phytoplankton are responsible for most of the ocean's primary productivity. Phytoplankton—and zooplankton, the small drifting or weakly swimming animals that consume them—are usually the first links in oceanic food webs. Plankton are most common along the coasts, in the upper sunlit layers of the temperate zone, in areas of equatorial upwelling, and in the southern subpolar ocean. Marine scientists have been inspired by the beauty and variety of plankton since first observing them under the microscope in the nineteenth century.

Actively swimming animals comprise the nekton. Nektonic organisms include invertebrates (such as the squid, nautiluses, and shrimps) and vertebrates (such as fishes, reptiles, birds, and mammals). Each organism has a continuously evolving suite of adaptations that has brought it through the rigors of food finding, predator avoidance, salt balance, and temperature regulation—all of the challenges of the marine environment—time and time again.

TERMS AND CONCEPTS TO REMEMBER

Arthropoda	krill
baleen	Mammalia
Carnivora	meroplankton
cartilage	mollusk
cephalopod	Mysticeti
Cetacea	nanoplankton
Chondrichthyes	nekton
coccolithophore	Odontoceti
cryptic coloration	Osteichthyes
diatom	pelagic
dinoflagellate	phytoplankton
drag	Pinnipedia
echolocation	plankton
exoskeleton	plankton net
Fissipedia	salt gland
flagella	schooling
foraminiferan	Sirenia
frustule	swim bladder
gas exchange	Teleostei
gill membrane	valve
holoplankton	vertebrate
invertebrate	zooplankton

STUDY QUESTIONS

1. What are plankton? How are plankton collected? How are zooplankton different from phytoplankton?

2. Describe the most important phytoplankters. Which are most efficient in converting solar energy to energy in chemical bonds? By what means is this conversion achieved?

3. Where in the ocean is plankton productivity the greatest? Why?

4. Describe five nektonic organisms that are *not* fish.

5. What are the major categories of fishes? What problems have fishes overcome to be successful in the pelagic world?

6. How does a marine bird differ from a terrestrial bird—a pigeon, for example?

7. What characteristics are shared by all marine mammals?

8. Compare and contrast the groups of living marine mammals.

9. How are odontocete (toothed) whales different from mysticete (baleen) whales? Which are the better known and studied? Why?

10. Without looking ahead to Chapter 15, can you list some of the effects human activities have had on pelagic communities?

FOR FURTHER STUDY

Berta, A. 1994. "What Is a Whale?" *Science* 263 (14 January): 180–1. Recent summary of whale evolution.

Ellis, R. 1998. *The Search for the Giant Squid.* New York: Lyons Press. Information on squid in general, and large ones in particular.

Falkowski, P. G., and J. A. Raven. 1997. *Aquatic Photosynthesis.* Haldon, MA: Blackwell Science. An up-to-date review of aquatic photosynthesis. Superb, thorough, technical.

Heyning, J. E. 1995. *Whales, Dolphins, and Porpoises: Masters of the Ocean Realm.* Seattle: University of Washington Press. Excellent general introduction, not overly technical.

Kozloff, E. N. 1990. *Invertebrates.* Philadelphia: W. B. Saunders. Outstanding text, very well written by an eminent marine biologist.

Lagler, K. F., et al. 1977. *Ichthyology.* 2d ed. New York: Wiley. An excellent and readable standard text.

Lohmann, K. J. 1992. "How Sea Turtles Navigate." *Scientific American,* January, 100–6. The most important cues are wave direction and Earth's magnetic field.

Love, R. M. 1995. *Probably More Than You Want to Know About Fishes of the Pacific Coast.* 2d ed. Santa Barbara: Really Big Press. An entertaining and informative book written by an expert.

Marshall, J. 1998. "Why Are Reef Fish So Colorful?" *Scientific American Presents,* Fall, 54–7.

Milne, D. H. 1995. *Marine Life and the Sea.* Belmont, CA: Wadsworth. A comprehensive treatment of marine organisms and their ecological setting.

Minasian, S. S. 1984. *The World's Whales: The Complete Illustrated Guide.* Washington, DC: Smithsonian Books. The title is not hyperbole. Extraordinary photographs and clear text, taxonomically arranged.

Oceanus 35, no. 3 (Fall 1992) is dedicated to the subject of biological oceanography and contains a number of important papers on the primary producers.

Pough, F. H., et al. 1989. *Vertebrate Life.* 3d ed. New York: Macmillan. Excellent and up-to-date general text containing a thorough discussion of the evolution of vertebrates.

Reeves, R. R., S. Leatherwood, and B. Stewart. 1992. *The Sierra Club Handbook of Seals and Sirenians.* San Francisco: Sierra Club Books.

Science 281, no. 5374 (10 July 1998) is dedicated to the chemistry and biology of the ocean.

Smith, D. L. 1977. *A Guide to Marine Coastal Plankton and Marine Invertebrate Larvae.* Dubuque, IA: Kendall-Hunt. A laboratory guide to plankton.

Stevens, J. D., ed. 1987. *Sharks.* New York: Facts on File. Perhaps the best general reference on sharks available to the interested amateur.

Thewissen, J.G.M., ed. 1998. *The Emergence of Whales— Evolutionary Patterns in the Origin of Cetacea.* New York: Plenum. The latest thinking on a complex topic.

Welty, J. C. 1982. *The Life of Birds.* 3d ed. Philadelphia: W. B. Saunders. Standard text.

Whitehead, H. 1984. "The Unknown Giants: A Rare Look at Sperm and Blue Whales." *National Geographic,* December, 772–89. Remarkable photographs!

Wilson, E. O. 1992. *The Diversity of Life.* New York: Harvard University Press, Belknap Press. A distinguished biologist discusses the unity and diversity of life, and its history on Earth.

For additional readings, go to InfoTrac College Edition, your online research library at:

http://infotrac.thomsonlearning.com/

Benthic Communities 14

THE RESOURCEFUL HERMIT

Benthic marine organisms live on or in the ocean floor. There are an enormous variety of benthic creatures, but let's spend a moment with one of the more entertaining ones, the hermit crab. These small, pleasant relatives of edible crabs and lobsters have engaged the attention of generations of seaside visitors. The hermits scurry around sand-swept rocks or the floors of tidal pools, withdrawing into their borrowed shells at the slightest sign of danger. Their activity appears random, but their fighting, snooping, hiding, probing, and scuffling is purposeful. Like all animals, hermit crabs must struggle to eat, avoid predators, and mate. Hundreds of structural and behavioral adaptations contribute to their success. Sensors on the hermit's antennae and mouth parts alert it to the presence of food; good eyesight, muscular coordination, and a tough form-fitting covering usually foil fast-moving predators; and brilliant blue bands around the tips of its legs may signal its availability for mating.

One unique behavioral adaptation shared by all hermit crabs involves the selection of a temporary home. The front parts of a hermit crab—mostly pincers, antennae, and mouth parts—are formidable, but its hindquarters are delicate and subject to attack. To protect its flank, a hermit searches for any enclosed portable object to climb into, usually an unoccupied snail shell. A hermit crab's borrowed or stolen shell seems a source of both inordinate pride and perpetual difficulty. No shell is ever completely satisfactory. A hermit crab will carefully inspect any substitute dwelling, occupied or not, and consider whether to abandon its current digs in favor of the new candidate. House hopping proceeds fairly smoothly until the supply of suitably sized shells is exceeded by the number of potential occupants (as might happen when the crabs grow rapidly in times of abundant food). Then things get serious. Snails can be evicted from their self-made homes even before they're through with them—the hermit may get a house *and* a meal in a single transaction. Two crabs may fight for hours or even days over one shell. Two others might simultaneously occupy the opposite ends of an abandoned worm tube, thus

A hermit crab surveys its domain.

spending most of their time pulling each other in different directions. Sometimes two crabs swap shells at a moment's notice. Renter's remorse sets in almost immediately, and they're off to see if they can find something even more suitable. At times human observers can't resist laughing at the all-too-human goings-on. 14-1

KEY CONCEPTS

1. Benthic organisms spend nearly all their lives on or in the ocean floor. Their distribution is rarely random; clumped distribution is the most common.

2. Nearshore temperate habitats often include multicellular marine algae: seaweeds. Unlike true plants, the photosynthetically efficient algae lack vessels to conduct sap.

3. Highly productive salt marshes and estuaries shelter a great variety of benthic life forms and serve as nurseries for some pelagic organisms.

4. Though among Earth's most rigorous habitats, high-energy rocky and sandy shores are heavily populated by diverse organisms able to take advantage of the abundant nutrients found there.

5. Coral reef communities are exceptions to the general rule that tropical oceans are unproductive. Closely cycled nutrients and zooxanthellae in coral organisms make high productivity, and high species diversity, possible.

6. The deep-sea floor is the ocean's most uniform habitat. It is lightly populated by a relatively few species of highly specialized organisms.

7. Recently discovered deep-vent communities, near black smokers and atop cold seeps, depend on chemosynthesis rather than photosynthesis for energy.

CHAPTER AT A GLANCE

U nlike the drifting or actively swimming pelagic organisms discussed in the last chapter, **benthic** organisms live on or in the ocean bottom. A benthic (*benthos* = bottom) habitat may be shallow or deep, warm or cold, brimming with food and life or nearly sterile. Some benthic creatures spend their lives buried in sediment, others rarely touch the solid seabed; most attach to, crawl over, swim next to, or otherwise interact with the ocean bottom continuously throughout their lives.

The diversity of benthic habitats (and the diversity of organisms within them) is astonishing. Kelp forests are benthic communities, as are rocky intertidal zones, sand beaches, salt marshes, and the strangely populated areas around deep vents. Coral reefs are also benthic habitats with communities that may include a greater number of species than any other habitat on our planet.

THE DISTRIBUTION OF BENTHIC ORGANISMS 14-2

Before looking at individual benthic communities, we will investigate ways that organisms are distributed within a community. Individual benthic organisms are almost never distributed randomly throughout their habitats. A **random distribution** implies that the position of *one* organism in a bottom community in no way influences the position of *other* organisms in the same community. Further, a truly random distribution (such as that shown in **Figure 14.1a**) indicates that conditions are precisely the same throughout the habitat, an extremely unlikely situation except possibly in the flat and unvarying benthic communities of abyssal plains.

The most common pattern for distribution of benthic organisms is small patchy aggregations, or clumps. **Clumped distribution** (**Figure 14.1b**) occurs when conditions for growth are optimal in small areas because of physical protection (in cracks in an intertidal rock), nutrient concentration (near a dead body lying on the bottom), initial dispersal (near the position of a parent), or social interaction.

Uniform distribution with equal space between individuals (**Figure 14.1c**), such as the arrangement of trees we see in orchards, is the rarest natural pattern of all. The distribution of some garden eels throughout their territories becomes almost uniform because each eel can extend from its

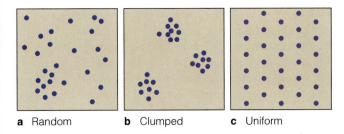

a Random   b Clumped   c Uniform

Figure 14.1 Random, clumped, and uniform population distribution patterns. The clumped pattern is most common in nature.

burrow just far enough to hassle neighbor eels spaced at an equal distance. There is eventually a break in the order of position—they don't line up row upon row like apple trees.

SEAWEEDS AND PLANTS 14-3

When benthic communities are mentioned, many people living near a temperate shore think first of tall, gently moving kelp forests. Although most of the ocean's primary productivity is generated by single-celled phytoplankton, between 2% and 10% is carried out by the large marine **multicellular algae** we informally call seaweeds.

Though photosynthetic, seaweeds are technically not plants. Structurally and biochemically they are different enough from vascular plants to be classified as protistans, a diverse taxonomic group of comparatively simple organisms that includes most phytoplankton. True marine plants include the sea grasses and mangroves discussed in **Box 14.1.**

Seaweeds occur in a great variety of sizes and shapes. The largest can reach 62 meters (205 feet) in length; the smallest appear as smears of cells on the surfaces of rocks. Some large algae form underwater forests, while others grow in isolation. Lifeless seaweeds drying on shore communicate none of their natural beauty and grace to the beachcomber, but to a diver the marine forest from which they came reveals extraordinary grace and form, discloses sheltering nurseries, conceals complex interrelationships, and even provides nourishment. None of these algae grow below the euphotic zone because all depend on photosynthesis to produce the energy-rich compounds necessary for life. Nearly 7,000 species of multicellular marine algae have been identified.

The Problems of Large Marine Autotrophs 14-4

At first glance marine algae and plants appear to have an easy life, but living near the ocean surface is not without hazards. Large intertidal autotrophs may find themselves exposed to the drying effects of air and sunlight when the tide is out and may be lashed against the rocks by waves when the tide is in. The physical nature of the organisms themselves provides some defense against these difficulties—their bodies are flexible, easily able to absorb shock, resistant to abrasion, streamlined to reduce water drag, and very strong.

Warmth and a lack of nutrients often limit the success of seaweeds. Higher temperatures lead to higher metabolic rates, and the oxygen level in warm seawater may not be high enough to support the respiratory needs of the algae at night. Warmer temperatures can also shatter the delicate accessory pigments and proteins required for photosynthesis and respiration. Large algae are rare in warm nutrient-poor waters, and divers visiting the tropics are usually surprised to find no sign of kelp forests. In contrast, chilly temperate and subpolar zones of nutrient upwelling often support thick algal mats and dense marine forests.

Yet another difficulty is the location of adequate anchorage or substrate. Attached seaweeds require a firm footing. A sandy or muddy bottom is unsuitable for colonization by most large algae, and less than 2% of the ocean floor is shallow enough and solid enough to permit their growth.

The marine lifestyle also offers some advantages. Marine algae and plants suffer no droughts and nearly always have enough carbon dioxide for photosynthesis. Assuming suitable nutrient levels and a good foothold, only sunlight is required for productivity. Being submerged in seawater brings the additional advantage of lightweight construction. A seaweed doesn't require strong support structures because it has nearly the same density as the surrounding seawater. More of its bulk can thus be dedicated to photosynthesis. Indeed, productivity in some seaweed beds may be the highest of *any* autotrophic community on Earth.

Structure of Seaweeds 14-5

Common terms like *leaf, stem,* and *root* are inappropriate for seaweeds because the definitions of those parts assume the presence of vessels. The structures that superficially resemble leaves are called **blades** (or fronds), the stemlike structures are termed **stipes,** and the root-shaped jumble at the base is appropriately named a **holdfast. Gas bladders** assist many species in reaching strongly illuminated surface water. Blades, stipes, and holdfast comprise the body of the organism, the **thallus.** These parts are labeled in **Figure 14.2b.**

An algal thallus may be large or small, branching or tufted, in sheet form or filamentous, encrusting or elongated, rounded or pointed—variety of form within the algae is tremendous. Algal blades are symmetrically equipped with photosynthesizing tissue, absorb gases across their entire surface, and even participate in reproduction. Stipes are strong, photosynthesizing, shock-absorbing links tying the blades (or sheets or filaments) into a unit. The holdfast does not take up water and nutrients from the substrate like the vascular root it superficially resembles, but it does anchor the plant in place and may provide incidental shelter for a rich variety of animal life.

Classification of Seaweeds 14-6

Seaweeds are classified in part by the presence of colored compounds in their tissues. These **accessory pigments** (or masking pigments) are light-absorbing compounds closely associated with chlorophyll molecules. They don't resemble chlorophyll chemically, but they bind loosely with it. Their presence in plants greatly enhances photosynthesis because they absorb the dim blue light at depth and transfer its energy to the adjacent chlorophyll molecules. Accessory pigments may be brown, tan, olive green, or red; they are what give most marine autotrophs, especially seaweeds, their characteristic color.

Multicellular marine algae are segregated into three divisions based on their observable color. The green algae, with their unmasked chlorophyll, are the Chlorophyta (*chloros* =

Seaweeds may look like plants, but they are actually a form of *multicellular algae.* The single-celled diatoms and dinoflagellates discussed in the last chapter are *unicellular algae. Algae* lack the vessels and other structural and chemical features of true plants. Most large land plants are vascular plants. On land, plumbing and sap are needed to transport water and nutrients from the ground to the leaves, where light and carbon dioxide are available for photosynthesis. Although this is generally unnecessary in the aquatic environment, a few species of vascular plants have recolonized the ocean. All have descended from land ancestors, and all live in shallow coastal water. The most conspicuous marine vascular plants are the sea grasses and the mangroves.

Sea Grasses

These plants are not true grasses, but they do have leaves and stems, as well as roots capable of extracting nutrients from the substrate. Extensive stands of sea grasses are found on the coasts of North America, on the Atlantic coast of Europe, in Eastern Asia, in temperate Australia, and in South Africa. They form broad gray or green submerged meadows, which support extraordinarily rich communities. About 45 species of sea grasses are known.

The most common sea grass is eelgrass, genus *Zostera,* a common inhabitant of the muddy shallows of calm bays and estuaries of the Atlantic and Pacific coasts of the United States. Similar habitats along the Gulf coast harbor stands of turtle grass (genus *Thalassia*) and manatee grass (genus *Syringodium*), angiosperms named after the animals that once shared their habitat and (in the case of the manatee) fed on them.

Perhaps the most beautiful sea grass is the vivid, emerald-green surf grass, genus *Phyllospadix,* with its seasonal flowers and fuzzy fruit (**Figure a**). These hardy plants survive in the turbulent, wave-swept intertidal and subtidal zones of temperate East Asia and western North America.

Mangroves

Low, muddy coasts in tropical and some subtropical areas are often home to tangled masses of trees known as **mangroves.** These large, flowering plants are never completely submerged, but because of their intimate association with the ocean we consider them to be marine plants (**Figure b**). They thrive in the sediment-rich lagoons, bays, and estuaries of the Indo-Pacific, tropical Africa, and the tropical Americas. Their distribution, shown in Figure 14.3, depends on temperature, currents, and rainfall.

The sediment in which mangrove trees live must be covered with brackish or salt water for part or all of the day. Many mangroves avoid taking up salt ions from seawater, or they selectively remove salt from sap with salt-excreting cells. The fine coastal mud they colonize doesn't provide firm footing for these substantial plants, so an intricate network of arching prop roots is required for support. The strutlike prop roots are supplemented by many smaller roots equipped with breathing pores and air passages. Atmospheric oxygen is conducted by these passages to the portions of the plant submerged in oxygen-deficient mud.

The root system also traps and holds sediments around the plant by interfering with the transport of suspended particles by currents. Mangrove forests thus assist in the stabilization and expansion of deltas and other coastal wetlands. The root complex also forms an impenetrable barrier and safe haven for organisms around the base of the trees.

The mangrove communities of south Florida, among the world's most widespread, consist primarily of the red mangrove (genus *Rhizophora*). The spreading leaves of the tree protect the residents from the tropical sun, and the tangled roots keep out large predators and also harbor huge numbers of fiddler crabs, worms, marine and terrestrial snails, fish, oysters, and red algae. Birds and insects inhabit the treetops, their droppings enriching the sediments below. **14-7**

b A mangrove (*Rhizophora*) growing in salt water in Everglades National Park, Florida. The tangled roots descending from the main branch are called prop roots or stilt roots. They provide anchorage for the mangrove, trap sediment, and provide protection for small organisms.

a The bright green sea grass *Phyllospadix* in a tide pool.

a

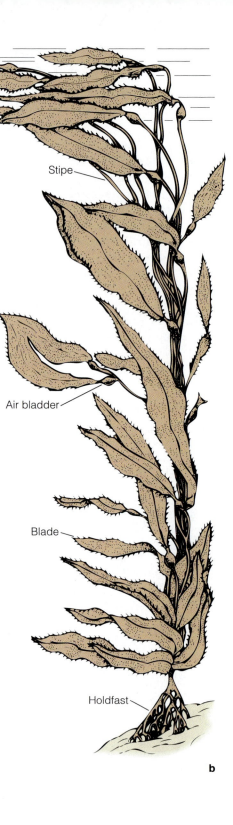

Stipe

Air bladder

Blade

Holdfast

b

Figure 14.2 (**a**) *Macrocystis,* a brown alga (phaeophyte), showing a close-up view of the blades with gas bladders at the base of each. This fast-growing, productive alga is one of the dominant species found in the spectacular kelp forest of western North America. (**b**) The thallus (body) of a typical multicellular alga. These organisms can grow at a rate of 50 centimeters (20 inches) per day, and reach a length of 40 meters (132 feet).

green + *phyton* = plant), the brown algae **Phaeophyta** (*phae* = tan, dusky), and the red algae **Rhodophyta** (*rhodon* = rosy red). Phaeophytes are most familiar to beachcombers, and rhodophytes the most numerous.

THE PHAEOPHYTES   Nearly all of the 1,500 living species of phaeophytes are marine. Some species of these largest of

algae, which include the **kelps,** can reach lengths of 40 meters (132 feet)—the record length exceeds 60 meters (200 feet). To attain these dimensions, the plant can grow at the spectacular rate of 50 centimeters (20 inches) per day! Some brown algae are annuals; others live for up to seven years. The tan or brown color of phaeophytes comes from an accessory pigment, which permits photosynthesis to proceed at greater

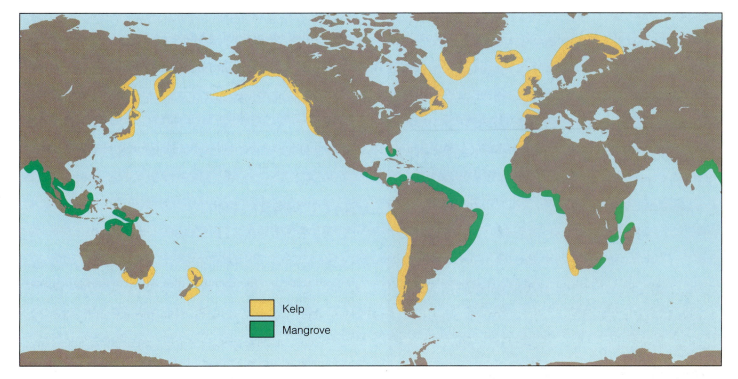

Figure 14.3 The distribution of kelp beds and mangrove communities. (Mangroves are discussed in Box 14.1.)

depths than is possible for the unmasked chlorophytes. In ideal circumstances, some larger brown algae can grow in water up to about 35 meters (115 feet) deep.

The Pacific's giant kelp forests, the world's largest, consist mostly of the magnificent genus *Macrocystis* (*macro* = large + *cyst* = bladder). The size and shape of the plant and its dense canopy are shown in **Figure 14.2.** Most brown algae live in temperate and polar habitats poleward of the 30° latitude lines, but a few live in the tropics. The worldwide distribution of kelp is shown in **Figure 14.3.**

THE RHODOPHYTES   Most of the world's seaweeds are red algae; there are more rhodophyte species than all other major groups of algae combined. Rhodophytes tend to be smaller and more complex than phaeophytes. Rhodophytes excel in dim light because of their sophisticated accessory pigments. These compounds absorb and transfer enough light energy to power photosynthetic activity at depths where human eyes cannot see light.

Paradoxically, many red algae also live on rocks right at the water's surface. The dark accessory pigments are believed to shade the productive machinery from brilliant light. Coralline, or calcareous, algae look like a brilliant ceramic glaze, a bit of melting raspberry sherbet, or a knobby plastic membrane thrown carelessly over the surface of the water.

One group of shallow rhodophytes is very important in the life of coral reefs. Like coral animals, encrusting coralline algae of genus *Lithothamnium* (**Figure 14.4**) also remove

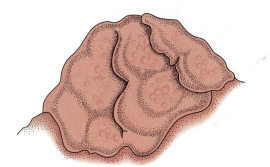

Figure 14.4 *Lithothamnium* are encrusting species of red algae (rhodophytes). Similar species deposit calcium and help to cement reefs into a mass capable of resisting heavy surf. The encrustation seen here is about 1.5 centimeters (¾ inch) across.

large quantities of dissolved calcium carbonate from seawater and deposit it within their tissues. This activity helps cement the reef into a mass capable of resisting heavy surf. Some reefs in the Indian Ocean are not, as was once thought, of coral origin, but instead were formed almost exclusively by the activity of coralline algae.

Seaweed Communities   14-8

The shelter and high productivity of a kelp forest can help provide a near-ideal environment for benthic animals. When light and nutrient conditions are optimal, large algae can make carbohydrate molecules so rapidly and in such quantity that

Figure 14.5 Sea urchins in a kelp bed. The urchins can absorb carbohydrates that leak from the algae, or gnaw the stipes and holdfasts with teeth.

sugars leak from their tissues like tea from a tea bag. Resident animals like sea urchins (**Figure 14.5**) are able to grow rapidly by collecting these molecules on their surfaces and transporting them directly into their bodies. As the algae weaken with age and productivity declines, the urchins' sharp teeth can gnaw at the thalluses. Other animals graze on the blades, nestle within the holdfasts, and consume kelp flakes and debris. Sea otters may eventually move in to eat the urchins. As with all other communities, the kelp forest changes as its inhabitants come and go and as time passes.

SALT MARSHES AND ESTUARIES

14-9

Muddy-bottomed salt marshes are among the most interesting and active benthic communities. Much of the high primary productivity of a salt marsh comes from sea grasses, mangroves, and other vascular plants that can prosper in a marine (or partly marine) environment.

As you may recall from Chapter 11, salt marshes often form in an **estuary,** a broad, shallow river mouth where fresh water and salt water mix (see Figures 11.24–26). Wave shock is usually reduced in estuaries—surf is blocked from estuaries by longshore bars or by twisting passages connecting to the ocean. The salinity of water within an estuary may vary with tidal fluctuations from seawater through brackish water (mixed salt and fresh water) to fresh water. Many of the organisms living in estuaries are able to tolerate varying salinities, but in areas near the river entrance the water may be almost fresh and near the outlet it may be of oceanic salinity. These different salinities often lead to distinct horizontal zonation of organisms. Temperature range is also potentially extreme, especially in the tropics or during the temperate zone summer when a receding tide abandons residents to the heat of the sun. Strong currents may move in estuaries as the tide rises and falls and the river flows.

Estuarine marshes (such as the one shown in **Figure 14.6**) are richer and exhibit greater species diversity than marshes

Figure 14.6 An estuarine marsh. Urban developers often destroy coastal marshes to build marinas and homes, but citizens near this marsh in Newport Beach, California, have recognized its natural value and have set it aside as a marine preserve.

Figure 14.7 Snails in an estuary search the surface mud for food.

exposed only to seawater. Primary productivity in estuaries is often extraordinarily high because of the availability of nutrients, the great variety of organisms present, strong sunlight, and the large number of niches. The mass of living matter per unit area in a typical estuary is among the highest per unit of surface area of any marine community.

Estuarine organisms show unique adaptations to their rich and variable environment. Some estuarine plants trap fine silt particles at their roots, thus countering the erosive action of current flow. Small plants are often filamentous, bristling with tiny projections anchoring them to the substratum. Larger plants have extensive root systems to hold themselves in place and to colonize new areas. Most of the resident animals burrow into the muck, scurry rapidly across the surface, or hide in the vegetation. Clams and snails work their way through the substratum, obtaining food and shelter at the same time (**Figure 14.7**). Polychaete worms dig for targets of opportunity, and crabs dart for any interesting morsels. Since planktonic larvae would be washed out to sea, most estuarine organisms produce nonplanktonic larvae, lay eggs on firm objects, or carry eggs on their bodies.

Estuaries are sometimes called marine nurseries because so many juvenile organisms are found there. This is especially true for fish. Many pelagic species spend their larval lives in the protective confines of an estuary taking advantage of the many feeding opportunities available. Most of the commercially exploited fish species on the Atlantic coast of the United States utilize estuaries as juvenile feeding grounds. The human pressures of development and pollution are thus doubly stressful in estuaries, affecting both permanent residents and the sensitive larval stages of open-water animals.

ROCKY INTERTIDAL COMMUNITIES

 14-10

Anyone who spends time at the shore, especially a rocky shore, is soon struck by a curious contradiction. Although the rocky shore looks like a very difficult place for organisms to make a living, the rocky **intertidal zone**—the band between the high-

est high-tide and lowest low-tide marks on a rocky shore—is one of Earth's most densely populated areas. Hundreds of species and individuals crowd this junction of land and sea.

The problems of living in this zone are formidable. The tide rises and falls, alternately drenching and drying out the animals and plants. **Wave shock,** the powerful force of crashing waves, tears at the structures and underpinnings of the residents. Temperature can change rapidly as cold water hits warm shells, or as the sun shines directly on exposed organisms. In high latitudes ice grinds against the shoreline, and in the tropics intense sunlight bakes the rocks. Predators and grazers from the ocean visit the area at high tide, and those from land have access at low tide. Freshwater runoff can osmotically shock the occupants during storms. Annual movement of sediment onshore and offshore can cover and uncover habitats. Yet, astonishingly, the richness, productivity, and diversity of the intertidal rocky community—especially in the world's temperate zones—is matched by very few other places. There is intense competition for space. Life abounds.

One reason for the great diversity and success of organisms in the rocky intertidal zone is the large quantity of food available. The junction between land and ocean is a natural sink for living and once-living material. The crashing of surf and strong tidal currents keep nutrients stirred and ensure a high concentration of dissolved gases to support a rich population of autotrophs. Minerals dissolved in water running off the land serve as nutrients for the inhabitants of the intertidal zone as well as for plankton in the area. Many of the larval forms and adult organisms of the intertidal community depend on plankton as their primary food source.

Another reason for the success of organisms here is the large number of habitats and niches available for occupation (see **Figure 14.8**). The habitats of intertidal animals and plants vary from hot, high, salty splash pools to cool, dark crevices. These spaces provide hiding places, quiet places to rest, attachment sites, jumping-off spots, cracks from which to peer to obtain a surprise meal, footing from which to launch a sneak attack, secluded mating nooks, or darkness to shield a retreat. The niches of the creatures in this community are varied and numerous—encrusting algae produce carbohydrates, snails scrape algae from rocks, hermit crabs scavenge for tidbits, and octopuses wait to surprise likely meals. Sea stars pry open mussels, and barnacles sweep bits of food from the water.

The most obvious and important physical factor in intertidal communities is the rise and fall of the tides (see Chapter 10). Organisms living between the high-tide and low-tide marks experience very different conditions from those residing below the low-tide line. Within the intertidal zone itself, organisms are exposed to varying amounts of emergence and submergence. For example, **Figure 14.9a** plots the number of hours of exposure to air in a California intertidal zone through six months' time versus the tidal height. Because some organisms can tolerate many hours of exposure while others are able to tolerate only a very few hours per week or month, the animals and plants sort themselves into three or four horizontal bands, or subzones, within the intertidal zone. Each distinct zone is an aggregation of animals and plants best adapted to

a

Figure 14.8 A Pacific coast tide pool and intertidal shore. (**a**) A diagrammatic view. (**b**) Key.

b

1 bushy red algae, *Endocladia*
2 sea lettuce, green algae, *Ulva*
3 rockweed, brown algae, *Fucus*
4 iridescent red algae, *Iridea*
5 encrusting green algae, *Codium*
6 bladderlike red algae, *Halosaccion*
7 kelp, brown algae, *Laminaria*
8 Western gull, *Larus*
9 intrepid marine biologist, *Homo*
10 California mussels, *Mytilus*
11 acorn barnacles, *Balanus*
12 red barnacles, *Tetraclita*
13 goose barnacles, *Pollicipes*
14 fixed snails, *Aletes*
15 periwinkles, *Littorina*
16 black turban snails, *Tegula*
17 lined chiton, *Tonicella*
18 shield limpets, *Collisella pelta*
19 ribbed limpet, *Collisella scabra*
20 volcano shell limpet, *Fissurella*
21 black abalone, *Haliotis*
22 nudibranch, *Diaulula*
23 solitary coral, *Balanophyllia*
24 giant green anemones, *Anthopleura*
25 coralline algae, *Corallina*
26 red encrusting sponges, *Plocamia*
27 brittle star, *Amphiodia*
28 commmon starfish, *Pisaster*
29 purple sea urchins, *Strongylocentrotus*
30 purple shore crab, *Hemigrapsus*
31 isopod or pil bug, *Ligia*
32 transparent shrimp, *Spirontocaris*
33 hermit crab, *Pagurus,* in turban snail shell
34 tide pool sculpin, *Clinocottus*

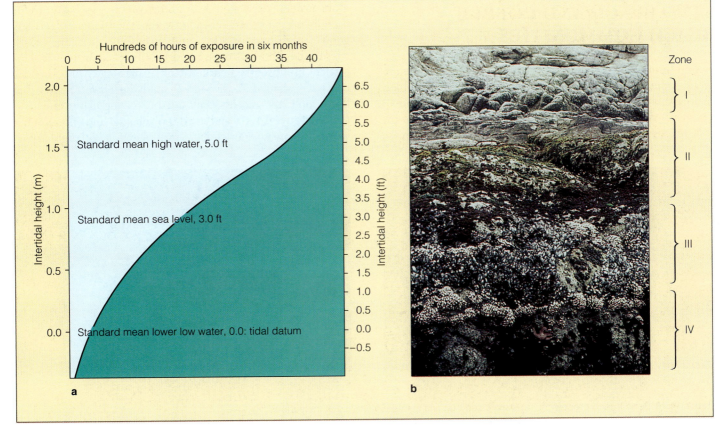

Figure 14.9 The relationship between amount of exposure and vertical zonation in a rocky intertidal community. (**a**) A graph showing intertidal height versus hours of exposure. The 0.0 point on the graph, the tidal datum, is the height of mean lower low water. (**b**) Vertical zonation, showing four distinct zones. The uppermost zone (I) is darkened by lichens and cyanobacteria; the middle zone (II) is dominated by a dark band of the red alga *Endocladia*; the low zone (III) contains mussels and gooseneck barnacles; and the bottom zone (IV) is home to sea stars (*Pisaster*) and anemones (*Anthopleura*). The bands in the photograph correspond approximately to the heights shown in the graph.

the conditions within that particular narrow habitat. The zones are often strikingly different in appearance, even to a person unfamiliar with shoreline characteristics. This zonation is clearly evident in the rocky shore of **Figure 14.9b.**

For intertidal areas exposed to the open sea, wave shock is a challenging physical factor. **Motile** (*motus* = moving) animals, like crabs, move to protective overhangs and crevices where they cower during intense wave activity. Attached, or **sessile** (*sessilis* = sitting), animals hang on tightly, often gaining assistance from rounded or very low profile shells, which deflect the violent forces of rushing water around their bodies. Some sessile animals have a flexible foot that wedges into small cracks to provide a good hold; others, like mussels, form shock-absorbing cables that attach to something solid.

Desiccation (drying) by exposure to air and sunlight is another source of intertidal stress. Again, motile organisms have an advantage because they can move toward water left in tidal pools or muddy depressions by the retreating ocean. Producers and consumers must await the water's return, huddled in low spots, moist pockets, or cracks in the rocks, or within tightly closed shells. Water trapped within a shell can keep gills moist for the needed exchange of gases. A protective mucous coating can retard evaporative water loss from exposed soft animal body parts or blades of seaweed.

SAND BEACH AND COBBLE BEACH COMMUNITIES   14-11

Not all intertidal areas are composed of firm rock; some are sandy, some are muddy, and others consist of gravel or cobbles. (A few shores combine these elements within a small area.) The usual rigors of the intertidal zone are intensified for organisms surviving on loose substrates. Indeed, it may surprise you to learn that in spite of its generally benign conditions, the ocean contains what may well be the most hostile, rigorous, and dangerous environments for small living things on Earth: high-energy sand and cobble beaches.

As environments go, sand beaches don't seem particularly nasty places to us humans; most of my students consider the beach to be about the finest habitat around. Seals and sea lions spend a lot of time at the beach and seem to enjoy the experience as much as people do. In short, for organisms of about our size, the problems of living on a beach are manageable.

For smaller organisms, however, a beach is a forbidding place. Sand itself is the key problem. Many sand grains have sharp, pointed edges, so rushing water turns the beach surface into a blizzard of abrasive particles. Jagged grit works its way into soft tissues and wears away protective shells. A small organism's only real protection is to burrow below the surface, but burrowing is difficult without a firm footing. When the grain size of the beach is small, capillary forces can pin down small animals and prevent them from moving at all. If these organisms are trapped near the sand surface, they may be exposed to predation, to overheating or freezing, to osmotic shock from rain, or to crushing as heavy animals walk or slide on the beach.

As if this weren't enough, those that survive must contend with the difficulty of separating food from swirling sand and the dangers of leaving telltale signs of their position for predators or being excavated by crashing waves. A few can run for their lives—some larger beach-dwelling crabs depend on their good eyesight and sprinting ability to outrace onrushing waves.

To these horrors must be added the usual problems of intertidal life discussed earlier. Not surprisingly, very few species have adapted to wave-swept sandy beaches! The few that have done so—mostly small fast-burrowing clams and sand crabs (**Figure 14.10**) and sturdy polychaetes and other minute worms—consume a rich harvest of plankton and organic par-

a   b

Figure 14.10 Sand beach organisms. (**a**) Dime-sized coquina clams (*Donax*) lie at the surface awaiting a ride up the beach on an incoming wave. They will bury themselves in the loose sediment, push up their siphons, and filter the water for food. When the tide retreats, they will again pop to the surface and allow the waves to take them back down the beach. (**b**) A sand crab (*Emerita*), beloved of all beach-going children and beginning lab students, attempts to bury itself in anticipation of an onrushing wave. It gleans food from passing water with its feathery antennae.

ticles washed onto the beach and filtered from the water by the uppermost layer of sand.

Cobble beaches are even more uninviting (and they're murder on bare feet). The rounded rocks clack and bump together as waves pound the shore; most small animals are crushed. Except for nimble insectlike "beach hoppers" and a few species of scavenging terrestrial insects, most loose rock-strewn shores are understandably sterile of anything much larger than microscopic organisms.

CORAL REEF COMMUNITIES

14-12

The tropical ocean is blue, brilliantly transparent, relatively high in salinity, and notable for its deep and abrupt thermocline and relatively low surface concentrations of dissolved nutrients and gases. It supports surprisingly little life.

But what of the travel-poster view of the tropical ocean? What about the thousands of brightly colored fish, strange invertebrates, and breathtaking scenes of divers swimming through living reef formations? Such scenes, photographed on reefs at the edges of islands and continents, can be found in less than 2% of the tropical ocean. The key to the difference between the open tropical ocean and the tropical reefs lies in the productivity of the **reefs** themselves, wave-resistant structures dominated by strong and rigid masses of living (or once-living) organisms. Not all reefs are built of coral—other reef builders include red and green algae, cyanobacteria, worms, even oysters—but we think first of coral reefs when the words *reef* and *tropics* are mentioned together.

Coral

14-13

Although they look like flowers, **corals** are related to sea anemones and jellyfish. Some corals are solitary animals with bodies up to 30 centimeters (12 inches) in diameter, but most of the more than 500 species are ant-sized organisms crowded into colonies called **coral reefs.** The coral animals themselves construct the reefs by secreting hard skeletons of a crystalline form of calcium carbonate. The matrix of cup-shaped individual skeletons secreted by coral animals gives the colony its characteristic shape.

An individual coral animal, or **polyp** (**Figure 14.11**), feeds by capturing and eating plankton that drift within reach of its tentacles. Victims are entrapped by stinging cells on each tentacle, transported to a central gastric cavity, and rapidly digested. Tropical corals feed at night; at dawn the polyps retract into their skeletal cups to withstand drying should the colony be exposed to air at low tide. The anatomy of a coral polyp is shown in **Figure 14.12.**

Tropical reef-building corals are **hermatypic** (*hermatos* = mound builder). Their bodies contain masses of tiny symbiotic dinoflagellates called **zooxanthellae** (*xanthos* = yellow). Coral's success in the nutrient-poor water of the tropics depends on its intimate biological partnership with zooxanthellae. The microscopic zooxanthellae carry on photosynthesis,

Figure 14.11 Close-up of hermatypic coral, showing expanded polyps.

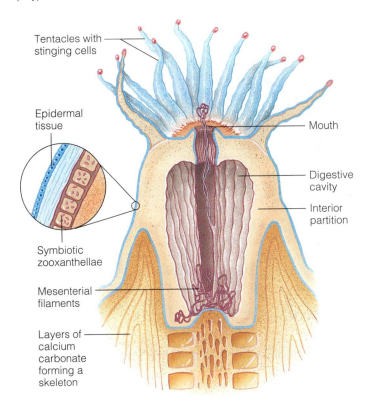

Figure 14.12 Anatomy of a coral polyp, with an enlargement showing a cross section of the outer covering and tissue.

a

Figure 14.14 (**a**) The coral reef habitat. (**b**) Key.

1 black-capped petrel
2 sea nettle
3 angelfish
4 lobed corals
5 sea whips and soft corals
6 triggerfish
7 sea fans
8 tube anemone
9 orange stone coral
10 bryozoans
11 brain coral
12 butterfly fish
13 moray eel
14 cleaner fish
15 tube corals

16 muricid snail
17 nudibranch
18 sponges
19 colonial tunicate
20 giant clam
21 purple pseudochromid fish
22 cobalt sea star
23 soft corals
24 barber pole shrimp
25 sea anemones
26 clown fish
27 worm tubes
28 cowry
29 sea fan

b

may live to be hundreds of years old. Two deep benthic representatives are seen in **Figure 14.18.**

The organisms within deep pelagic and benthic communities share some curious adaptations. Gigantism is a common characteristic—individuals of representative families in deep water often tend to be much larger than related individuals in the shallow ocean. Fragility is also common in the depths. Not only are heavy support structures unnecessary in the calm deep environment, but the low water pH and deficiency of dissolved calcium discourage skeletal development. Some animals have slender legs or stalks to raise them above the sediment, and some come apart like warm gelatin at the slightest touch. Except for its influence on enzyme activity, hydrostatic pressure is not a problem for these animals. Because they lack gas-filled internal spaces, their internal pressure is precisely the same as that outside their bodies.

VENT COMMUNITIES

 14-17

The oceanographic world was excited in 1977 when scientists in Woods Hole Oceanographic Institution's submersible *Alvin* (Figure b, Box 4.1; Appendix V) discovered an entirely new type of marine community more than 3,000 meters (10,000 feet) below the surface. They were searching the near-freezing bottom for the source of some unusually warm water, which had been detected by remote probes. What they found were jets of superheated water (to 350°C, 650°F) blasting from rift vents in the young oceanic ridge 350 kilometers

(220 miles) north and east of the Galápagos Islands. Clustered around the vents were dense aggregations of large, previously unknown animals. Bottom water in the area was laden with hydrogen sulfide, carbon dioxide, and oxygen, upon which specialized bacteria were found to live. These bacteria form the base of a food chain that extends to the unique animals. Large crabs, clams, sea anemones, shrimp, and unusual worms were found in this warm oasis.

Some of the "tube worms," contained in their own long parchmentlike tubes, measured 3 to 4 meters (10 to 13 feet) in length and were the diameter of a human arm. These strange animals are pogonophorans, members of a small phylum of invertebrates also found in fairly shallow water. Three species of the appropriately named genus *Riftia* (**Figure 14.19a**) have been identified so far. The tubes of these pogonophorans are flexible and capable of housing the length of the animal when it retracts. The animals extend tufts of tentacles from the openings of their tubes. Feeding was something of a puzzle because these animals have no mouth, digestive tract, or anus. The trunks of the worms were found to contain large "feeding bodies" tightly packed with bacteria similar to those seen in the water and on the bottom near the geothermal vents. The worms' tentacles absorb hydrogen sulfide from the water and transport it to the bacteria, which then use the hydrogen sulfide as an energy source to convert carbon dioxide to organic molecules. The ultimate source of the worms' energy (and the energy of most other residents in this community) is this energy-binding process, called chemosynthesis, which replaces photosynthesis in the world of darkness. Puzzle solved.

Figure 14.15 The three types of coral reefs. (**a**) The structure of a fringing reef like that on Moorea, in the Society Islands. (**b**) The structure of a barrier reef, an example of which is Bora Bora, also in the Society Islands. (**c**) An atoll, such as Aratika, in the Tuamotu Archipelago.

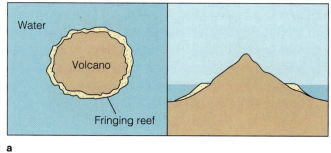

a

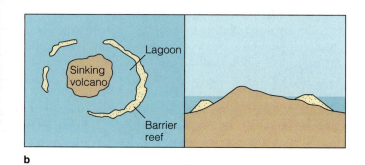

b

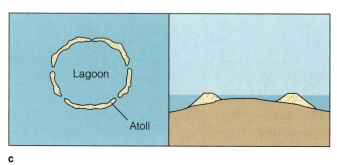

c

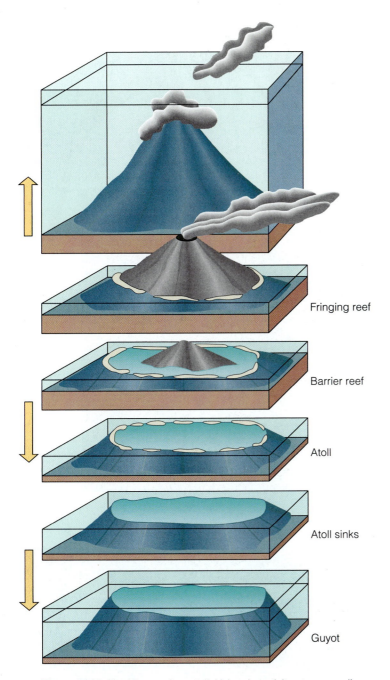

Figure 14.16 The history of an atoll. Volcanic activity at a spreading center builds the island, which acquires a fringing reef. As the island moves away from the spreading center, the volcano becomes inactive, the island slowly subsides, and the coral animals continue to build, forming first a barrier reef and then an atoll. If the subsidence rate increases above about 1 centimeter (½ inch) a year, the coral dies and the atoll becomes a guyot.

Labels on figure: Fringing reef / Barrier reef / Atoll / Atoll sinks / Guyot

Figure 14.17 A tripod fish, an abyssal benthic species. The long, curved projections are thought to aid in sensing the distant vibrations of prospective prey.

The clams and shrimp of the vent communities are equally unusual. For example, the large white clam *Calyptogena* grows among uneven basaltic mounds (**Figure 14.19b**). Each the size of a shoe, the clams shelter the same kinds of bacteria as *Riftia*. Though the clam retains its filter-feeding structures, it too derives nutrition from the bacteria. Small shrimp discovered at the vents in 1985 have been found to possess special organs that may allow them to see heat from the vents. Such an adaptation would permit them to range away from the vents for food, yet return to the warmth and richness of the home community.

Similar vent communities have now been found off Florida, California, and Oregon, and in several other locations along oceanic ridges. Cold seeps—and their attendant

a

b

c

Figure 14.18 Abyssal benthic animals. (**a**) *Oneirophanta,* a 10-centimeter (4-inch) holothuran found on the abyssal plains of the North Atlantic. (**b**) *Apseudes galatheae,* a blind, thumb-sized crustacean found in the Kermadec Trench, north of New Zealand. (**c**) *Oneirophanta*—as in (a)—and brittle stars (foreground) search for food on a continental slope at a depth of about 1,000 meters (3,300 feet). Brittle stars like these are among the world's most cosmopolitan organisms, found on nearly all deep sediments.

a

b

Figure 14.19 Some large organisms of hydrothermal vents. (**a**) *Riftia,* large tube worms (pogonophorans) that contain masses of chemosynthetic bacteria in special interior pouches. (**b**) A vent field dominated by the giant white clam *Calyptogena magnifica.* Each clam is about the size of a man's shoe, and contains chemosynthetic bacteria within its gill filaments.

communities—have also been located, and these areas usually support a larger number of species than the hot-vent areas. Could vent communities occupy the active central rift valleys of a significant percentage of the 65,000 kilometers (41,000 miles) of Earth's oceanic ridges? Perhaps the deep-vent communities will prove to be more important in marine biology than has been previously supposed. Marine biologists are eager to continue their explorations.

QUESTIONS FROM STUDENTS

1. If the rocky, sandy, or muddy intertidal zones represent such a challenging mix of environmental factors, why do so many organisms live there?

Difficulty in biology is a relative term. It may seem a circular argument, but wherever organisms live, conditions for life at that place are biologically tolerable, food is available, and environmental conditions are not so extreme as to preclude success. Organisms live in abundance where energy is available. Where there is food, or sunlight, or biodegradable compounds, there is life. Natural selection has sorted out the ways that work in this zone from the ways that do not, and the adaptations that work give the organisms living in the intertidal zone's many niches access to a rich harvest of nutrients.

2. Could there be any huge undiscovered Godzilla-type sea monsters in the deep ocean?

Probably not, unless they can extract energy directly from water molecules! The deep pelagic feeding situation is simply not rich enough to support the energy needs of an active population of violent, aggressive, city-eating (metrophagous?) reptiles. Scientists never say never, but classic science fiction films aside, it doesn't look promising.

3. Does the great diversity of marine species seen at the surface of some tropical ocean areas occur in the deep sea as well?

No. The cold, unchanging regions of the deep ocean are populated by the same kinds of specialized organisms all over the world. Down there, below the pycnocline, water is cold, food is scarce, and only a few species have adapted. A deep-bottom sample off Tahiti would yield the same sorts of brittle stars, sea cucumbers, and unusual cnidarians as a sample from similar depth in the Arctic Ocean. Conditions—and species—below about 3,500 meters (12,000 feet) are similar anywhere in the world ocean.

4. Can zooxanthellae live outside of a coral animal?

In laboratory cultures, yes. They change from the spherical shape seen within the coral to the typical biflagellate form characteristic of motile dinoflagellates. Researchers are uncertain whether all corals are host to the same species of zooxanthellae or whether several species exist. As far as we know, they do not normally live free in the ocean.

CHAPTER SUMMARY

Benthic organisms live on or in the ocean bottom. They may be distributed through their habitats randomly, uniformly, or (most commonly) in clumped distributions.

Nearshore benthic habitats in the temperate zones often contain multicellular algae, large nonvascular plants known as seaweeds. Carbohydrates produced by these highly productive large algae (and the vascular plants) can provide much of the energy needed by animals of the benthic communities.

Salt marshes and estuaries are among the ocean's most productive habitats, and estuaries shelter a remarkable variety of juvenile forms, some of which are refugees from the forbidding and competitive open-ocean environment. Rocky intertidal communities are among the ocean's richest and most diverse. Although the problems of rocky shore living are formidable, hundreds of organisms have adapted to its rigors because of the wealth of food available there. Sand and cobble beaches may seem more benign, but the difficulty of maintaining a dependable foothold and separating food from inedible particles severely limits the number of organisms able to live there. Except for vent communities associated with mid-ocean ridges, the deep seabed is the most sparsely populated benthic habitat due largely to a limited food supply. It stands in remarkable contrast to the world of tropical coral reefs, places of overwhelming productivity, diversity, and beauty.

TERMS AND CONCEPTS TO REMEMBER

accessory pigments
ahermatypic
atoll
barrier reef
benthic
blade
clumped distribution
coral reef
corals
desiccation
estuary
fringing reef
gas bladder
hermatypic
holdfast
intertidal zone

kelps
mangroves
motile
multicellular algae
Phaeophyta
polyp
random distribution
reef
Rhodophyta
sessile
stipe
thallus
uniform distribution
wave shock
zooxanthellae

STUDY QUESTIONS

1. What factors influence the distribution of organisms within a benthic community? How are these distributions described? Why is random distribution so rare?

2. What are algae, and how are they different from plants? Are all algae seaweeds? How are seaweeds classified? Which seaweeds live at the greatest depths? Why?

3. What problems confront the inhabitants of the intertidal zone? How do you explain the richness of the intertidal zone in spite of these rigors? Which intertidal area has larger numbers of species and individuals: sand beach or rocky beach? Why?

4. Which benthic marine habitat is the most sparsely populated? Why?

5. If tropical oceans generally support very little life, why do coral reefs contain such astonishing biological diversity and density?

6. Explain Charles Darwin's classification scheme for coral reefs. Is the classification still in use?

7. What is the primary source of biological energy in rift-vent and cold-seep communities?

FOR FURTHER STUDY

Ellis, R. 1996. *Deep Atlantic: Life, Death, and Exploration in the Abyss.* New York: Knopf. Photos, paintings, information, stories.

Erikson, J. 1996. *Marine Geology: Undersea Landforms and Life Forms.* New York: Facts on File. The combination equals more than the sum of its parts.

Ghiorse, W. C. 1997. "Subterranean Life." *Science* 275 (7 February): 789–90. Extremophiles triumphant!

Heezen, B. C., and C. D. Hollister. 1971. *The Face of the Deep.* New York: Oxford University Press. An outstanding compendium of photos.

Lemonick, M. D. 1995. "The Last Frontier." *Time,* 14 August, 52–60. An excellent overview, well written and thoroughly illustrated.

MacDonald, I. R., and C. Fisher. 1996. "Life Without Light." *National Geographic,* October, 88–97. Extraordinary photos, more information on the distribution of bacteria around vents.

Miller, G. T. 2000. *Living in the Environment.* 11th ed. Belmont, CA: Wadsworth. Part 2, on basic ecological concepts, is especially recommended.

Milne, D. H. 1995. *Marine Life and the Sea.* Belmont, CA: Wadsworth. An excellent ecological approach, readable and complete.

Morell, V. 1995. "Life on a Grain of Sand." *Discover,* April, 78–86. Extraordinary scanning electron micrographs of some highly specialized organisms.

National Research Council, eds. 1995. *Understanding Marine Biodiversity.* Washington, DC: National Academy Press. The diversity of marine life is being dramatically affected by human activity.

Nybakken, J. W. 1993. *Marine Biology: An Ecological Approach.* 3d ed. New York: Harper & Row. Excellent general marine biology text, with an ecological emphasis.

Oceanography 9, no. 1 (1996) contains a variety of papers on marine biodiversity.

Oceanus 38, no. 2 (Fall/Winter 1995) is devoted to marine biodiversity.

Ricketts, E. F., J. Calvin, J. Hedgpeth, and D. W. Phillips. 1985. *Between Pacific Tides.* 5th ed. Palo Alto, CA: Stanford University Press. A gracefully written classic applicable especially to the Pacific coast of the United States.

Starr, C., and R. Taggart. 1996. *Biology: The Unity and Diversity of Life.* 8th ed. Belmont, CA: Wadsworth. Chapters 45 and 46 provide a concise general introduction to the principles of ecology.

For additional readings, go to InfoTrac College Edition, your online research library at:

http://infotrac.thomsonlearning.com/

Uses and Abuses of the Ocean

15

An Endless Supply?

World economies began a period of unprecedented growth in the decade following World War II. Leaders began to look to the sea to provide a greater proportion of the mineral and biological resources needed to sustain their economies and feed their citizens. The great ocean seemed an inexhaustible source of food, oil, and ore. It was comforting to think that as we learned more about the ocean, we would surely discover vast treasure troves available for the taking.

But memories were short. Resources that seemed limitless can be depleted in a few generations. Consider, for example, the history of the North Atlantic's most valuable resources, its fisheries. For hundreds of years, Georges Bank, one of the world's most productive fishing grounds, yielded seemingly endless supplies of cod, haddock, flounder, and halibut. American colonists considered halibut a trash fish, unfit for human consumption. By the 1830s, however, halibut had become fashionable and demand skyrocketed. Individual halibut boats fishing Georges Bank brought in ten tons of halibut per day! As the reproducing stock was removed, the fishery was rapidly and catastrophically depleted. Today, catches of halibut are so rare in the northwestern Atlantic that regulators don't bother to keep statistics on them.

More recently, on the other side of the North Atlantic, fishers of at least seven European nations compete for cod and haddock whose historic abundance will soon be little more than a memory. Fishing pressure is now so

A haul of Alaskan pollock from the Bering Sea. (Pollock is a codlike fish popular in fish sticks and fast food.) About 60% of the fish landed in the United States are pulled from the Bering Sea, a resource worth $1.1 billion before processing. Once thought inexhaustible, pollock stocks are plunging—about 25% of the pollock in the Bering Sea are caught each year.

great that less than 1% of one-year-old cod remain in the ocean long enough to spawn. Each year 60% of all the cod present are swept into nets. The hunt on the Dutch continental shelf is so intense that every square meter of seabed is trawled, on average, once to twice each year! The 1992 collapse of the northwestern cod fishery cost 35,000 jobs (18,000 of them in Newfoundland alone). The northeastern fishery is now teetering on the same brink.

Marine fish are the only wildlife still hunted on a large scale. Properly managed, fish can be a renewable resource, but minerals, oil, natural gas, and most other physical resources are nonrenewable. The most important physical resources are fluid resources such as petroleum (oil) and natural gas. These substances are probably as abundant in the continental shelves as on dry land, but obtaining them is very expensive and often dangerous.

With very few exceptions, the persistent prediction that the continued exploitation of the ocean will provide an increasing percentage of the material needs of a growing human population—of oceanic riches for everyone—cannot be realized. As we will see, marine resources grow more valuable as they become scarcer. 15-1

KEY CONCEPTS

1. Marine resources can be categorized as physical, biological, and nonextractive. They can be renewable or nonrenewable. The contribution of marine resources to the world economy has become so large that international laws now govern their allocation.

2. Pollution can be a side effect of resource extraction and use.

3. Though expensive to obtain, offshore oil and natural gas presently provide nearly a third and a quarter (respectively) of world needs.

4. Though fishery output continues to increase, *per capita* use of oceanic biological resources has fallen since about 1970. Biological resources are often taken without regard to replenishment time.

5. Worldwide bykill may amount to one-third of total biological resource utilization.

6. Aquaculture accounts for more than one-quarter of all fish consumed by humans.

7. Sewage, industrial waste, and air pollution are significant contributors to ocean pollution.

8. Changes in the composition of the atmosphere can influence global climate, ocean temperatures, and sea level.

9. The solution to environmental problems, if one exists, lies in education and action.

CHAPTER AT A GLANCE

There are two sides to humanity's use of the ocean. On the one hand, we find the ocean's resources useful, convenient, essential. On the other, we find we cannot exploit those resources without damaging their source. Through the last few generations our increasingly anxious efforts at resource extraction and utilization have affected the ocean and the atmosphere on a global scale. World economies are now dependent on oceanic materials, and we are unwilling to abandon or diminish their use until we see clear signs of severe environmental damage. Humanity has embarked on an unintentional global experiment in marine resource exploitation and waste management. We adjust course only with hesitation as we rush into the unknown.

MARINE RESOURCES 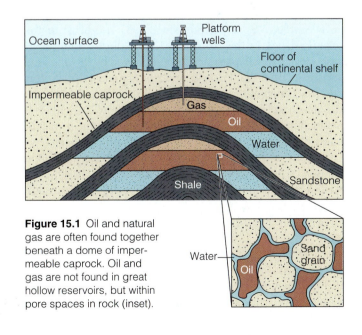 15-2

We will discuss three groups of marine resources in this chapter:

- **Physical resources** result from the deposition, precipitation, or accumulation of useful substances in the ocean or seabed. Most physical resources are mineral deposits, but petroleum and natural gas, mostly remnants of once-living organisms, are included in this category. Fresh water obtained from the ocean is also a physical resource.
- **Biological resources** are living animals and plants collected for human use.
- **Nonextractive resources** are uses of the ocean in place; transportation of people and commodities by sea, recreation, and waste disposal are examples.

Marine resources can be classified as either renewable or nonrenewable.

- **Renewable resources** are naturally replaced by the growth of marine organisms or by other natural processes.
- **Nonrenewable resources** such as oil, gas, and solid mineral deposits are present in the ocean in fixed amounts and cannot be replenished over time spans as short as human lifetimes.

PHYSICAL RESOURCES 15-3
Petroleum and Natural Gas 15-4

Offshore petroleum and natural gas generated nearly $220 billion in worldwide revenues in 1998. The ocean makes a significant contribution to present world needs: About 32% of the crude oil and 24% of the natural gas produced in 1998 came from the seabed. About a third of known world reserves of oil and natural gas lie along the continental margins. Major U.S. marine reserves are located on the continental shelf of southern California, off the Texas and Louisiana Gulf coast, and along the North Slope of Alaska.

Oil is a complex chemical soup containing perhaps a thousand compounds, mostly hydrocarbons. Petroleum is almost always associated with marine sediments, suggesting that the organic substances from which it was formed were once marine. Planktonic organisms or soft-bodied benthic marine animals are the most likely candidates. Their bodies apparently accumulated in quiet basins where the supply of oxygen was low and there were few bottom scavengers. The action of anaerobic bacteria converted the original tissues into simpler, relatively insoluble organic compounds that were probably buried—possibly first by turbidity currents, then later by the continuous fall of sediments from the ocean above. Further conversion of the hydrocarbons by high temperatures and pressures must have taken place at considerable depth, probably 2 kilometers (1.2 miles) or more beneath the surface of the ocean floor. Slow cooking under this thick sedimentary blanket for millions of years completed the chemical changes that produce oil.

If the organic material cooked too long, or at too high a temperature, the mixture turned to methane, the dominant component of natural gas. Deep sedimentary layers are older and hotter than shallow ones and have higher proportions of natural gas to oil. Very few oil deposits have been found below a depth of 3 kilometers (1.8 miles). Below about 7 kilometers (4.4 miles), only natural gas is found.

Oil is less dense than the surrounding sediments, so it can migrate from its source rock through porous overlying formations. It collects in the pore spaces of reservoir rocks when an impermeable overlying layer prevents further upward migration of the oil (**Figure 15.1**). When searching for oil, geologists use sound reflected off subsurface structures to look for the signature combination of layered sediments, depth, and reservoir structure before they drill.

Drilling for oil offshore is far more costly than drilling on land, because special drilling equipment and transport systems are required. Most marine oil deposits are tapped from offshore platforms resting in water less than 100 meters (330 feet) deep. As oil demand (and therefore price) continues

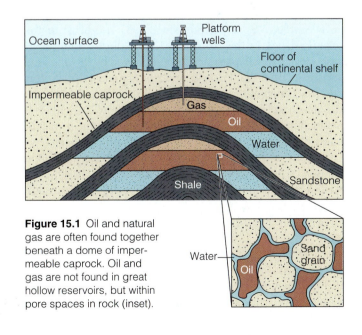

Figure 15.1 Oil and natural gas are often found together beneath a dome of impermeable caprock. Oil and gas are not found in great hollow reservoirs, but within pore spaces in rock (inset).

a

Figure 15.2 (**a**) Shell Oil Company's tension-leg platform *Ursa,* deployed in 1,160 meters (3,800 feet) of water 108 kilometers (130 miles) southeast of New Orleans. (**b**) Platform *Mars,* deployed off Louisiana in 912 meters (2,933 feet) of water in 1996, is pictured here in relation to the Houston skyline. Tension-leg platforms are held in position by steel cables anchored in the seabed, pulling against the semisubmerged platform's buoyancy. Twelve lateral cables anchored to the seabed (not shown) prevent sideways movement. Both platforms are designed to withstand hurricane-force waves of 22 meters (71 feet) and winds of 225 kilometers (140 miles) per hour.

to rise, however, deeper deposits farther offshore will be exploited from larger platforms. The largest and heaviest platform is *Statfjord-B,* in position since 1981 northeast of the Shetland Islands in the North Sea. The tallest drilling platform is *Ursa,* a tension-leg structure deployed by the Shell Oil Company in 1998 that floats in water more than 1,160 meters (3,800 feet) deep (**Figure 15.2**). It is held in position off the Louisiana coast by 16 lengths of 81-centimeter (32-inch) steel cable anchored in the seabed and pulling against the semisubmerged platform's buoyancy. The 14 wells on platform *Ursa* are projected to produce 30,000 barrels of oil and 80 million cubic feet of natural gas per day. Total cost of platform *Ursa* exceeded $1.45 billion. Even larger tension-leg platforms are being planned.

Sand and Gravel

 15-5

Sand and gravel are used in cement and concrete for construction. They may not be very glamorous marine resources, but sand and gravel are second in dollar value only to oil and nat-

b

Figure 15.3 Salt evaporation ponds at the southern end of San Francisco Bay in California. Operators can segregate the various salts from one another by shifting the residual brine from pond to pond at just the right time during the evaporation process. The colors in the ponds are imparted by algae and other microorganisms that thrive at varying levels of salinity. In general, the ponds with the highest salinity have a reddish cast.

ural gas. More than 500 million metric tons (550 million tons) of sand and gravel, valued at more than half a billion dollars, were mined offshore in 1998. The seafloor supplies about 20% of the sand and gravel used in the island nations of Japan and the United Kingdom.

Most of the exploitable U.S. deposits of marine sand and gravel are found off the coasts of Alaska, California, Washington, and the East Coast states from Virginia to Maine. Nearshore deposits are widespread, easily accessible, and used extensively in buildings and highways. Offshore oil wells in Alaska are built on huge man-made gravel platforms; the large quantities of gravel available at those locations make offshore drilling practical there.

Salts

15-6

As you may remember from Chapter 6, the ocean's salinity varies from about 3.3% to 3.7% by weight. When seawater evaporates, the remaining major constituent ions (see Figure 6.4) combine to form various salts, including table salt (NaCl), calcium carbonate ($CaCO_3$), and gypsum ($CaSO_4$).

Seawater is evaporated in large salt ponds in arid parts of the world (**Figure 15.3**). Operators can segregate the various salts from one another by shifting the residual brine from pond to pond at just the right time during the evaporation process. The magnesium salts are used as a source of magnesium metal and magnesium compounds. The potassium salts are processed into chemicals and fertilizers. Bromine (a useful component of certain medicines, chemical processes, and antiknock gasolines) is extracted from the residue. About a third of the world's table salt is currently produced from seawater by evaporation. In North America, some of this salt is used for snow and ice removal. Salt is also used in water softeners, agriculture, and food processing. In 1998 the United States produced by evaporation about 4.7 million metric tons (5.2 million tons) of table salt with a value of about $145 million.

Fresh Water

15-7

Only 0.017% of Earth's water is liquid, fresh, and available at the surface for easy use by humans. Another 0.6% is available as groundwater within half a mile of the surface. Unfortunately,

Figure 15.7 Fish jam the rearmost part of a trawl net before it is hoisted from the water.

tons per unit of effort and are ranging farther afield in their urgent search for food. We may be perilously close to the catastrophic collapse of more fisheries. As of 1997, 30% of recognized marine fisheries were overexploited or already depleted, and 50% more were at their presumed limit of exploitation. The U.S. National Marine Fisheries Service estimates that 45% of the fish stocks whose status is known are now **overfished**—so many fish have been harvested that there is not enough breeding stock left to replenish the species (**Figure 15.7**). The fishing industry's dominant motivating force is often quick financial return, even if it means depleting a stock and disrupting the equilibrium of a fragile ecosystem. Long-term stability is forsaken for short-term profit. When the catch begins to drop, the industry increases the number of boats and develops more efficient techniques for capturing the remaining animals. When the impending catastrophe is obvious, governments will sometimes intervene to set limits or close a fishery altogether. In 1999 New England officials approved a plan to close a large section of the Gulf of Maine to fishing in an attempt to replenish once-abundant cod stocks. As many as 2,500 fishers and 700 boats were idled, and losses may exceed $21 million a year. Given the choice between immediate profit and a long-term sustainable fishery, fishers made their priorities clear by initiating a recall campaign against officials who had approved the ban.

The intended organism is not the only victim. In some fisheries, **bykill**—animals unintentionally killed when desirable organisms are collected—exceeds target catch. In 1992 in the Bering Sea, fishers discarded 16 million red king crabs and kept only 3 million. Discarded creatures outnumber target shrimp by between 125% and 830%. In 1993 in the Gulf of Mexico, fishing for shrimp resulted in a bykill of 34 million red snappers and 2,800 metric tons (3,090 tons) of sharks. (The annual commercial catch of red snapper in the Gulf is typically 3 million fish.) At the end of the 1990s commercial fisheries in the United States discarded about 9 million metric tons (10 million tons) of non-target fish each year—twice the catch of desired commercial and recreational fishing combined! Worldwide bykill approached 30 million metric tons (33 million tons) in 1998, *a quantity nearly one-third of total landings*!

Some progress has been made to minimize bykill. Turtle exclusion devices (TEDs), chutes through which turtles are ejected from nets, have been mandated for shrimp fishing in U.S. territorial waters. As we shall see in our discussion of whaling, by forcing the redesign of tuna nets and the adoption of special ship maneuvers during netting, the "dolphin safe" tuna campaign, an adjunct to the 1972 Marine Mammals Protection Act, has spared the lives of hundreds of thousands of dolphins.

There are other disruptive, mistargeted fishing techniques. One employs **drift nets**—fine, vertically suspended nylon nets as much as 7 meters (25 feet) high and 80 kilometers (50 miles) long. A deployed drift net is shown in **Figure 15.8.** Until 1993, Taiwanese, Korean, and Japanese vessels deployed some 48,000 kilometers (30,000 miles) of these "walls of death" each night—more than enough to encircle Earth! Drift nets caught the fish and squid for which they were designed, but they also entangled everything else that touched them, including turtles, birds, and marine mammals (**Figure 15.9**). The process has been compared to stripmining—the ocean is literally sieved of its contents. An estimated 30 kilometers (18 miles) of these fine, nearly invisible nets were lost each night—about 1,600 kilometers (1,000 miles) per season. These remnants, made of nonbiodegradable plastic, become *ghost nets*, continuing to entangle fish and other animals for decades.

Though large-scale drift net fishing is now banned, pirate netting continues in the Pacific and has begun in the Atlantic. Completely unrestrained by regulation, these outlaw fishermen operate where their activities cause maximum damage to valuable reproductive stock.

Whaling   15-12

Since the 1880s, whales have been hunted to provide meat for human and animal consumption; oil for lubrication, illumination, industrial products, cosmetics, and margarine; bones for fertilizers and food supplements; and baleen for corset stays.

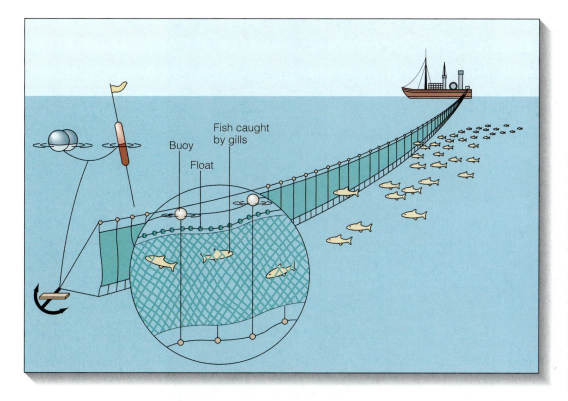

Figure 15.8 Drift net fishing. Typically, a fleet of huge factory ships uses sophisticated electronic detection devices and aircraft to find large schools of fish. Each ship then launches 20 to 50 fast, small catcher boats to set hundreds to thousands of miles of drift nets, weighted to stay at the desired depth. After drifting overnight the nets are hauled in by the factory ships, and the catch is processed on board by canning or freezing. Large-scale drift net fishing was officially banned in 1993, but it continues in the Pacific and has begun in the Atlantic.

Figure 15.9 An unintentional but unavoidable consequence of drift net fishing.

Figure 15.10 The whaling industry has pushed most of the dozen or so species of great whales to the brink of extinction through overharvesting. This photograph shows a blue whale being butchered at the Coal Harbor Whaling Station, British Columbia, Canada, in 1963. Commercial whaling of great whales was banned in 1985 but began again in early 1993.

Figure 15.10 shows a whale being butchered for food and oil. An estimated 4.4 million large whales existed in 1900; today slightly more than 1 million remain. Eight of the 11 species of large whales once hunted by the whaling industry are commercially extinct. The industry pursued immediate profits despite obvious signs that most of the "fishery" was exhausted.

Substitutes exist for all whale products, but the harvest of most commercial species did not stop until whaling became uneconomical. In 1986 the International Whaling Commission, an organization of whaling countries established to manage whale stocks, placed a moratorium on the slaughter of large whales. Except for a suspiciously large harvest of whales taken by the Japanese for "scientific purposes," commercial whaling ceased in 1987. Fewer than 700 large whales were taken in 1988. Now, however, the number being taken is rising. What protection there is may have come too late to save some species from extinction—and protection may only be temporary. Under intense pressure from its major fishing industry, Norway resumed whaling in 1993. Iceland was scheduled to begin in 2000. Japan never stopped. Their target, the small minke whale, is the smallest and most numerous of the great whale species (see Figure 13.22). The meat of this whale is prized in Japan as an expensive delicacy. The minke whale population has been estimated at 1.2 million, a number that may withstand the present level of harvest. A decision to increase minke whaling, however, may doom the minke to the same fate as most other whale species.

There is a glimmer of hope. In 1994 the International Whaling Commission voted overwhelmingly to ban whaling in about 21 million square kilometers (8 million square miles) around Antarctica, thus protecting most of the remaining large whales, which feed in those waters. The sanctuary is often ignored. In their quest for minke (and other whales), Japanese and Norwegian whalers have entered the area and harpooned many animals. The Japanese took about 600 minkes in the region in 1999; the Norwegians took 500 in 1998. Chile, Peru, and North Korea are considering joining the hunt. Pirate whalers based in other countries can also catch whales in the Antarctic, and sell the flesh on the Japanese market. Conservation efforts can do some good, however. Although it was hunted nearly to extinction, the California gray whale has long been off limits to all but aboriginal hunters. Its numbers have grown, and it was removed from the endangered list in 1993.

Many more small whales have been killed than large ones, but not for food or raw materials. For reasons that are not well understood, dolphins (which are small whales) gather above schools of yellowfin tuna in the open sea. Fishermen have learned to find the tuna by spotting the dolphins. Nets cast to catch the tuna also entangle the air-breathing dolphins, and the mammals drown. The dead dolphins are simply pitched over the side as waste. More than 6 million small whales have been killed in association with tuna fishing.

Passage of the U.S. Marine Mammals Protection Act in 1972 brought a drop in the number of dolphin deaths. However, from 1977 to 1987 the number of tuna boats in the U.S. fleet declined from more than 100 boats to 34 boats, while the foreign fleet rose from fewer than 10 to more than 70 boats. Foreign fishermen do not always abide by the dolphin-saving provisions of the act, but the U.S. Congress voted in 1988 to

Figure 15.11 A kelp cutter harvesting kelp. Metal shears about a meter below the surface trim the kelp, and a moving ramp brings it aboard.

ban importation of tuna not caught in accordance with new methods designed to reduce the kill. In 1990, American tuna processing companies agreed to buy tuna only from fishermen whose methods do not result in the deaths of dolphins. American commercial fishermen have agreed to comply, and conservationists hope that the foreign fleets will follow their lead.

Botanical Resources 15-13

Marine plants are also commercially exploited. The most important commercial product is **algin,** made from the mucus that slickens seaweeds. When separated and purified, algin's long, intertwining molecules are used to stiffen fabrics; to form emulsions such as salad dressings and cheese spreads, paint, and printer's ink; to prevent the formation of large crystals in ice cream; to clarify beer and wine; and to suspend abrasives. The U.S. seaweed gel industry produces more than $220 million worth of algin each year, and the annual worldwide value of products containing algin (and other seaweed substances) was estimated to be more than $41 billion in 1998. A harvesting barge is shown in **Figure 15.11.**

Seaweeds are also eaten directly. The Japanese consume 150,000 metric tons (165,000 tons) of *nori* each year; seaweed and seaweed extracts are also eaten in the United States, Britain, Ireland, New Zealand, and Australia. Their mineral content and fiber are useful in human nutrition.

Aquaculture 15-14

Aquaculture is the growing or farming of plants and animals in any water environment under controlled conditions. Aquaculture production currently accounts for more than one-quarter of all fish consumed by humans. Most aquaculture production occurs in China and the other countries of Asia.

Mariculture is the farming of *marine* organisms, usually in estuaries, bays, or nearshore environments (**Figure 15.12**), or in specially designed structures using circulated seawater.

Worldwide mariculture production is thought to be about one-eighth that of freshwater aquaculture. Several species of fish, including plaice and salmon, have been grown commercially, and marine and brackish-water fish account for 67% of the total production. Shrimp mariculture is the fastest growing and most profitable segment, with an annual global value exceeding $6 billion in 1998.

Ranching, or open-ocean mariculture, is an interesting variation. Juveniles are grown in a certain area, released into the ocean, and expected to return when mature. The natural homing instinct of salmon makes salmon ranching quite successful—about 20% of the world supply of salmon is ranched. Yet only about 1 in 50 of ranched juveniles return to the area of release. In Japan, recent attempts to extend ranching techniques to yellowtail and tuna have met with only limited success.

Drugs 15-15

Modern medical researchers estimate that perhaps 10% of all marine organisms are likely to yield clinically useful compounds. One such medicine is derived from a Caribbean sponge and is already in use: Acyclovir, the first antiviral compound approved for humans, has been fighting herpes infections of the skin and nervous system since 1982. A class of anti-inflammatory drugs known as pseudopterosins, developed by researchers at the University of California, has also been successful.

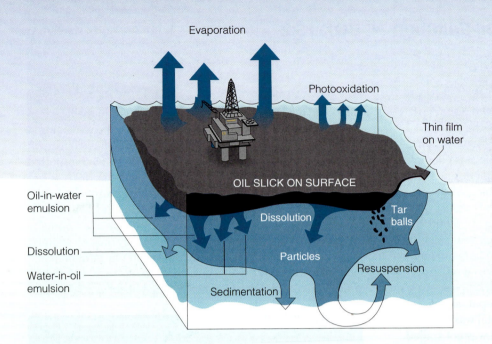

Evaporation

Photooxidation

OIL SLICK ON SURFACE

Thin film on water

Oil-in-water emulsion

Dissolution

Water-in-oil emulsion

Dissolution

Particles

Tar balls

Sedimentation

Resuspension

a The fate of oil spilled at sea. Smaller molecules evaporate or dissolve into the water beneath the slick. Sunlight and atmospheric oxygen cause oxidation of the surface layer. Within a few days, water motion coalesces the oil into tar balls and semisolid emulsions of water-in-oil and oil-in-water. Tar balls and emulsions may persist for months after formation. Although bonding with sediment particles may sink oil droplets, suspended oil does not enter benthic subtidal sediments in large quantities. If crude oil is left undisturbed, bacterial activity will eventually consume it. Refined oil, however, can be more toxic, and natural cleaning processes take a proportionally longer time to complete.

waters. This oil residue—especially if derived from refined oil—can have long-lasting effects on seafloor communities. The fate of spilled oil is summarized in **Figure a.**

The methods used to contain and clean up an oil spill sometimes cause more damage than the oil itself. When the supertanker *Exxon Valdez* ran aground in Alaska's Prince William Sound on 24 March 1989, more than 40 million liters (almost 11 million gallons; 29,000 metric tons) of Alaskan crude oil—about 22% of her cargo—escaped from the crippled hull. Only about 17% of this oil was recovered by a work crew of more than 10,000 people using containment booms, skimmer ships, bottom scrapers, and absorbent sheets. About 35% of the oil evaporated, 8% was burned, 5% was dispersed by strong detergents, and 5% biodegraded in the first five months. The rest of the oil, some 30% of the spill, formed oil slicks on Prince William Sound and fouled more than 450 kilometers (300 miles) of coastline.

Recent analysis of the affected parts of Prince William Sound shows the cleaned areas to be in generally worse shape than areas left alone. Most of the small animals that make up the base of the food chain in these areas were cooked by the 65°C (150°F) water used to blast oil from between the rocks (**Figure b**). Others were smothered when the high-pressure jets rearranged sand and mud. It appears that an overambitious cleanup program can be counterproductive. The biological cost of the spill will not be known for a decade. The cost of the cleanup has exceeded $3.5 billion. The legal costs could be even higher: In 1994

a federal jury awarded $5 billion in punitive damages to 14,000 plaintiffs including fishermen, native villagers, and fish processors. Exxon has appealed the verdict; litigation continues.

The best way to deal with oil pollution is to prevent it from happening in the first place. Tanker design is being modified to limit the amount of oil intentionally released in transport. Laws now limit new U.S.-flagged tanker construction to stronger, double-hull designs (although shipping experts are uncertain that even a double-hull design could have prevented the *Exxon Valdez* spill). Perhaps most important, crew testing and training are being upgraded.

15-18

b Cleaning up the 1989 *Exxon Valdez* oil spill, Prince William Sound, Alaska.

Figure 15.13 *Regina Maersk,* one of the world's largest container-ships, enters the Port of New York.

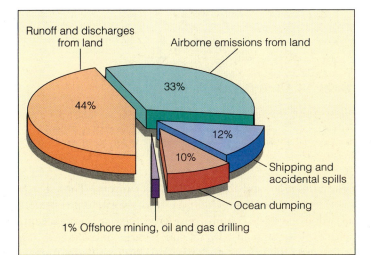

Runoff and discharges from land

Airborne emissions from land

33%

44%

12%

10%

Shipping and accidental spills

Ocean dumping

1% Offshore mining, oil and gas drilling

Source: Joint Group of Experts on the Scientific Aspects of the Marine Environment, *The State of the Marine Environment,* UNEP Regional Seas Reports and Studies No. 115 (Nairobi: U.N. Environment Programme, 1990)

Figure 15.14 Sources of marine pollution.

spontaneously or through physical processes (like the shattering of large molecules by sunlight). Sometimes pollutants are removed from the environment through biological activity. Indeed, many pollutants are ultimately **biodegradable**—that is, able to be broken down by natural processes into simpler compounds. Many pollutants resist attack by water, air, sunlight, or living organisms, however, because the synthetic compounds of which they are composed resemble nothing in nature.

The ways in which pollutants are changing the ocean and the atmosphere are often difficult for researchers to determine. Environmental impact cannot always be predicted or explained. As a result, marine scientists vary widely in their opinions about what pollutants are doing to the ocean and the atmosphere and what to do about it. Environmental issues are frequently emotional, and media reports tend to sensationalize short-term incidents (like oil spills—see Box 15.2) rather than more serious long-term problems (like atmospheric changes or the effects of long-lived chlorinated hydrocarbon compounds).

Pollution is costly. In 1995 government and industry in the United States spent about $200 billion on the control of atmospheric, terrestrial, and marine pollution—an average of $735 for each American. This figure was equivalent to about 1.6% of the gross national product, or 2.7% of capital expenditures by U.S. business. That same year the United States lost 4% of its gross national product through environmental damage.

Failure to control the pollution associated with exploitation of marine resources will eventually threaten our food supply (marine and terrestrial), destroy whole industries, produce a greater disparity between have and have-not nations, and cause a decline in the health of all the planet's citizens. To these costs must be added the aesthetic costs of an ocean despoiled by pollution; none of us look forward to sharing the beach with oiled birds, jettisoned diapers, or clumps of medical waste.

THE OCEAN AS WASTE DUMP

 15-19

Each year 7.7 million metric tons (8.5 million tons) of sewage and industrial waste flow into U. S. rivers and empty into the ocean. Air pollution is responsible for almost one-third of the toxic contaminants and nutrients that enter coastal areas and the ocean.

Heavy Metals and Synthetic Organic Chemicals

15-20

Among the dangerous heavy metals being introduced into the ocean are mercury and lead. Human activity releases about 5 times as much mercury and 17 times as much lead as is derived from natural sources, and incidents of mercury and lead poisoning, major causes of brain damage and behavioral disturbances in children, have increased dramatically over the last two decades. Lead particles from industrial wastes, landfills, and gasoline residue reach the ocean through runoff from land during rains, and the lead concentration in some shallow-water bottom-feeding species is increasing at an alarming rate.

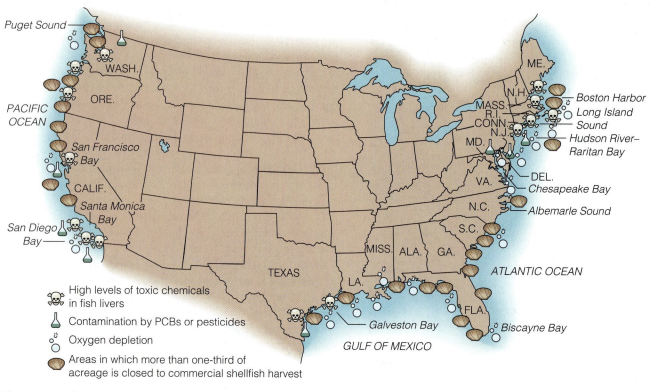

Figure 15.15 Chemical assault on major U.S. coastal areas.

Wellness-conscious consumers see fish as a safe and healthful food, but with the ocean receiving heavy-metal-contaminated runoff from the land, a rain of pollutants from the air, and the fallout from shipwrecks, we can only wonder how much longer seafood will be safe to eat. Consumers should be wary of seafoods taken near shore in industrialized regions.

Halogenated hydrocarbons—a class of synthetic hydrocarbon compounds that contain chlorine, bromine, or iodine—are used in pesticides, flame retardants, industrial solvents, and cleaning fluids. The concentration of **chlorinated hydrocarbons**—the most abundant and dangerous halogenated hydrocarbons—is so high in the water off New York State that officials have warned women of childbearing age and children under 15 who live there not to consume more than half a pound of local bluefish a week. (They are told *never* to eat striped bass caught in the area.) One administrator of the U.S. Environmental Protection Agency has written, "Anyone who eats the liver from a lobster taken from an urban area is living dangerously."

The level of synthetic organic chemicals in seawater is usually very low, some organisms at higher levels in the food chain can concentrate these toxic substances in their flesh. This **biological amplification** is especially hazardous to top carnivores in a food web. Biologists were alarmed by the recent discovery that many of the nearshore dolphins off U.S. coasts are intensely contaminated. The concentration of chlorinated hydrocarbons in these animals exceeds 6,900 parts per million (ppm), concentrations high enough to disrupt the dolphins' immune systems, hormone production, reproductive success,

Figure 15.16 Rachel Carson, author of the influential 1962 book *Silent Spring* on the misuse of synthetic pesticides. This work is generally credited with beginning the modern environmental movement in the United States.

neural function, and ability to stave off cancers.[2] These levels vastly exceed the 50 ppm limit the U.S. government considers hazardous, and the 5 ppm considered the maximum acceptable level for humans! Investigations are continuing, as are probes into the effects of dioxin and other synthetic organic poisons accumulating in the oceanic sink.

---

[2] If you dragged a beached porpoise into the ocean, you could theoretically receive a $10,000 fine for improper disposal of polluted materials!

Figure 15.15 indicates the locations of U.S. coastal areas degraded by synthetic organic chemicals. Production of such chemicals subject to biological amplification in food chains presently exceeds 91 million metric tons (100 million tons) each year. Even vaster volumes of ocean will be affected if the predicted 1% of that production reaches the sea.

Solid Waste

 15-21

Not all pollutants enter the ocean in a dissolved state; much of the burden arrives in solid form. Some solid waste is ultimately biodegradable, but plastic, which now makes up almost 8% of the solid waste stream, is not. Scientists estimate that some kinds of synthetic materials—plastic six-pack holders, for example—will not decompose for about 400 years!

Americans generate 120 million metric tons of plastic waste, about 500 kilograms (1,100 pounds) per person, each year. Much of this waste plastic finds its way to the sea. In 1997 more than 4,500 volunteers spent two weekends scouring the New Jersey–New York coast to collect debris. More than 75% of the 209 tons recovered was plastic. In isolated areas the dunes of plastic debris can build to surreal proportions (see **Figure 15.17**).

A staggering 100,000 marine mammals and 2 million seabirds die *each year* after ingesting or being caught in plastic debris! Sea turtles mistake plastic bags for their jellyfish prey and die from intestinal blockages. Seals and sea lions starve after becoming entangled in nets (see **Figure 15.18**) or muzzled by six-pack rings. The same kinds of rings strangle fish and seabirds. Adding ingredients to plastics that would hasten their decomposition would add only 5% to 7% to their cost.

Figure 15.18 Sea lions (seen here) and seals die by the tens of thousands each year after becoming entangled in plastic debris, especially discarded and broken fishing nets.

Figure 15.17 Plastic mounds on an isolated shore of Niihau, one of the Hawaiian Islands.

What should we do with plastic and other solid wastes such as glass and paper, disposable diapers, scrap metal, building debris, and all the rest? Dumping it into the ocean is clearly unacceptable, yet places to deposit this material are becoming scarce. In 1998 the average New Yorker threw away 1.6 tons of waste annually. California's Los Angeles and Orange Counties generate enough solid waste to fill Dodger Stadium every eight days. Transportation of waste to sanitary landfills becomes more expensive as nearby landfills reach capacity.

Is incineration the answer? We currently burn about 10% of our trash. By contrast, Sweden burns half its solid wastes. Unfortunately, even with air pollution control devices, incinerators still emit great quantities of tiny particles, heavy metal residue, and immense amounts of carbon dioxide. Dumping of the residual ash is also a problem—ash from Philadelphia's garbage incinerators has been turned away by states as far away as South Carolina, and one shipload of the stuff was refused entry at an African port.

Is recycling the answer? The Japanese currently recycle about 50% of their solid waste and are importing even more; scrap metal and waste paper headed for Japan are the two biggest exports from the Port of New York. Americans are buying back their own refuse in the form of appliances, automobiles, and the cardboard boxes that hold their televisions and compact disc players. Massachusetts has set a goal of recycling 25% of its waste. The direct savings to consumers, as well as the environmental rewards to ocean and air, will be significant.

The *best* solution is a combination of recycling and reducing the amount of debris we generate by our daily activities. We will soon have no other choice.

Sewage 15-22

About 98% of sewage is water. Sewage treatment separates the fluid component from the solids, treats the water to kill disease organisms and reduce the levels of nutrients, and releases it into a river or the ocean. The remaining **sewage sludge** is a semisolid mixture of organic matter containing bacteria and viruses, toxic metal compounds, synthetic organic chemicals, and other debris. It is digested, thickened, dried, and shipped to landfills, burned to generate electricity, or dumped into the ocean. The liquid effluent wanders from its outlet and circulates with currents, but sludge and other insoluble residues may stay near the outfall or dump site for years. The amount of wastewater and sewage sludge released into the ocean from the United States increased by 60% in the 1990s.

Not only human sewage is involved. In June 1995, 26 million gallons of hog feces and urine spilled into a narrow stream feeding North Carolina's New River estuary. Thousands of fish died from direct effects, and hundreds of thousands more perished as algae grew and consumed oxygen.

Sludge may be an even greater problem than raw sewage. The water just south of Long Island, New York, has been one of the most intensely used ocean dumping sites in the world. The area is covered with sludge, which creates an oxygen-poor environment in which few animals can survive. During storms some of this material washes up on local beaches, contaminates shellfish beds, and routinely causes disease outbreaks among people consuming raw oysters and clams from the area. After a public outcry, a new dumping site was selected beyond the edge of the continental shelf. Ten million tons of wet sludge produced by treatment plants in New York and New Jersey since March 1986 have been dropped there by huge barges. The plume from this sludge now contaminates the Gulf Stream.

ATMOSPHERIC CHANGES 15-23

The ocean and the atmosphere are extensions of each other, and human activity has changed the atmosphere as it has changed the ocean. Pollutants injected into the air can have global consequences for the ocean and for all Earth's inhabitants. Two of the most potentially destructive atmospheric problems are depletion of the ozone layer and global warming.

Ozone Layer Depletion 15-24

Ozone is a molecule formed of three atoms of oxygen. Ozone occurs naturally in the atmosphere. A diffuse layer of ozone mixed with other gases—the **ozone layer**—surrounds the world at a height of about 20 to 40 kilometers (12 to 25 miles).

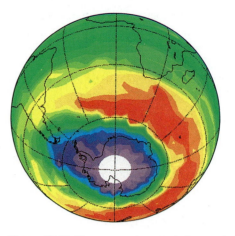

Figure 15.19 The seasonal ozone hole over Antarctica, as recorded by the *Nimbus 7* satellite in 1993. The lowest ozone values (the "hole") are indicated by magenta and purple.

Seemingly harmless synthetic chemicals released into the atmosphere—primarily **chlorofluorocarbons** (**CFCs**) used as cleaning agents, refrigerants, fire-extinguishing fluids, spray-can propellants, and insulating foams—are converted by the energy of sunlight into compounds that attack and partially deplete Earth's atmospheric ozone. Ozone levels in the stratosphere have decreased by at least 3% over most of the United States since 1969. A 4% drop has been noted over Australia and New Zealand, and a 50% decrease has been observed near the North and South Poles. The amount of depletion varies with latitude (and with the seasons) because of variations in the intensity of sunlight (see **Figure 15.19**).

This decline in ozone alarms scientists because stratospheric ozone intercepts some of the high-energy ultraviolet radiation coming from the sun. Ultraviolet radiation injures living things by breaking strands of DNA and unfolding protein molecules. Species normally exposed to sunlight have

Figure 15.20 If ozone depletion continues, the "perfect tan" may have to be none at all. (© The New Yorker Collection 1989 Stuart Leeds from cartoonbank.com. All rights reserved.)

evolved defenses against average amounts of ultraviolet radiation, but increased amounts could overwhelm those defenses. Land plants such as soybeans and rice would be subjected to sunburn that decreases their yields. Even plankton in the uppermost 2 meters (6.5 feet) of ocean would be affected: Recent research indicates an alarming decrease in phytoplankton primary productivity of between 6% and 12% in the coastal waters around Antarctica.

Our own species would not escape: A 1% decrease in atmospheric ozone would probably be accompanied by a 5% to 7% increase in human skin cancer (see **Figure 15.21**). According to medical researchers' estimates, the 4% ozone depletion over Australia and New Zealand will cause at least a 20% increase in human skin cancers over the next two decades. Strong ultraviolet light can also suppress the immune system and cause eye cataracts.

In June 1990, representatives of 53 nations agreed to ban major production and use of ozone-destroying chemicals by the year 2000. A sense of great urgency surrounds ongoing research to find safe substitutes for these substances. Recent data indicate these measures are having an effect. CFC concentrations peaked near the beginning of 1994. Researchers believe we could see a decrease in the sizes of polar ozone holes by 2005 or 2010.

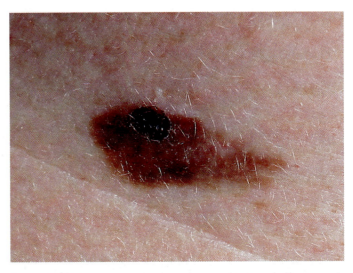

Figure 15.21 A malignant melanoma in an early stage. Left untreated, this most dangerous form of skin cancer could spread and cause the death of the patient. Skin cancer appears to be associated with exposure to the sun. If you spend much time in the sun, you increase your risk of skin cancer. Be alert for changes in moles or for any pigmented lesion that is asymmetrical, has an irregular border, is an unusual color, grows in size, or begins to itch. If you notice any of these characteristics, see a physician.

Global Warming   15-25

The surface temperature of Earth fluctuates slowly over time. The global temperature trend has been generally upward in the 18,000 years since the last ice age, but the *rate* of increase has recently accelerated. This rapid warming may be the result of an enhanced **greenhouse effect,** the trapping of heat by the atmosphere. Glass in a greenhouse is transparent to light but not to heat. The light is absorbed by objects inside the greenhouse, and its energy is converted into heat. The temperature inside a greenhouse rises because the heat is unable to escape. On Earth **greenhouse gases**—carbon dioxide, water vapor, methane, CFCs, and others—take the place of glass. Heat that would otherwise radiate away from the planet is absorbed and trapped by these gases, causing a rise in surface temperature. **Figure 15.22** shows this mechanism.

The greenhouse effect is necessary for life; without it, Earth's average atmospheric temperature would be about

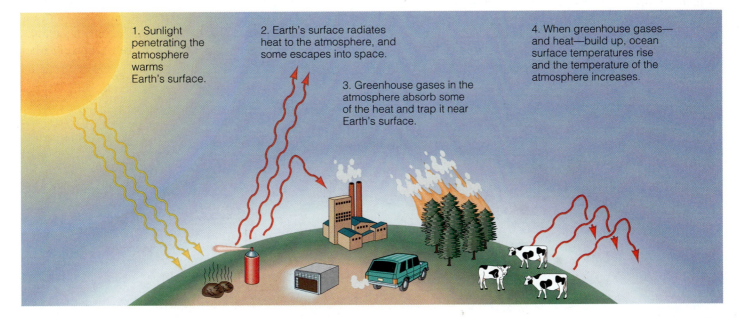

1. Sunlight penetrating the atmosphere warms Earth's surface.

2. Earth's surface radiates heat to the atmosphere, and some escapes into space.

3. Greenhouse gases in the atmosphere absorb some of the heat and trap it near Earth's surface.

4. When greenhouse gases—and heat—build up, ocean surface temperatures rise and the temperature of the atmosphere increases.

Figure 15.22 How the greenhouse effect works. Since the beginning of the Industrial Revolution, emissions of CO_2, methane, and other greenhouse gases have increased. Most researchers believe these gases have contributed to a general warming of Earth's atmosphere and ocean.

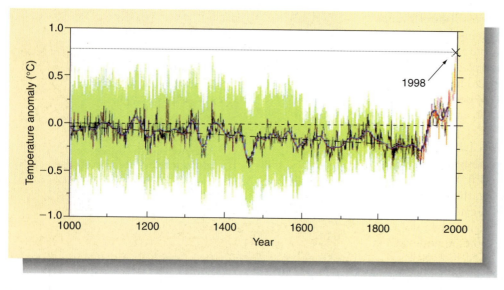

Figure 15.23 Three of the last 5 years of the 20th century were the warmest on record. Temperature records preserved in tree rings and glacial ice have revealed a large and abrupt warming unique in the past 1,000 years. In April, 2000, a report by the United Nations Panel on Climate change concludes ". . . there has been a discernible human influence on global climate." (Mann *et al.* Copyright © 1999 American Geophysical Union.)

−18°C (0°F). Earth has been kept warm by natural greenhouse gases. The sources of these gases are volcanic and geothermal processes, the decay and burning of organic matter, and respiration and other biological sources. The removal of these gases by photosynthesis and absorption by seawater appears to prevent the planet from overheating. But the human demand for quick energy to fuel industrial growth, especially since the beginning of the Industrial Revolution, has injected unnatural amounts of new carbon dioxide into the atmosphere from the combustion of fossil fuels. Carbon dioxide is now being produced at a greater rate than it can be absorbed by the ocean. The atmosphere's carbon dioxide content now rises at the rate of 0.4% each year. Though CFCs are now declining, methane levels continue to increase. (The primary sources of methane include intestinal gases from cows and termites, and the decay

of refuse and vegetation cleared from land.)

There has been a 4°C (7°F) rise in global temperature from the end of the last ice age until today. Carbon dioxide and other human-generated greenhouse gases produced since 1880 are thought to be responsible for about 1°C (1.8°F) of that increase. A new survey of 300 borehole temperatures on four continents has confirmed that Earth is getting warmer and that the rate of warming has been increasing since 1900. Subsurface temperatures confirm that the average global temperature has increased 0.4 to 0.8°C (0.7 to 1.4°F) over the last century. (see **Figure 15.23**). If greenhouse emissions continue to rise, the average should increase another 1°C by 2050. (This estimate is based on projections of data, not on models.)

The south polar ice sheets are already shrinking in the warmth. Imagine the effect of a significant rise in sea level on the harbors, coastal cities, river deltas, and wetlands where one-third of the world's people now live. As **Figure 15.24** suggests, costs to society would be enormous.

Scientists are still uncertain about the magnitude of human-induced global warming. Unpredictable natural processes, such as fluctuations in the energy output of the sun and the 1991 eruption of Mount Pinatubo in the Philippines, can greatly influence temperature predictions. It's difficult to gather and interpret data fast enough to answer the important questions. Most researchers, however, are convinced that global warming is or will become a reality. In January 2000, a panel of scientists convened by the National Research Council concluded that warming of Earth's surface is "undoubtedly real." Experts suggest a reduction in the generation of carbon

Figure 15.24 For island nations such as the Maldives, even a small rise in sea level could spell disaster. Strung out across 880 kilometers (550 miles) of the Indian Ocean, 80% of this island nation of 263,000 people lies less than a meter (3.3 feet) above sea level. Of the country's 1,180 islands, only a handful would survive the median estimate of sea level rise by 2100. Most of the population lives in fishing villages on low islands like the one shown here, where the effects of this century's sea level rise of 10–25 centimeters (4–10 inches) have already been felt.

dioxide, methane, and other greenhouse gases until more data can be accumulated.

Unfortunately, it will be exceedingly difficult to curtail our production of carbon dioxide and methane. In the last hundred years, industrial production has increased fiftyfold; we have burned roughly 1 billion barrels of oil, 1 billion metric tons of coal, and 10 billion cubic meters of natural gas. Carbon dioxide is a major product of combustion for these hydrocarbon compounds. The world's energy demand is projected to increase 3.5 times between now and the year 2025, with carbon dioxide emissions 65% higher than today. Some environmentalists have proposed the planting of millions of trees to take up the extra carbon dioxide, but these efforts would be futile in the face of such massive increases.

Will citizens of industrial countries (and countries wishing to become industrialized) agree to lessen the danger of increased global warming by slowing their economic growth, decreasing their dependence on fossil fuel combustion, and developing safe alternate sources of energy? Some insight may be gained from the behavior of ranchers and industrialists in the rain forests of New Guinea, the Philippines, and Brazil. The Amazon rain forest of Brazil is being burned at a rate of about 12 square kilometers (almost 5 square miles) *each hour,* acreage equivalent to the area of West Virginia every year. Huge stands of trees that should be nurtured to absorb excess carbon dioxide are being destroyed. The cleared land is used for farms, cattle ranches, roads, and cities. The priorities are clear.

A CAUTIONARY TALE   15-26

Let me tell you a story.

Easter Island was home to a culture that rose to greatness amid abundant resources, attained extraordinary levels of achievement, and then died suddenly and alone, terrified, in the empty vastness of the Pacific. The inhabitants had destroyed their world.

Europeans first saw Easter Island (Rapa Nui) in 1722. The Dutch explorer Jacob Roggeveen encountered the small volcanic speck on Easter morning while on a scouting voyage. The island, dotted with withered grasses and scorched vegetation, was populated by a few hundred skittish, hungry, ill-clad Polynesians who lived in caves. During his one-day visit, Roggeveen was amazed to see more than 200 massive stone statues standing on platforms along the coast (**Figure 15.25**). At least 700 more statues were later found partially completed, lying in the quarries as if they had just been abandoned by workers. Roggeveen immediately recognized a problem: "We could not comprehend how it was possible that these people, who are devoid of thick heavy timber for making machines, as well as strong ropes, nevertheless had been able to erect such images." The islanders had no wheels, no powerful animals, and no resources to accomplish this artistic, technical, and organizational feat. And there were too few of them—600 to 700 men and about 30 women.

When Captain James Cook visited in 1774, the islanders paddled to his ships in canoes "put together with manifold small planks . . . cleverly stitched together with very fine twisted threads." Cook noted the natives "lacked the knowledge and materials for making tight the seams of the canoes; they are accordingly very leaky, for which reason they are compelled to spend half the time in bailing." The canoes held only one or two people, and they were less than 3 meters (10 feet) long. Only three or four canoes were seen on the entire island, and Cook estimated the human population at less than 200. By the time of Cook's visit, nearly all the statues had been overthrown, tipped into pits dug for them, often onto a spike placed to shatter their faces upon impact.

Archaeological research on Easter Island has revealed a chilling story. Only 165 square kilometers in extent (64 square miles), Easter Island is one of the most isolated places on Earth, the easternmost outpost of Polynesia. Voyagers from the Marquesas first reached Easter Island about A.D. 350, possibly after being blown off course by a series of storms. The place was a miniature paradise. In its fertile volcanic soils grew dense

Figure 15.25 Giant statues look to the horizon on Easter Island (Rapa Nui). The civilization that built the statues had all but destroyed its world when the first Europeans arrived in 1722.

forests of palms, daisies, grasses, hauhau trees, and toromino shrubs. Large oceangoing canoes could be built from the long, straight, buoyant Easter Island palm, and strengthened with rope made from the hauhau trees. Toromino firewood cooked the fish and dolphins caught by the newly arrived fishermen, and forest tracts were cleared to plant crops of taro, bananas, sugarcane, and sweet potatoes. The human population thrived.

By the year 1400 the population had blossomed to between 10,000 and 15,000 people. Gathering, cultivating, and distributing the rich bounty for so many inhabitants required complex political control. As numbers grew, however, stresses began to be felt—the overuse and erosion of agricultural land caused crop yields to decline, and the nearby ocean was stripped of benthic organisms so the fishermen had to sail greater distances in larger canoes. As resources became inadequate to support the growing population, those in power appealed to the gods. They redirected community resources to carve worshipful images of unprecedented size and power. The people cut down more large trees for ropes and rolling logs to place the heads on huge carved platforms.

By this time the island's seabirds had been consumed, and no new birds came to nest. The rats that had hitchhiked to the island on the first canoes were raised for food. Palm seeds were prized as a delicacy. All available land was under cultivation. As resources continued to shrink, wars broke out over dwindling food and space. By about 1550 no one could venture offshore to harpoon dolphins or fishes because the palms needed to construct seagoing canoes had all been cut down. The trees used to provide rope lashings were extinct, their wood burned to cook what food remained. The once-lush forest was gone. Soon the only remaining ready source of animal protein was being utilized: the people began to hunt and eat each other.

Central authority was lost and gangs arose. As tribal wars raged, the remaining grasses were burned to destroy hiding places. The rapidly shrinking population retreated into caves from which raids were launched against enemy forces. The vanquished were consumed, their statues tipped into pits and destroyed. Around 1700 the human population crashed to fewer than a tenth of its peak numbers. No statues stood upright when Cook arrived.

Even if the survivors had wanted to leave the island, they could not have done so. No suitable canoes existed; none could be made. What might the people have been thinking as they chopped down the last palm? Generations later, their successors died, wondering what the huge stone statues had been looking for.

As you read this chapter, you may have been struck by similarities between Easter Island and Earth. But the Easter Islanders had no books, and no histories of other doomed societies. Might we be able to learn from their mistakes?

WHAT CAN BE DONE?

 15-27

We think of ourselves as more advanced, more intelligent, more foresighted than the Easter Islanders. Yet our births exceed deaths by about three people per second. The six-

billionth human was born in October 1999. *Each year* there are 90 million more of us, a total equal to the combined populations of Mexico City, Tokyo and Yokohama, São Paulo, New York, Shanghai, and Calcutta. The number of people has tripled in this century and is expected to double again before reaching a plateau sometime in the next century. Another billion humans will join the world population in the next ten years, 92% of them in third-world countries.[3]

This exploding population will not be content with using the same proportion of resources used today. Citizens of the world's least-developed countries are influenced by education and advertising to demand a developed-world standard of living. They look with justifiable envy on the United States, a country with 4.5% of the planet's population that consumes 55% of its raw material resources and 25% of its energy, while generating 30% of industry-related carbon dioxide. Can the world support a population whose expectations are rising as rapidly as their numbers? The burgeoning human population is the greatest environmental problem of all.

We cannot expect science to solve the problem for us. Most of the decisions and necessary actions fall outside pure science in the areas of values, ethics, morality, and philosophy. The solution to environmental problems, if one exists, *lies in education and action.* Each of us is obliged to become informed on issues that affect Earth, its ocean, and its air—to learn the arguments and weigh the evidence. Once informed, we must act in rational ways. Chaining yourself to an oil tanker is not rational, but selecting well-designed, long-lasting, and recyclable products made by responsible companies with minimal impact on the environment (and encouraging others to do so) certainly is.

The present trade-off between financial and ecological considerations is often strongly tilted in the direction of immediate gain, of short-term profit, of immediate convenience. Education may be the only way to modify these destructive behaviors.

Humanity is part of the natural world, not its master. We may be able to learn to live in harmony with this small, beautiful, blue world. True convenience and true progress depend on the preservation of open space, serious and sustained attention to population control, conversion to a steady-state economy instead of one that must grow to stay alive, business incentives for preservation, the use of renewable resources, and, above all, *public education on environmental issues.* We must ask ourselves difficult questions: What is the optimal quality of life? How can I achieve balance between my material needs and the needs of Earth? What do I want to leave for my children? How can I reserve the quiet, renewing ocean for myself, for all species, and for the future? Our cities are crowded and our tempers are short. Times of turbulent change lie before us. The trials ahead will be severe.

Each of us, individually, needs to take a stand. We must preserve the sunsets and fog, the waves to ride, the wind-

---

[3] In the United States alone, the population is growing by the equivalent of four Washington D.C.s every year, another New Jersey every 3 years, another California every 12.

Figure 15.26 An abandoned plastic fishing net drapes the intertidal zone at remote Clarion Island, Revillagigedo Islands, Mexico. Artifacts like this on isolated islands suggest the furious assault to which the ocean is being subjected. The last Easter Islanders would have understood its significance.

blown spray on our faces. Margaret Mead summarized our potential for making a difference: "Never doubt that a small group of thoughtful, committed citizens can change the world. Indeed, it is the only thing that ever has." We need to start now.

QUESTIONS FROM STUDENTS

1. When will we run out of oil?

By the time the world's crude oil supplies have been substantially depleted, the total amount of oil extracted is expected to range from 1.6 to 2.4 trillion barrels. If this estimate is correct, consumption could continue at the present level until some time between 2025 and 2040, at which time it would drop quite rapidly. But, in fact, we will never run *completely* out of oil—there will always be some oil within Earth to reward great effort at extraction. The days of unlimited burning of so valuable a commodity are nearing an end, however. Future civilizations will surely look back in horror at the fact that their ancestors actually burned something as valuable as lubricating oil.

2. The problems of overfishing and bykill concern me. What seafood should I eat, and what should I avoid?

Eat sparingly, if at all:
Sharks, especially shark-fin soup
Swordfish
Bluefin tuna (most canned tuna is OK)
Orange roughy
Grouper
Toothfish

OK to eat:
Pacific cod
Atlantic striped bass (taken offshore)
Atlantic mackerel
Atlantic herring
Farm-raised shrimp
Ranched salmon
Canned tuna with "dolphin safe" labels

(*Sources*: Nash, 1997, p. 67; National Marine Fisheries Statistics fax-on-demand service)

3. What can an individual do to minimize his or her impact on the ocean and the atmosphere?

Remember that Earth and all life forms are interconnected. There are no true consumers, only users: Nothing can truly be thrown away (there is no "away"). We must abandon the pollute-and-move-on ethic that has guided the actions of most humans for thousands of years, and we must work toward a society more in harmony with the fundamental rhythms of life that sustain us. *Our task is not to multiply and subdue Earth*. It may not be too late to change our ways. We need to act individually to effect change. We should think globally and act locally.

So drive fewer miles. Eat lower on the food chain. Turn off lights when they're not being used. Take shorter showers. Recycle newspapers, glass, cans, and plastics. Participate in river or coast cleanups and plant trees. Have fewer children or none at all.

4. Is pollution always a bad thing?

Some forms of pollution bring temporary benefits. For example, some of Florida's once-endangered manatees are thriving at the warm outfalls of coastal power stations. On a larger scale, if increased greenhouse warming does develop, some computer models indicate increased rainfall, longer growing seasons, and increased crop yields over broad latitude bands in the temperate zones of both hemispheres. On the whole, though, the less human intervention in complex natural systems, the better.

5. What are the most dangerous threats to the environment overall?

The underlying causes of the problems discussed in this chapter are (1) human population growth and (2) a growth-dependent economy. Stanford professor Paul Ehrlich has said, "Arresting global population growth should be second in importance on humanity's agenda only to avoiding nuclear war." The present world population, now above 6 billion, seems doomed to reach at least 12 billion before leveling off.

6. What role does public perception play in pollution issues?

A large role, indeed! Until recently, relatively small-scale but highly visible incidents have claimed most of the public's attention and have driven us to action. The messy breakup of the oil tanker *Torrey Canyon* off the southern coast of England in 1967 galvanized world opinion and added impetus to the present environmental movement. More recently, in the summer of 1988 beachgoers were horrified to discover that more than 80 kilometers (50 miles) of northern New Jersey and Long Island beaches had been temporarily closed because of medical debris littering the shore. Some of the dozens of vials of blood, syringes, stained bandages, and surgical sutures tested positive for the viruses that cause AIDS and hepatitis B. Similar incidents occurred in Rhode Island and Massachusetts.

Appalling and visible though such incidents are, their long-term effects on the ocean as a whole are negligible. Public attention has recently turned to issues of larger consequence. The drought and heat of the past few summers brought terms like *greenhouse effect* and *ozone layer* to local newspapers and dinner table conversation. Weekend fishermen worry about eating their catch. They wonder about the unseen threats as much as the obvious ones. Perceptions are changing.

CHAPTER SUMMARY

Marine resources include physical resources such as oil, natural gas, building materials, and chemicals; biological resources such as seafood and kelp; and nonextractive resources like transportation and recreation. The contribution of marine resources to the world economy has become so large that international laws now govern their allocation.

In spite of their abundance, marine resources provide only a fraction of the worldwide demand for raw materials, human food, and energy. Similar resources on land can usually be obtained more safely and at lower cost.

Our species has always exercised its capacity to consume resources and pollute its surroundings, but only in the last few generations have our efforts affected the ocean and the atmosphere on a planetary scale. The introduction into the biosphere of unnatural compounds (or natural compounds in unnatural quantities) has had, and will continue to have, unexpected detrimental effects. The destruction of marine habitats and the uncontrolled harvesting of the ocean's living resources have also disturbed delicate ecological balances. We find ourselves in difficult situations for which solutions do not come easily.

TERMS AND CONCEPTS TO REMEMBER

algin	mariculture
aquaculture	marine pollution
biodegradable	maximum sustainable yield
biological amplification	nonextractive resource
biological resource	nonrenewable resource
bykill	overfishing
chlorinated hydrocarbons	ozone
chlorofluorocarbons (CFCs)	ozone layer
desalination	physical resource
drift net	potable water
exclusive economic zone (EEZ)	renewable resource
	sewage sludge
greenhouse effect	territorial waters
greenhouse gases	United States Exclusive
high seas	Economic Zone

STUDY QUESTIONS

1. Distinguish between physical and biological resources, and between renewable and nonrenewable resources.

2. What are the three most valuable physical resources from the ocean? How does the contribution of each to the world economy compare to the contribution of that resource derived from land?

3. Does the ocean provide a substantial percentage of all protein needed in human nutrition? Of all animal protein? What is the most valuable biological resource? The fastest growing fishery?

4. What is a nonextractive resource? Give some examples.

5. What is pollution? What factors determine how dangerous a pollutant is?

6. Why is refined oil more hazardous to the marine environment than crude oil? Which is spilled more often?

7. What heavy metals are most toxic? How do these substances enter the ocean? How do they move from the ocean to marine organisms and people?

8. Few synthetic organic chemicals are dangerous in the very low concentrations in which they enter the ocean. How are these concentrations increased? What can be

the outcome when these substances are ingested by organisms in a marine food chain?

9. What are the signs of overfishing? How does the fishing industry often respond to these signs? What is the result?

10. What synthetic chemicals appear to be causing depletion of Earth's protective ozone layer? What is the likely result?

11. What is the greenhouse effect? Is it always detrimental? What gases contribute to the greenhouse effect? Why do most scientists believe Earth's average surface temperature will increase over the next few decades? What may result?

12. What, if anything, can be done to minimize the environmental disturbances that result from the exploitation of marine resources?

FOR FURTHER STUDY

Avery, W. H., and C. Wu. 1994. *Renewable Energy from the Ocean.* New York: Oxford University Press.

Bakun, A. 1996. *Patterns in the Ocean: Ocean Processes and Marine Population Dynamics.* San Diego: University of California, California Sea Grant. A senior fisheries resource officer looks at the underlying dynamics of marine ecosystems, fishing, and overfishing.

Broecker, W. S. 1995. "Chaotic Climate." *Scientific American,* November, 62–7. The near-random state of climate makes long-term prediction very difficult.

Brown, L. 2000. *State of the World.* New York: Norton. An annual updating of facts about the state of our relationship with the environment.

Caldeira, K., and P. B. Duffy. 2000. "The Role of the Southern Ocean in Uptake and Storage of Anthropogenic Carbon Dioxide." *Science* 287 (no. 5453, 28 January): 620–1.

Cane, M. A., et al. 1997. "Twentieth-Century Sea Surface Temperature Trends." *Science* 275 (14 February): 957–60. Evidence for global warming is presented.

Carson, R. 1962. *Silent Spring.* Boston: Houghton Mifflin. This elegantly written book is said to have begun the modern era of environmental awareness in the United States.

Cohen, J. E. 1995. *How Many People Can the Earth Support?* New York: Norton. The author doubts a further doubling will occur. About 14% of the world's population presently live in the 30 countries with zero or negative population growth.

Costanza, R., et al. 1998. "Principles for Sustainable Governance of the Oceans." *Science* 281 (no. 5374, 10 July): 198–9. Key article outlining six core principles to guide governance.

Cramer, D. 1995. "Troubled Waters." *Atlantic Monthly,* June, 22–6. The troubled state of the North Atlantic fishery.

Hasselmann, K. 1997. "Are We Seeing Global Warming?" *Science* 276 (no. 5314, 9 May): 914–5. Maybe.

Hatvell, C. D., et al. 1999. "Emerging Marine Diseases—Climate Links and Anthropogenic Factors." *Science* 285 (no. 5433, 3 September): 1505–10. Mass mortalities due to disease have recently affected major groups of oceanic organisms. Is it something we did?

Holmes, B. 1994. "Biologists Sort the Lessons of Fisheries Collapse." *Science* 264 (27 May): 1252–3.

Jacobson, J. L., and A. Rieser. 1998. "The Evolution of Ocean Law." *Scientific American Presents,* Fall, 100–5.

Junger, S. 1997. *The Perfect Storm.* New York: Norton. A perfectly riveting, perfectly horrifying view of the worst possibilities in the lives of fishermen. Exceptionally well written, highly recommended.

Kaiser, J. 1999. "The Exxon Valdez's Scientific Gold Rush." *Science* 284 (no. 5412, 9 April): 247–9. Scientists are learning valuable lessons from research done in the disaster's wake.

Kerr, R. A. 1991. "A Lesson Learned, Again, at Valdez." *Science* 252 (no. 5004, 19 April): 371. The cleanup at Prince William Sound was more damaging than the effects of the crude oil spilled.

Kerr, R. 2000. "Draft Report Affirms Human Influence." *Science* 288 (28 April), 589–90. Twentieth century warming was unique in the millennium. Human influence is thought responsible.

Kerr, R. 1998. "The Next Oil Crisis Looms Large—and Perhaps Close." *Science* 281 (no. 5380, 21 August): 1128–31. Most economists see another 50 years of relatively cheap oil, but some geologists believe oil will begin to run out in only 10 years.

Malakoff, D. 1997. "Extinction on the High Seas." *Science* 277 (no. 5325, 25 July): 446–8. Biologists are realizing the ocean is not too vast to prevent extinctions.

Miller, G. T. 2000. *Living in the Environment.* 11th ed. Belmont, CA: Wadsworth. Excellent reference and general text on environmental science.

Nash, M. 1997. "The Fish Crisis." *Time,* 11 August, 65–7. A compact and eye-opening survey.

National Geographic, November 1995, is largely devoted to oceanic issues.

Oceanus 33, no. 2 (Fall/Winter 1990) deals with ocean disposal.

Oceanus 38, no. 2 (Fall/Winter 1995) is devoted to marine biodiversity.

Parfit, M. 1995. "Diminishing Returns." *National Geographic,* November, 2–39. A major article, magnificently illustrated, on the state of fisheries.

Safina, Carl. 1995. "The World's Imperiled Fish." *Scientific American,* November, 46–52. Pivotal article by an expert. Outlines a sad tale of overexploitation.

Schmidt, K. 1998. "Ecology's Catch of the Day." *Science* 281 (no. 5374, 10 July): 192–3. A massive federal undertaking to learn where marine fish live could lead to major changes in the management of U.S. fisheries.

Science 281, no. 5374 (10 July 1998) is subtitled "Chemistry and Biology of the Oceans."

Scientific American, June 1992, is subtitled "Meeting the Challenge of Sustainable Development."

Scientific American Presents, Fall 1998, is devoted to "The Oceans."

Suess, E., et al. 1999. "Flammable Ice." *Scientific American,* November, 76–83. Methane-laced ice crystals in the seabed store more energy than all the world's fossil fuel reserves combined.

Taylor, K. 1999. "Rapid Climate Change." *American Scientists* 87 (no. 4): 320–7. New evidence suggests Earth's climate can change dramatically in only a decade. Greenhouse gases might be involved.

Wiley, J. P. 1999. "We're Scraping Bottom." *Smithsonian,* April, 25–8. Drag nets and dredges destroy the biome.

• For additional readings, go to InfoTrac College Edition, your online research library at:

http://infotrac.thomsonlearning.com/

Afterword

The marine sciences are at the threshold of a new age. The recent revolutions in biology and geology are being assimilated, and the road ahead seems clearer. A revolution in the design of sampling devices, robot submersible vehicles, and data processing has brought new vigor to oceanography. Satellite-borne sensors can provide data in an instant that would have taken years to collect using surface ships. Shipboard technology has become so sophisticated that Wyville Thomson or Fridtjof Nansen would hardly recognize our sensors or sampling devices.

The tools may be different, but the spirits of those who use them remain the same. Today's marine scientists are like all the men and women who have gone before: *We want to know about the ocean.* We eagerly search our mailboxes for journals bearing the latest research news, watch television listings for any new ocean shows, inspect new samples with the enthusiasm of little kids, and share our insights with anyone at the drop of a hat. I am personally delighted that you have traveled with me this far. Those of us who enjoy an oceanographic background (and this now includes you) look at Earth with greater understanding than we did before we began. The whole concept of an ocean world appeals to us, gives us profound pleasure, and sobers us with a deep sense of responsibility. In no other field of science do so many ideas interweave to form so rich a tapestry.

Our journey together is over, but before we go our separate ways, I have three last ideas to share:

- *Change* has been a recurrent theme of this book. Earth's climate has changed with time, as has its atmospheric composition, its ocean chemistry, the size and positions of its continents, and its life forms. Earth may seem a calm and stable home, but it is really a violent place for inhabitation by such seemingly delicate objects as living things. Even so, life and the ocean have grown old together. The story of Earth is the story of change and chance; its history is written in the rocks, the water, and the genes of the millions of organisms that have evolved here. We are survivors.

  That survival may now be in question. Change is now progressing at an unnatural rate, and these human-induced changes are imposing stress on natural systems. What we do *with* and *to* the ocean is literally of planetary consequence. In the last century we have developed the physical, chemical, and biological machinery to destroy or rejuvenate the world ocean and all of its life. A painful time of inadvertent global experimentation lies just ahead.

  All of us who love the colors and textures of this small wet world need to act to moderate the negative effects of the looming environmental crisis. In Chinese, the written character for the word *crisis* has two components: danger and opportunity. Informed citizens will express their concern, discuss this concern with others, and act whenever possible to minimize the threats and take advantage of new opportunities. Intelligence and beauty must triumph; we have no other rational alternative.

- Appreciation of the ocean doesn't come exclusively from the realm of science. Philosophers, artists, composers, and poets have had much to say about the sea. Read Homer's description of the ocean in *The Odyssey* (try Books IV, X, and XI). See how Lord Byron's feeling for the ocean colors his poetry (see, for instance, *Childe*

Conversion Factors

Area

1 square inch (in^2)	6.45 square centimeters
1 square foot (ft^2)	144 square inches
	0.0929 square meter
1 square centimeter (cm^2)	0.155 square inch
	100 square millimeters
1 square meter (m^2)	10^4 square centimeters
	10.8 square feet
1 square kilometer (km^2)	247.1 acres
	0.386 square mile
	0.292 square nautical mile

Mass

1 kilogram (kg)	2.2 pounds
	1,000 grams
1 metric ton	2,205 pounds
	1,000 kilograms
	1.1 tons
1 pound	16 ounces
	454 grams
	0.45 kilogram
1 ton	2,000 pounds
	907.2 kilograms
	0.91 metric ton

Volume

1 cubic inch (in^3)	16.4 cubic centimeters
1 cubic foot (ft^3)	1,728 cubic inches
	28.32 liters
	7.48 gallons
1 cubic centimeter (cc; cm^3) 0.61 cubic inch	1 milliliter
1 liter	1,000 cubic centimeters
	61 cubic inches
	1.06 quarts
	0.264 gallon
1 cubic meter (m^3)	10^6 cubic centimeters
	264.2 gallons
	1,000 liters
1 cubic kilometer (km^3)	10^9 cubic meters
	10^{15} cubic centimeters
	0.24 cubic mile

Length

1 micrometer (μm)	0.001 millimeter
	0.0000349 inch
1 millimeter (mm)	1,000 micrometers
	0.1 centimeter
	0.001 meter
	0.03937 inch
1 centimeter (cm)	10 millimeters
	0.394 inch
	10,000 micrometers
1 meter (m)	100 centimeters
	39.4 inches
	3.28 feet
	1.09 yards
1 kilometer (km)	1,000 meters
	1,093 yards
	3,281 feet
	0.62 statute mile
	0.54 nautical mile
1 inch (in)	25.4 millimeters
	2.54 centimeters
1 foot (ft)	30.5 centimters
	0.305 meter
1 yard	3 feet
	0.91 meter
1 fathom	6 feet
	2 yards
	1.83 meters
1 statute mile	5,280 feet
	1,760 yards
	1,609 meters
	1.609 kilometers
	0.87 nautical mile
1 nautical mile	6,076 feet
	2,025 yards
	1,852 meters
	1.15 statute miles
1 league	15,840 feet
	5,280 yards
	4,804.8 meters
	3 statute miles
	2.61 nautical miles

Speed

1 statute mile per hour	1.61 kilometers per hour
	0.87 knot
1 knot (nautical mile per hour)	51.5 centimeters per second
	1.15 miles per hour
	1.85 kilometers per hour
1 kilometer per hour	27.8 centimeters per second
	0.62 mile per hour
	0.54 knot

Temperature

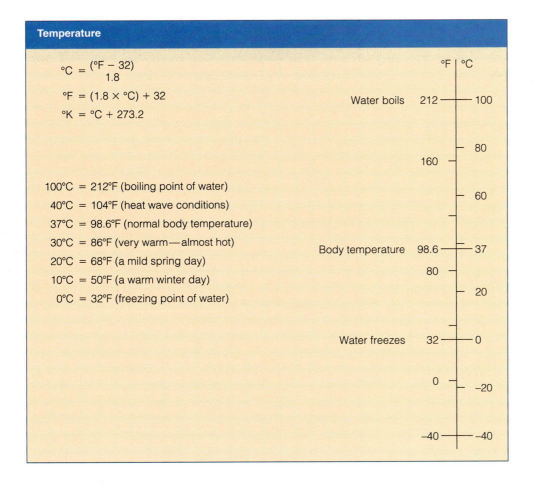

$$°C = \frac{(°F - 32)}{1.8}$$

$$°F = (1.8 \times °C) + 32$$

$$°K = °C + 273.2$$

100°C = 212°F (boiling point of water)

40°C = 104°F (heat wave conditions)

37°C = 98.6°F (normal body temperature)

30°C = 86°F (very warm—almost hot)

20°C = 68°F (a mild spring day)

10°C = 50°F (a warm winter day)

0°C = 32°F (freezing point of water)

	°F	°C
Water boils	212	100
		80
	160	60
Body temperature	98.6	37
	80	20
Water freezes	32	0
	0	−20
	−40	−40

Some Familiar Metric Approximations

Measurement	Metric Unit	Approximate Size of Unit
Length	millimeter	diameter of a paper clip wire
	centimeter	a little more than the width of a paper clip (about 0.4 inch)
	meter	a little longer than a yard (about 1.1 yards)
	kilometer	somewhat farther than ½ mile (about 0.6 mile)
Mass (Weight)	gram	a little more than the mass (weight) of a paper clip
	kilogram	a little more than 2 pounds (about 2.2 pounds)
	metric ton	a little more than a short ton (about 2,200 pounds)
Volume	milliliter	a fifth of a teaspoon
	liter	a little larger than a quart (about 1.06 quarts)
Pressure	kilopascal	atmospheric pressure is about 100 kilopascals

Source: U.S. Metric Board Report.

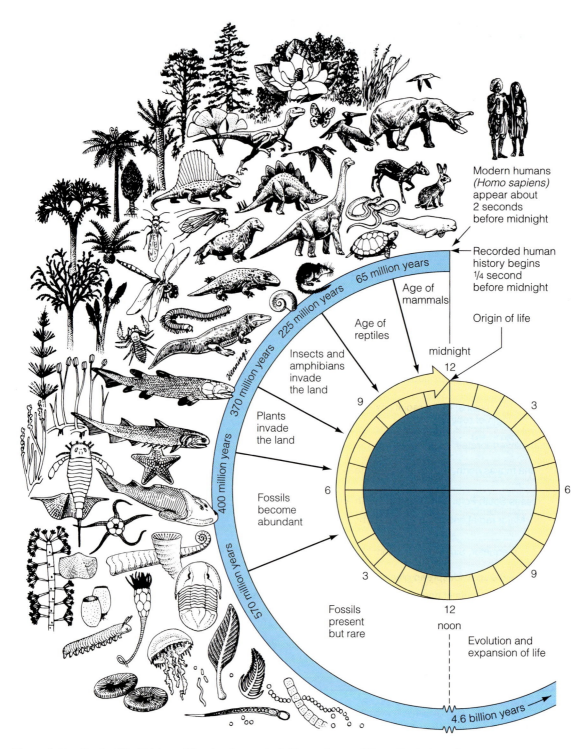

Figure 2 Greatly simplified history of the development of different forms of life on Earth through biological evolution, compressed to a 24-hour time scale.

APPENDIX III

Latitude and Longitude, Time, and Navigation

The ocean is large and easy to get lost in. A backyard, like that shown in **Figure 1,** is smaller, but we can still be lost in it if we don't have a frame of reference. Note that the yard is framed by a fence. We can refer to this frame to establish our position—in this case, at the intersection of perpendicular lines drawn from fence posts 2 and C. Many towns are arranged in this way: Fourth and D streets intersect at a precise spot based on the municipal frame of reference.

But the World Is Round: Spherical Coordinates

If the world were flat, a simple scheme of rectangular coordinates would serve all mapping purposes—a rectangle, like the yard in Figure 1, has four sides from which to measure. A sphere has no edges, no beginnings or ends, so what shall we use as a frame of reference for Earth? Since Earth turns, the poles, the axis of rotation, are the only absolute points of reference. We can draw an imaginary line equidistant from the North and South Poles, a line that *equates* the globe into northern and southern halves: the equator. Other lines, drawn parallel to the equator, further divide the sphere north and south of the equator. These lines, or parallels, are lines of **latitude** (**Figure 2**).

We can further subdivide Earth by drawing lines at regular intervals through both poles. Note that unlike the parallels, these lines, called meridians, are all equally long. Meridians are lines of **longitude** (**Figure 3**).

If you travel north from the equator, you can count the parallels (lines of latitude) that cross your path to find out how far you have gone. Likewise, if you travel east from a reference meridian, you can count the meridians (lines of longitude) that cross your path to find out how far you have gone. Just as a football player on the field knows his distance from the goal by the yard lines that cross his run, so you know how far north or east you have gone by the lines that have crossed your path.

Since there are no continuous lines of fence posts on the spherical Earth, our reference frame for latitude and

Figure 1

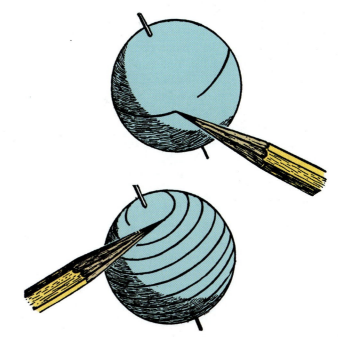

Figure 2

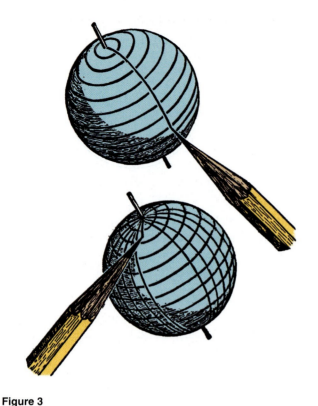

Figure 3

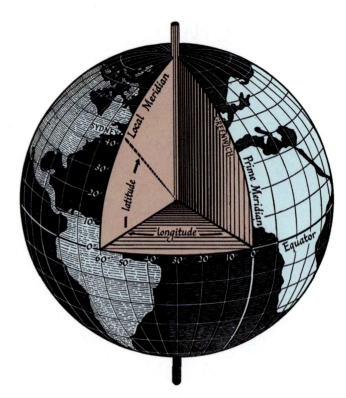

Figure 4 The latitude and longitude of Sydney, Canada.

longitude must be marked from the equator and poles by some other means. This is done by degrees.

Why Degrees?

Degrees measure fractions of a circle. We need to know what fraction of Earth's circumference separates us from the equator and from the reference meridian to have a definite idea of our location.

Babylonian astronomers first divided the circle into 360 degrees (°). Why 360? The moon cycles around Earth every 30 days. It takes about 12 months ("moonths") to make a year. Thus, 30 × 12 = 360, the number of days they supposed was in a year. Circles were divided the same way. As we saw in Chapter 1, the Greek librarian Hipparchus applied this division to the surface of Earth.

In **Figure 4,** we have marked the position of Sydney, Canada. A line drawn from Sydney to the center of Earth intersects the plane of the equator at an angle of 46° to the north. That is its latitude.

The reference meridian, the meridian from which all others are marked, is known as the prime meridian. Unlike the equator, there is no earthly reason why the prime meridian should pass through any particular place. It passes through Greenwich, England, because an international agreement signed in 1884 decreed it so. The meridian on which Sydney, Canada, lies intersects the plane of the prime meridian at an angle of 60°. The angular distance of Sydney from the prime meridian is 60° to the west. That is its longitude. So its position is 46°N 60°W.

We can do this for each hemisphere. A line drawn to the center of Earth from Sydney, *Australia*, intersects the plane of the equator at an angle of 34° south latitude. Sydney, Australia, lies 151° east of the prime meridian. Thus, its position is 34°S 151°E. (Note that the greatest possible longitude is 180°; once you pass 180°, the line opposite the prime meridian, you begin to come around the other side of Earth, and the angle to Greenwich decreases.)

What Does Time Have to Do with This?

Meridians are often numbered from prime meridian in 15° increments. Earth takes 24 hours to complete a 360° rotation. Divide 360° by 24 hours and you get 15, the number of degrees the sun moves across the the sky in 1 hour. Meridians on a globe are often spaced to represent 1 hour's turning of Earth toward or away from the sun, toward or away from the moon.

You can use this fact to find your east-west position, your longitude. Imagine that you have a radio that can tell you the precise time of noon at Greenwich.[1] If your local noon comes *before* Greenwich noon, you are east of Greenwich. For instance, if the sun is highest in your sky at 10 A.M. Greenwich time, you are 2 hours before—30° east of—Greenwich. Earth must turn 2 more hours before the sun will shine directly above

---

[1]Any shortwave radio will do. Tune it to 2.5, 5, 10, 15, or 20 mHz for radio stations WWV (Colorado) or WWVH (Hawaii). These stations broadcast time signals giving a measure of coordinated universal time, an international time standard based on the time at Greenwich. For a telephone report of coordinated universal time, call WWV at (303) 499-7111, or go to www.time.gov.

the Greenwich meridian. If your local noon is *after* Greenwich noon, you are west of Greenwich. Suppose that the sun is at high noon and your chronometer, set at Greenwich time, says 6 P.M. That means that Earth has been turning 6 hours since noon at Greenwich, and 6 hours times 15° per hour is 90°. That's your longitude relative to Greenwich: 90°W.

Navigation

Longitude is half the problem. To find latitude and obtain a position, we need to measure the angle north or south of the equator. But we can't use the time difference between local noon and Greenwich noon to determine longitude because the sun moves from east to west and we want to measure north-south position. Instead, we use the angle of the North Star above the horizon. Polaris, the current North Star, lies almost exactly above the North Pole. If we were standing at the North Pole, the North Star would appear almost directly overhead; ideally, the angle from the horizon to the star would be 90°, the same as the latitude of the North Pole. At Sydney, Canada, the angle from the horizon to the star would be about 46°—again, the same as the latitude. What would the angle of Polaris be at the equator, 0° latitude? (If you enjoyed your high school geometry course, you might try to prove that the angle from the horizon to Polaris is equal to the latitude at any position in the Northern Hemisphere.)

Polaris is not visible in the Southern Hemisphere, so how can we find south latitude? By finding the angle above the horizon of other stars. In practice, navigators in both hemispheres use a sextant to measure angles from the horizon to selected stars, planets, the moon, and the sun. The time of the observation is carefully noted. The navigator takes these readings to his or her stateroom, consults a series of mathematical tables, does some relatively simple calculations to compensate for observational errors, and comes up to the pilothouse with the vessel's latitude and longitude, accurate (in the best of circumstances) to within ½ mile, marked on a small slip of paper. The daily results are always entered into the ship's log.

New Tricks

Discovering position by measuring the singular positions of heavenly bodies—celestial navigation—is a dying art. Global positioning satellites, loran-C, inertial platforms, radar, and other electronic wonders have largely replaced the romance of a navigator standing on the bridge squinting through a sextant. The slip of paper has been supplanted by the glow of back-lit liquid-crystal readouts or a chart with an X marking the ship's position, accurate to within 50 feet, feeding out of a slot. Still when the power fails, the human navigator becomes the most popular person on board.

APPENDIX IV

Maps and Charts

It is easier to draw a diagram to show someone how to get to a place than to describe the process in words. For centuries, travelers have made special diagrams—maps and charts—to jog their own memories and to show others how to reach distant destinations. A **map** is a representation of some part of Earth's surface, showing political boundaries, physical features, cities and towns, and other geographical information. A **chart** is also a representation of Earth's surface, but it has been specially designed for convenient use in navigation. It is intended to be worked on, not merely looked at. A **nautical chart** is primarily concerned with navigable water areas. It includes information such as coastlines and harbors, channels, obstructions, currents, depths of water, and the positions of aids to navigation.

Any flat map or chart is necessarily a distortion of the spherical Earth. If we roll a flat sheet of paper around a globe to form a cylinder, the paper will contact the globe only along one curve. Let's assume that it's the equator. If the lines of latitude and longitude on the globe are covered with ink, only the equator will contact the paper and print an exact replica of it-self. Unroll the cylinder, and that part of the new map will be a perfect representation of Earth. To include areas north and south of the equator, we will have to "throw them forward" onto the paper; we need to *project* them in some way.

Now imagine our globe to be a translucent sphere. If we place a bright light at its center, we can project the lines of latitude and longitude onto the rolled paper cylinder (**Figure 1**). Careful tracing of these lines will result in a map, but the areas away from the equator will be distorted: The farther from the equator, the greater the distortion. A useful modification of this projection—one that does not distort high latitudes as dramatically—was devised by Gerhardus Mercator, a Flemish cartographer who published a map of the world in 1569. Though landmasses and ocean areas are not depicted as accurately in a Mercator projection as they would be on a globe, such a map is still useful because it enables mariners to steer a course over long distances by plotting straight lines.

The distortion in Mercator projections has led generations of schoolchildren to believe that Greenland is the same size as South America (**Figure 2**). Mercator charts can distort

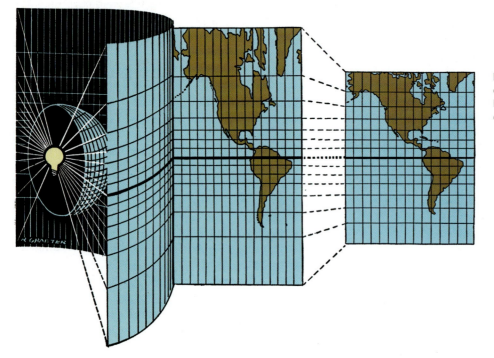

Figure 1 Central projection of a globe upon a cylinder, and a modified map structure, the Mercator, made to the same scale along the equator.

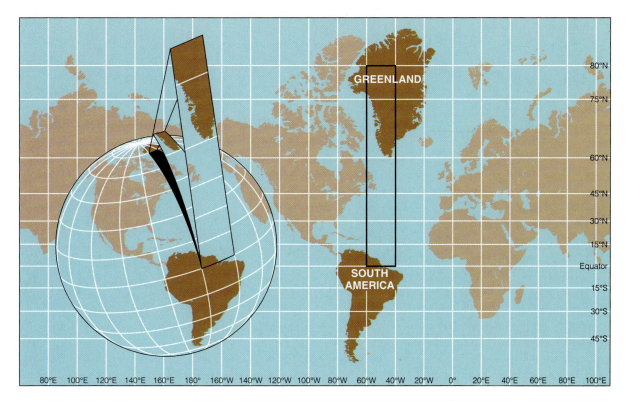

Figure 2 A gore of the globe peeled and protected according to the scheme devised by Gerhardus Mercator. This is the protection used on modern sailing charts. Note that this projection's distortion at high latitudes makes Greenland and South America appear about the same size. Next time you're near a globe, check their real sizes.

our perceptions of the ocean as well: The area of the continental shelves at high latitudes, the amount of primary productivity in the polar regions, and the importance of ocean currents at the northerly or southerly extremes of an ocean basin may be exaggerated if presented in Mercator projection. The projection used in this book—a further modification of the Mercator projection known as the Miller projection—was chosen for its more accurate representation of surface area at high latitudes.

Mapmakers have invented other projections, each with advantages and disadvantages for particular uses. Some are conical projections: a flat sheet of paper wrapped into a cone with its edge touching the globe at a line of latitude north (or south) of the equator and the point of the cone above the North (or South) Pole. Conical projections do not distort high-latitude areas in the same way a Mercator projection does, and if drawn for the ocean area in which a mariner is sailing, can be used to draw great circle routes as straight lines. However, the distortions inherent in a conical projection prevent it from being used to represent more than about one-third of the globe on a single sheet of paper. Other projections try to minimize

distortion around a specific location. All map and chart projections are distorted in some way; a sphere cannot be flattened onto a plane without deformation. Marine scientists necessarily become familiar with various chart projections and are careful to use the proper chart for its intended purpose.

FOR FURTHER STUDY

Herring, T. 1996. "The Global Positioning System." *Scientific American,* February, 44–50.

Krause, G., and M. Tomczak. 1995. "Do Marine Scientists Have a Scientific View of the Earth?" *Oceanography* 8 (no. 1): 11–16. Chart distortions often distort our interpretation of data, as this well-illustrated paper demonstrates.

Wilford, J. N. 1998. "Revolution in Mapping." *National Geographic,* February, 6–39. The usual thorough treatment of a rapidly changing topic.

APPENDIX V

Working in Marine Science

Working in the marine sciences is wonderfully appealing to many people. They sometimes envision a life of diving in warm, clear water surrounded by tropical fish, or descending to the seabed in an exotic submersible outfitted like Captain Nemo's fictional submarine in *20,000 Leagues Under the Sea*, or living with intelligent dolphins in a marine life park. Then reality sets in. There are rewards from working in the marine sciences, but they tend to be less spectacular than the first dreams of students looking to the ocean for a life's work.

A marine science worker is paid to bring a specific skill to a problem. If that problem lies in warm, tropical water or in a marine park, fine. But more likely, the problem will yield only to prolonged study in an uncomfortable, cold, or dangerous environment. The intangible rewards can be great; the physical rewards are often slim. Having said that, let me add that no endeavor is more interesting or exciting, and few are more intellectually stimulating. Doing marine science is its own reward.

Training for a Job in Marine Science

Marine science is, of course, science. And science requires mathematics—you need math to do the chemistry, physics, measurements, and statistics that lie at the heart of science. Your first step in college should be to take a math placement test, enroll in an appropriate math class, and spend time doing math. *Math is the key to further progress in any area of marine science.*

With your math skills polished, start classes in chemistry, physics, and basic biology. Surprisingly, except for one or two introductory marine science classes, you probably won't take many marine science courses until your junior year. These introductory classes will be especially valuable because a balanced survey of the marine sciences can aid you in selecting an appealing specialty. Then, with a good foundation in basic science, you can begin to concentrate in that specialty.

Other skills are important too. The ability to write and speak well is crucial in any science job. Also critical is computer literacy. Expertise in photography or foreign languages or the ability to field-strip and rebuild a diesel engine or hydraulic winch will put you a step above the competition at hiring time. Certification as a scuba diver is almost mandatory; you can never have too much diving experience. (Remember, though, diving is only a tool, a way to deliver an informed set of eyes and an educated brain to a work site.) You should be in

good health. Indeed, good aerobic fitness is essential in most marine science jobs; stamina is often a crucial factor in long experiments under difficult conditions at sea. It is also desirable to be physically strong—marine equipment is heavy and often bunglesome. And it helps greatly if you are not prone to seasickness.

Deciding what school to attend will depend on your skills. Readers of this book will probably be enrolled in a general oceanography course in a college or university. The first step would be to discuss your interests with your professor (or his or her teaching assistants). You'll need to attend a four-year college or university to complete the first phase of your training. If you're attending a two-year institution, picking a specific transfer institution can come later, but keep a few things in mind: No matter where you take your first two years of training, you need thorough preparation in basic science. You should attend an institution with strengths in the area of your specialty (such as geology, biology, and marine chemistry). And you should be reasonable in your expectations of acceptance if you're a transfer student (that is, don't try for Stanford or Yale with a B average).

Another thing: Most marine scientists have completed a graduate degree (a master's degree or doctorate). Most graduate students hold teaching or research assistantships (that is, they get paid for being grad students). In all, progress to a final degree is a long road, but the journey is itself a pleasure.

If the thought of four or more years of higher education doesn't appeal, does that mean there's no hope? Not at all. Many students begin a program with the goal of becoming a marine technician, animal trainer at a marine life park, marina or boatyard employee or manager, or crew member on a private yacht. Those jobs don't always require a bachelor's degree. Jobs at Sea World and other marine theme parks do require athletic ability, extreme patience, public speaking skills, a love of animals, and, usually, diving experience. Few positions are available, but there is some turnover in the ranks of junior trainers, and being hired is certainly possible.

Becoming a marine technician is an especially attractive alternative to the all-out chemistry-physics-math academic route. For every highly trained marine scientist, there are perhaps five technical assistants who actually do the experiments, maintain the equipment, work daily with organisms, and build special apparatus. Marine technicians tend to spend more time at hands-on tasks than marine scientists. Most of these folks (including the author of the letter that ends this appendix) have the equivalent of a two-year technical degree, usually from a community college.

Don't quit your job, burn your bridges, leave your family, sell your possessions, and dedicate yourself monklike to marine science. Do some investigation. Nothing is as valuable as *actually going out and talking to people who do things that you'd like to do.* Ask them if they enjoy their work. Is the pay OK? Would they start down the same road if they had it to do all over again? You may decide to expand your involvement in marine science in a more informal way, by becoming a volunteer; joining the Sierra Club, Audubon Society, Greenpeace, or other environmental group; working for your state's fish and game office as a seasonal aide; or attending lectures at local colleges and universities.

If you decide to continue your education, don't be discouraged by the time it will take. Have a general view of the big picture, but proceed one semester at a time. Again, remember that the educational journey is itself a great pleasure. Don Quixote reminds us of the joys of the road, not the inn.

The Job Market

Marine science is very attractive to the general public. People are naturally drawn to thoughts of working in the field. Unfortunately, there aren't a great many jobs in the marine sciences. But there will always be some jobs, and people will fill them. Those people will be the best prepared, most versatile, and most highly motivated of those who apply. Perhaps not surprisingly, marine biology is the most popular marine science specialty. Unfortunately, it is also the area with the smallest number of nonacademic jobs. Museums, aquariums, and marine theme parks employ biologists to care for animals and oversee interpretive programs for the public. A few marine biologists are employed as monitoring specialists by water management agencies like sanitation districts, which discharge waste into the ocean. Electrical utilities that use seawater to cool the condensers in power-generating plants almost always have a handful of marine biologists on staff to watch the effects of discharged heat on local marine life and to write the reports required by watchdog agencies. State and federal agencies employ marine biologists to read and interpret those documents and to set standards. Relatively small businesses, like private shipyards, agricultural concerns, and chemical plants, can't afford their own staff biologists, so private consulting firms staffed by marine biologists and other specialists have arisen to assist in the preparation of the environmental impact reports required of businesses under various legislation.

There are more jobs in physical oceanography: marine geology, ocean engineering, and marine chemistry and physics. Thousands of marine geologists work for oil and mineral companies; indeed, with the increasing emphasis on offshore resources, the market for these people may be increasing. Marine engineers are needed to design, construct, and maintain offshore oil rigs, ships, and harbor structures. Marine chemists are hard at work figuring ways to stop corrosion and to extract chemicals from seawater. Physicists are vitally interested in the transmission of underwater sound and light, in the movement of the ocean, and in the role the ocean plays in global weather and climate. Economists, lawyers, writers, and mathematicians also work in the marine science field.

Many biological and physical oceanographers are teachers and professors. Indeed, there are nearly as many marine scientists employed in the academic world as there are in private industry and government. If you like the idea of teaching, you might consider this avenue. The demand for science teachers at all educational levels is already great and is expected to increase.

Four factors will be significant in influencing your employability:

1. *Experience.* Employers are favorably impressed by experience, especially work experience related to the duties of the position for which you are applying. Volunteer work counts.
2. *Grades.* Good grades are important, especially for positions in government agencies. A grade point average of 3.0 or higher in all college work increases your chances of employment and should give you a higher starting salary.
3. *Geographical availability.* Don't restrict yourself geographically. Not everyone can work in Hawaii or California, but four out of ten marine scientists work in just three states: California, Maryland, and Virginia.
4. *Diversification.* Again, mastery of more than one specialty gives you an employment edge. Being a plankton connoisseur and being able to repair a balky computer while ordering in-port supplies over a radiotelephone in Spanish makes a lasting impression.

Report from a Student

Students in marine science programs graduate, get jobs, and move on. One of the pleasures of being a professor is hearing from them. One of our former students, an employee of the Marine Science Institute at the University of California, Santa Barbara, recently reported his activities as part of a team using the submersible *Alvin* to investigate plumes of warm water issuing from hydrothermal vents along the southern Juan de Fuca Ridge. The nature of his work—and his enthusiasm for it—is clearly evident in this excerpt. Dan Dion writes:

> The buoyant plume experiment wasn't going very well. The chemistry dives were pushed back because of technical difficulties and poor weather (rough seas cut two dives). The first two buoyant plume dives ended in failure. The first one because of mechanical/electrical problems, the second because of a computer crash. Everyone worked around the clock to get things in order for dive 2440. I was scheduled to go down with John Trefrey, from the Florida Institute of Technology. Cindy Van Dover was our pilot. We launched *Alvin* right on schedule at 0800, and descended from the glacier blue water into the bioluminescent snowstorm of the euphotic zone. During the hour and a half descent we listened to the music of Enya in the soft light of the sub as we busily prepared ourselves for the experiment: booting up the computers, loading film in the cameras, tapes in the recorders, etc. We had

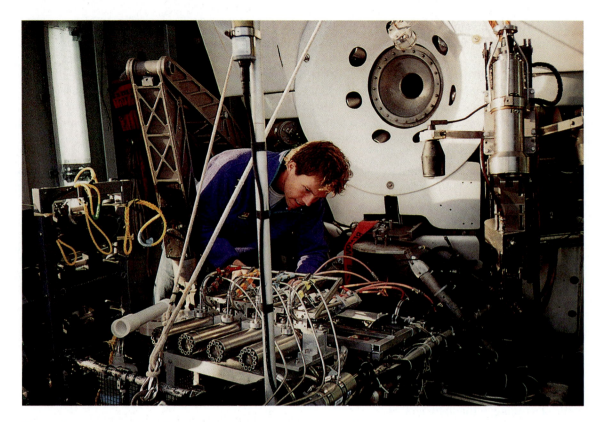

Figure 1 Dan Dion attaching hydraulic actuators to water-sampling bottles, in preparation for a dive by *Alvin*. The bottles were part of a sampling program that included measurements of water conductivity, transmissivity, temperature, and iron and manganese ion content near a hydrothermal vent. This information was later merged with data from transponder navigation to obtain a three-dimensional map of plume structure and chemistry.

three laptop computers to deal with in the cramped spaces of the sub. I was in charge of two of them, one that plotted our in-sub navigation (from transponders), and the other that controlled and recorded data from the [continuous temperature-depth-conductivity probe]. The third laptop was connected to the chemical analyzer and John was in control of that. All of the instruments were operating perfectly. I periodically saved the computer file in the event of another crash. We reached the bottom right on target; Monolith Vent was in sight, 2261 meters below the surface. We did a video survey of the vent, especially a chimney that was rapidly growing back after the geologists had decapitated it just a few days earlier. We ascended to 55 meters-off-bottom and began our drive-throughs. To me, the navigator, it was the ultimate video game. From the computer screen I would guide *Alvin* through a dark abyss, calling out headings that would maneuver us into a "lawn mower" pattern crisscrossing the plume. It was quite visible, and even beautiful; wispy, intricate patterns of "smoke" which seemed to dance like graceful ghosts. We completed passes at 35, 20, 10, and 5 meters above the bottom, then one last one at 45 meters. Eight hours of sub time went by so quickly! Our dive was a huge success; in addition to all the samples we obtained, we generated over 25 megabytes of data. I used everything I learned . . . from computer skills to navigation and marlinspike seamanship (and, of course, chemistry!).

Marine science is equipment training sessions, long cruises, seminars and lectures, visiting experts, hot sand vol-

Figure 2 Dan Dion enters the hatch of *Alvin* to visit hydrothermal vents 2,261 meters (7,416 feet) below the surface off the coast of Oregon.

People in Marine Science

The last two decades have seen a significant change in the profile of people involved in oceanographic research in the United States. Some of these changes are summarized here.

1. The number of oceanographers employed in all sectors increased from 1,130 in 1973 to 2,460 in 1989.
2. The percent of oceanographers who consider teaching as their primary activity decreased from 21% in 1973 to 11% in 1989.
3. The percent of oceanographers who consider research as their primary activity remained stable at 40%.
4. In 1989, a majority of Ph.D. oceanographers (60%) were employed at educational institutions.
5. The distribution of oceanographers by discipline (biological, chemical, geological, physical) has remained almost constant.
6. The percentage of women oceanographers has increased from 4% in 1973 to 11% in 1989. In 1992, 30% of students at major oceanographic institutions were women.
7. The two largest oceanographic institutions are the Scripps Institution of Oceanography (affiliated with the University of California, San Diego) and the Woods Hole Oceanographic Institution. In 1970 the faculty at these two institutions constituted 40% of the field. In 1990 they constituted only 25%.

Source: "Workshop Report, National Science Foundation Conference on Undergraduate Programs in Ocean Sciences," Galveston, Texas, 1992

leyball games, and chilly labs with classical music. Marine science is a long and demanding road; but it is, quite honestly, great fun. Captain Nemo and his sub never had it this good!

FOR MORE INFORMATION

Organizations

1. Write to the following organizations:

American Society of Limnology and Oceanography A nonprofit, professional scientific society that seeks to promote the interests of limnology and oceanography and related sciences and to further the exchange of information across the range of aquatic science disciplines. C. Susan Weiler, Whitman College, Walla Walla, WA 99362

Association for Women in Science A nonprofit association dedicated to increasing the educational and employment opportunities for both girls and women in all fields of science. 1522 K St., Suite 820, Washington, DC 20005, 202/408-0742, fax: 202/408-8321, e-mail: awis@access.digex.net

Center for Marine Conservation A nonprofit membership organization dedicated to protecting marine wildlife and its habitats and to conserving coastal and ocean resources. 1725 DeSales St., NW, Washington, DC 20036

The Cousteau Society A nonprofit environmental education organization dedicated to the protection and improvement of the quality of life for present and future generations. 870 Greenbrier Circle, Suite 402, Chesapeake, VA 23320

Marine Advanced Technology Training Center For training of marine technicians—specialists in the deployment and maintenance of tools used in marine science. Marine Advanced Technology Training Center, Monterey Peninsula College, 980 Fremont Street, Monterey, CA 93940, www.marinetech.org

Marine Technology Society An international, interdisciplinary society devoted to ocean and marine engineering

Oceanography Programs in North America

Here is a list of the largest and most comprehensive marine science teaching and research institutions in North America.

Bedford Institute of Oceanography, Dartmouth, Nova Scotia, Canada
Dalhousie University, Halifax, Nova Scotia, Canada
Louisiana State University, Baton Rouge, LA 70803
Massachusetts Institute of Technology, Cambridge, MA 02139
McGill University, Montreal, Canada
Mississippi State University, Starkville, MS 39759
North Carolina State University, Raleigh, NC 27695
Old Dominion University, Norfolk, VA 23529
Oregon State University, Corvallis, OR 97331-2134
State University of New York, Stony Brook, NY 11794-5001
Texas A & M University, Galveston, TX 77553-1675
University of Alaska, Fairbanks, AK 99775-5040
University of British Columbia, Vancouver BC, Canada
University of California, San Diego, La Jolla, CA 92093-0232[1]
University of Connecticut, Groton, CT 06340
University of Delaware, Newark, DE 19716
University of Florida, Gainesville, FL 32611-0409
University of Georgia, Athens, GA 30602
University of Hawaii, Honolulu, HI 96822
University of Maine, Orono, ME 04469-5715
University of Maryland, College Park, MD 20742
University of Michigan, Ann Arbor, MI 48109-2099
University of New Hampshire, Durham, NH 03824
University of North Carolina, Wilmington, NC 28403
University of Rhode Island, Narragansett, RI 02882-1197
University of South Carolina, Columbia, SC 29208
University of Southern California, Los Angeles, CA 90089-1231
University of Virginia, Charlottesville, VA 22903
University of Washington, Seattle, WA 98105
Woods Hole Oceanographic Institution, Woods Hole, MA 02543

[1]The Scripps Institution of Oceanography is affiliated with the University of California, San Diego.

science and policy. 1828 L St., NW, Suite 906, Washington, DC 20036-5104, 202/775-5966, fax: 202/429-9417

National Marine Educators Association An organization for those interested in the study and enjoyment of the world of water—both fresh and salt. PO Box 51215, Pacific Grove, CA 93950, 831/648-4837, e-mail: mrigsby@mbayaq.org

National Oceanic and Atmospheric Administration A government agency that guides use and protection of our oceans and coastal resources, warns of dangerous weather, charts the seas and skies, and conducts research to improve our understanding and stewardship of the environment. NOAA Correspondence Unit, 1305 East-West Highway, #8624, Silver Spring, MD 20910

National Science Foundation An independent agency of the federal government established in 1950 to promote and advance scientific progress in the United States. NSF Office of Legislative and Public Affairs, Rm. 1245, 4201 Wilson Blvd., Arlington, VA 22230, 703/306-1070, fax: 703/306-0157

Oceanic Engineering Society An organization that promotes the use of electronic and electrical engineers for instrumentation and measurement work in the ocean environment and the ocean/atmosphere interface. Norman D. Miller, 2644 NW Esplanada Dr., Seattle, WA 98117-2527, phone/fax: 206/784-7154, e-mail: n.miller@ieee.org

The Oceanography Society A professional society for scientists in the field of oceanography. 4052 Timber Ridge Drive, Virginia Beach, VA 23455, 804/464-0131, fax: 804/464-1759, e-mail: jrhodes@ccpo.odu.edu

Office of Naval Research A component of the U.S. Navy, ONR plans, fosters, and encourages scientific research and technology development in recognition of their paramount importance to national security. ONR Public Affairs Office, 800 North Quincy St., Arlington, VA 22217, 703/696-7034, fax: 703/696-5940

The Society for Marine Mammology A professional organization that supports the conservation of marine mammals and the educational, scientific, and managerial advancement of marine mammal science. Daniel K. Odell, SMM Education Committee, Sea World, 7007 Sea World Drive, Orlando, FL 32821-8097, 407/363-2662, fax: 407/345-5397, e-mail: odell@pegasus.cc.ucf._edu

Publications

2. Read one or more of the following publications:

   Careers in Oceanography and Marine-Related Fields— available from The Oceanography Society

   Education and Training Programs in Oceanography and Related Fields—$6, available from the Marine Technology Society

   Strategies for Pursuing a Career in Marine Mammal Science ($1.50 for students, $3 for others)—Allen Press, P.O. Box 1897, Lawrence, KS 66044-8897, 800/627-0629

   Taking the Initiative: Report on a Leadership Conference for Women in Science and Technology— available from the Assoc. for Women in Science

   You Can Teach Science (high school version)—available from the Office of Public Information, National Science Teachers Assoc., 1840 Wilson Blvd., Arlington, VA 22201, 703/243-7100

3. Consult the catalog of any college or university offering a marine science curriculum.

4. Send for a copy of *Ocean Opportunities: A Guide to What the Oceans Have to Offer.* The pamphlet is available from The Marine Technology Society, 1828 L Street, NW, Suite 906, Washington, DC 20036-5104, 202/775-5966.

5. See Wunsch, Carl. 1993. "Marine Science in the Coming Decades." *Science* 259 (no. 5093, 15 January): 296–7. An expert oceanographer's glimpse into the future.

6. See Yentsch, C. M., and C. J. Snyderman. 1992. *The Woman Scientist: Meeting the Challenges for a Successful Career.* New York: Plenum. Two marine scientists examine the facets of surviving and succeeding in academia and research science.

7. See American Meteorological Society. 1992. *Curricula in the Atmospheric, Oceanic, and Related Sciences.* Boston: American Meteorological Society. Source of current information on marine science programs in the United States, Canada, and Puerto Rico.

Note: The organizations and publications listed here are but a sampling of the resources that are available to help you learn more about careers in the marine sciences. Each contact you make in your search for information will lead you to more contacts and more information.

Glossary

absorption Conversion of sound or light energy into heat.

abyssal hill Small sediment-covered inactive volcano or intrusion of molten rock less than 200 meters (650 feet) high, thought to be associated with seafloor spreading. Abyssal hills punctuate the otherwise flat abyssal plain.

abyssal plain Flat, cold, sediment-covered ocean floor between the continental rise and the oceanic ridge at a depth of 3,700 to 5,500 meters (12,000 to 18,000 feet). Abyssal plains are more extensive in the Atlantic and Indian Oceans than in the Pacific.

abyssal zone The ocean between about 4,000 and 5,000 meters (13,000 and 16,500 feet) deep.

accessory pigment One of a class of pigments (such as fucoxanthin, phycobilin, and xanthophyll) present in various photosynthetic plants and that assist in the absorption of light and the transfer of its energy to chlorophyll. Also called *masking pigment*.

accretion An increase in the mass of a body by accumulation or clumping of smaller particles.

acid A substance that releases a hydrogen ion (H+) in solution.

active margin Continental margin near an area of lithospheric plate convergence. Also called *Pacific-type margin*.

adhesion Attachment of water molecules to other substances by hydrogen bonds. Wetting.

ahermatypic Describing coral species lacking symbiotic zooxanthellae and incapable of secreting calcium carbonate at a rate suitable for reef production.

air mass A large mass of air with nearly uniform temperature, humidity, and density throughout.

algin A mucilaginous commercial product of multicellular marine algae. Widely used as a thickening and emulsifying agent.

alkaline Basic. See *base*.

amphidromic point A "no-tide" point in an ocean caused by basin resonances, friction, and other factors around which tidal crests rotate. About a dozen amphidromic points exist in the world ocean. Sometimes called a *node*.

Antarctic Bottom Water The densest ocean water (1.0279 g/cm^3), formed primarily in Antarctica's Weddell Sea during Southern Hemisphere winters.

Antarctic Circumpolar Current See *West Wind Drift*.

aphelion The point in the orbit of a satellite where it is farthest from the sun; opposite of *perihelion*.

aphotic zone The dark ocean below the depth to which light can penetrate.

apogee The point in the orbit of a satellite farthest from the main body; opposite of *perigee*.

aquaculture The growing or farming of plants and animals in a water environment under controlled conditions. Compare *mariculture*.

Arthropoda The phylum of animals that includes shrimp, lobsters, krill, barnacles, and insects. The phylum Arthropoda is the world's most successful.

artificial system of classification A method of classifying an object based on attributes other than its reason for existence, its ancestry, or its origin. Compare *natural system of classification*.

asthenosphere The hot, plastic layer of the upper mantle below the lithosphere, extending some 300 kilometers (187 miles) below the surface. Convection currents within the asthenosphere power plate tectonics.

atmospheric circulation cell Large circuit of air driven by uneven solar heating and the Coriolis effect. Three circulation cells form in each hemisphere. See also *Hadley cell*; *Ferrel cell*; *polar cell*.

atoll A ring-shaped island of coral reefs and coral debris enclosing, or almost enclosing, a shallow lagoon from which no land protrudes. Atolls often form over sinking, inactive volcanoes.

authigenic sediment Sediment formed directly by precipitation from seawater. Also called *hydrogenous sediment*.

autotroph An organism that makes its own food by photosynthesis or chemosynthesis.

backshore Sand on the shoreward side of the berm crest, sloping away from the ocean.

backwash Water returning to the ocean from waves washing onto a beach.

baleen The interleaved, hard, fibrous, hornlike filters within the mouth of baleen whales.

barrier island A long, narrow, wave-built island lying parallel to the mainland and separated from it by a lagoon or bay. Compare *sea island*.

barrier reef A coral reef surrounding an island or lying parallel to the shore of a continent, separated from land by a deep lagoon. Coral debris islands may form along the reef.

barycenter The center of mass and of rotation of the Earth–moon system, 1,700 kilometers (1,060 miles) inside Earth.

basalt The relatively heavy crustal rock that forms the seabeds, composed mostly of oxygen, silicon, magnesium, and iron. Its density is about 2.9 g/cm^3.

base A substance that combines with a hydrogen ion (H+) in solution.

bathyal zone The ocean between about 200 and 4,000 meters (700 and 13,000 feet) deep.

bathymetry The discovery and study of ocean floor contours.

bathyscaphe Deep-diving submersible designed like a blimp, which uses gasoline for buoyancy and can reach the bottom of the deepest ocean trenches. From the Greek *batheos* ("depth") and *skaphidion* ("a small ship").

bay mouth bar An exposed sandbar attached to a headland adjacent to a bay and extending across the mouth of the bay.

beach A zone of unconsolidated (loose) particles extending from below water level to the edge of the coastal zone.

beach scarp Vertical wall of variable height marking the landward limit of the most recent high tides. Corresponds with the berm at extreme high tides.

benthic zone The zone of the ocean bottom. See also *pelagic zone.*

berm A nearly horizontal accumulation of sediment parallel to shore. Marks the normal limit of sand deposition by wave action.

berm crest The top of the berm; the highest point on most beaches. Corresponds to the shoreward limit of wave action during most high tides.

big bang The hypothetical event that started the expansion of the universe from a geometric point. The beginning of time.

biodegradable Able to be broken by natural processes into simpler compounds.

biogenous sediment Sediment of biological origin. Organisms can deposit calcareous (calcium-containing) or siliceous (silicon-containing) residue.

biological amplification Increase in concentration of certain fat-soluble chemicals such as DDT or heavy-metal compounds, in successively higher trophic levels within a food web.

biological resource A living animal or plant collected for human use. Also called a *living resource.*

biomass The mass of living material in a given area or volume of habitat.

biosynthesis The initial formation of life on Earth.

blade Algal equivalent of a vascular plant's leaf. Also called a *frond.*

breakwater An artificial structure of durable material that interrupts the progress of waves to shore. Harbors are often shielded by a breakwater.

buffer A group of substances that tend to work against change in the pH of a solution by combining with free ions.

buoyancy The ability of an object to float in a fluid by displacement of a volume of fluid equal to it in mass.

bykill Animals unintentionally killed when desirable organisms are collected.

calcareous ooze Ooze composed mostly of the hard remains of organisms containing calcium carbonate.

calorie The amount of heat needed to raise the temperature of 1 gram (0.035 ounce) of pure water by 1°C (1.8°F).

capillary wave A tiny wave with a wavelength of less than 1.73 centimeters (0.68 inch), whose restoring force is surface tension; the first type of wave to form when the wind blows.

Carnivora The order of mammals that includes seals, sea lions, walruses, and sea otters.

cartilage A tough, elastic tissue that stiffens or supports.

cartographer A person who makes maps and charts.

cephalopod The group of marine predators that includes squid, octopuses, and nautiluses.

Cetacea The order of mammals that includes porpoises, dolphins, and whales.

***Challenger* expedition** The first wholly scientific oceanographic expedition, 1872–76. Named for the steam corvette used in the voyage.

chart A map that depicts mostly water and the adjoining land areas.

chemical bond An energy relationship that holds two atoms together as a result of changes in their electron distribution.

chemical equilibrium In seawater, the condition in which the proportion and amounts of dissolved salts per unit volume of ocean are nearly constant.

chemosynthesis The synthesis of organic compounds from inorganic compounds using energy stored in inorganic substances such as sulfur, ammonia, and hydrogen. Energy is released when these substances are oxidized by certain organisms.

chlorinated hydrocarbons The most abundant and dangerous class of halogenated hydrocarbons, synthetic organic chemicals hazardous to the marine environment.

chlorinity A measure of the total weight of chloride, bromide, and iodide ions in seawater. We may derive salinity from chlorinity by multiplying by 1.80655.

chlorofluorocarbons (CFCs) A class of halogenated hydrocarbons thought to be depleting Earth's atmospheric ozone. CFCs are used as cleaning agents, refrigerants, fire-extinguishing fluids, spray-can propellants, and insulating foams.

Chondrichthyes The class of fishes with cartilaginous skeletons: the sharks, skates, rays, and chimaeras.

clamshell sampler Sampling device used to take shallow samples of the ocean bottom.

clay Sediment particle smaller than 0.004 millimeter in diameter; the smallest sediment size category.

climate The long-term average of weather in an area.

climax community A stable, long-established community of self-perpetuating organisms that tends not to change with time.

clumped distribution Distribution of organisms within a community in small, patchy aggregations, or clumps; the most common distribution pattern.

coast The zone extending from the ocean inland as far as the environment is immediately affected by marine processes.

coastal cell The natural sector of a coastline in which sand input and sand outflow are balanced.

coastal upwelling Upwelling adjacent to a coast, usually induced by wind.

coccolithophore A very small planktonic alga carrying discs of calcium carbonate, which contributes to biogenous sediments.

cohesion Attachment of water molecules to each other by hydrogen bonds.

Columbus, Christopher (1451–1506) Italian explorer in the service of Spain who discovered islands in the Caribbean in 1492. Although traditionally credited as the discoverer of America, he never actually sighted the North American continent.

community The populations of all species that occupy a particular habitat and interact within that habitat.

compass An instrument for showing direction by means of a magnetic needle swinging freely on a pivot and pointing to magnetic north.

conduction The transfer of heat through matter by the collision of one atom with another.

constructive interference The addition of wave energy as waves interact, producing larger waves.

continental crust The solid masses of the continents, composed primarily of granite.

continental drift The theory that the continents move slowly across the surface of Earth.

continental margin The submerged outer edge of a continent, made of granitic crust. Includes the continental shelf and the continental slope. Compare *ocean basin.*

continental rise The wedge of sediment forming the gentle transition from the outer (lower) edge of the continental slope to the abyssal plain. Usually associated with passive margins.

continental shelf Gradually sloping, submerged extension of a continent, composed of granitic rock overlain by sediments. Has features similar to the edge of the nearby continent.

continental slope The sloping transition between the granite of the continent and the basalt of the seabed. The true edge of a continent.

convection Movement within a fluid resulting from differential heating and cooling of the fluid. Convection produces mass transport or mixing of the fluid.

convection current A single closed-flow circuit of rising warm material and falling cool material.

convergence zone The line along which waters of different density converge. Convergence zones form the boundaries of tropical, subtropical, temperate, and polar areas.

convergent plate boundary A region where plates are pushing together and where a mountain range, island arc, and/or trench will eventually form. Often a site of much seismic and volcanic activity.

Cook, James (1728–1779) Officer in the British Royal Navy who led the first European voyages of scientific discovery.

coral Any of more than 6,000 species of small cnidarians, many of which are capable of generating hard calcareous (aragonite, $CaCO_3$) skeletons.

coral reef A linear mass of calcium carbonate (aragonite and calcite) assembled from coral organisms, algae, mollusks, worms, and so on. Coral may contribute less than half of the reef material.

core The innermost layer of Earth, composed primarily of iron, with nickel and heavy elements. The inner core is thought to be solid, the outer core liquid. The average density of the outer core is about 11.8 g/cm^3, and that of the inner core is about 16 g/cm^3.

Coriolis, Gaspard Gustave de (1792–1843) The French scientist who in 1835 worked out the mathematics of the motion of bodies on a rotating surface. See *Coriolis effect*.

Coriolis effect The apparent deflection of a moving object from its initial course when its speed and direction are measured in reference to the surface of the rotating Earth. The object is deflected to the right of its anticipated course in the Northern Hemisphere and to the left in the Southern Hemisphere. The deflection occurs for any horizontal movement of objects with mass and has no effect at the equator.

cosmogenous sediment Sediment of extraterrestrial origin.

crust The outermost solid layer of Earth, composed mostly of granite and basalt; the top of the lithosphere. The crust has a density of 2.7–2.9 g/cm^3 and accounts for 0.4% of Earth's mass.

cryptic coloration Camouflage. May be active (under control of the animal) or passive (an unalterable color or shape).

current Mass flow of water. (The term is usually reserved for horizontal movement.)

cyclone A weather system with a low-pressure area in the center around which winds blow counterclockwise in the Northern Hemisphere and clockwise in the Southern Hemisphere. Not to be confused with a tornado, a much smaller weather phenomenon associated with severe thunderstorms. See also *extratropical cyclone; tropical cyclone*

deep-water wave A wave in water deeper than half its wavelength.

deep zone The zone of the ocean below the pycnocline, in which there is little additional change of density with increasing depth. Contains about 80% of the world's water.

degree An arbitrary measure of temperature. One degree Celsius (°C) = 1.8 degrees Fahrenheit (°F).

delta The deposit of sediments found at a river mouth, sometimes triangular in shape (hence the name after the Greek letter).

density The mass per unit of volume of a substance, usually expressed in grams per cubic centimeter (g/cm^3).

density curve A graph showing the relationship between a fluid's temperature or salinity and its density.

density stratification The formation of layers in a material, with each deeper layer being denser (weighing more per unit of volume) than the layer above.

desalination The process of removing salt from seawater or brackish water.

desiccation Drying.

destructive interference The subtraction of wave energy as waves interact, producing smaller waves.

diatom Earth's most abundant, successful, and efficient single-celled phytoplankton. Diatoms possess two interlocking valves made primarily of silica. The valves contribute to biogenous sediments.

dinoflagellate One of a class of microscopic single-celled flagellates, not all of which are autotrophic. The outer covering is often of stiff cellulose. Planktonic dinoflagellates are responsible for "red tides."

disphotic zone The lower part of the photic zone, where illumination is not sufficient for photosynthetic productivity.

dissolution The dissolving by water of minerals in rocks.

disturbing force The energy that causes a wave to form.

diurnal tide A tidal cycle of one high tide and one low tide per day.

divergent plate boundary A region where plates are moving apart and where new ocean or rift valley will eventually form. A spreading center forms the junction.

doldrums The zone of rising air near the equator known for sultry air and variable breezes. See also *intertropical convergence zone (ITCZ)*.

downwelling Circulation pattern in which surface water moves vertically downward.

drag The resistance to movement of an organism induced by the fluid through which it swims.

drift net Fine, vertically suspended net that may be 7 meters (25 feet) high and 80 kilometers (50 miles) long.

dynamic theory of tides Model of tides that takes into account the effects of finite ocean depth, basin resonances, and the interference of continents on tide waves.

eastern boundary current Weak, cold, diffuse, slow-moving current at the eastern boundary of an ocean (off the west coast of a continent). Examples include the Canary Current and the Humboldt Current.

ebb current Water rushing out of an enclosed harbor or bay because of the fall in sea level as a tide trough approaches.

echolocation The use of reflected sound to detect environmental objects. Cetaceans use echolocation to detect prey and avoid obstacles.

echo sounder A device that reflects sound off the ocean bottom to sense water depth. Its accuracy is affected by the variability of the speed of sound through water.

ectotherm An organism incapable of generating and maintaining steady internal temperature from metabolic heat and therefore whose internal body temperature is approximately the same as that of the surrounding environment. A cold-blooded organism.

eddy A circular movement of water usually formed where currents pass obstructions, or between two adjacent currents flowing in opposite directions, or along the edge of a permanent current.

Ekman transport Net water transport, the sum of layer movement due to the Ekman spiral. Theoretical Ekman transport in the Northern Hemisphere is 90° to the right of the wind direction.

electron A tiny negatively charged particle in an atom responsible for chemical bonding.

El Niño A southward-flowing nutrient-poor current of warm water off the coast of western South America, caused by a breakdown of trade wind circulation.

endotherm An organism capable of generating and regulating metabolic heat to maintain a steady internal temperature. Birds and mammals are the only animals capable of true endothermy. A warm-blooded organism.

ENSO Acronym for the coupled phenomena of El Niño and the Southern Oscillation. See also *El Niño; Southern Oscillation.*

epicenter The point on Earth's surface directly above the focus of an earthquake.

equatorial upwelling Upwelling in which water moving westward on either side of the geographical equator tends to be deflected slightly poleward and replaced by deep water often rich in nutrients. See also *upwelling.*

equilibrium theory of tides Idealized model of tides that considers Earth to be covered by an ocean of great and uniform depth capable of instantaneous response to the gravitational and inertial forces of the sun and the moon.

Eratosthenes of Cyrene (276–192 B.C.) Greek scholar and librarian at Alexandria who first calculated the circumference of Earth about 230 B.C.

erosion A process of being gradually worn away.

estuary A body of water partially surrounded by land where fresh water from a river mixes with ocean water, creating an area of remarkable biological productivity.

euphotic zone The upper layer of the photic zone in which net photosynthetic gain occurs. Compare *photic zone.*

eustatic change A worldwide change in sea level, as distinct from local changes.

evaporite Deposit formed by the evaporation of ocean water.

excess volatiles Compounds found in the ocean and the atmosphere in quantities greater than can be accounted for by the weathering of surface rock. Such compounds probably entered the ocean and the atmosphere from deep crustal and upper mantle sources through volcanism.

exclusive economic zone (EEZ) The offshore zone claimed by signatories to the 1982 United Nations Draft Convention on the Law of the Sea. The EEZ extends 200 nautical miles (370 kilometers) from a contiguous shoreline. See also *United States Exclusive Economic Zone.*

exoskeleton A strong, lightweight, form-fitted external covering and support common to animals of the phylum Arthropoda. The exoskeleton is made partly of chitin and may be strengthened by calcium carbonate.

experiment A test that simplifies observation in nature or in the laboratory by manipulating or controlling the conditions under which observations are made.

extratropical cyclone A low-pressure mid-latitude weather system characterized by converging winds and ascending air rotating counterclockwise in the Northern Hemisphere and clockwise in the Southern Hemisphere. An extratropical cyclone forms at the front between the polar and Ferrel cells.

fault A fracture in a rock mass along which movement has occurred.

Ferrel cell The middle atmospheric circulation cell in each hemisphere. Air in these cells rises at 60° latitude and falls at 30° latitude. See also *westerlies.*

fetch The uninterrupted distance over which the wind blows without a significant change in direction, a factor in wind wave development.

Fissipedia The carnivoran suborder that includes sea otters.

fjord A deep, narrow estuary in a valley originally cut by a glacier.

flagellum (plural *flagella*) A whiplike structure used by some small organisms and gametes to move through the environment.

flood current Water rushing into an enclosed harbor or bay because of the rise in sea level as a tidal crest approaches.

food General term for organic molecules capable of providing energy to heterotrophs when combined with oxygen during biochemical respiration.

food web A group of organisms associated by a complex set of feeding relationships in which the flow of food energy can be followed from primary producers through consumers.

foraminiferan One of a group of planktonic amoeba-like animals with a calcareous shell, which contributes to biogenous sediments.

Forchhammer's principle See *principle of constant proportions.*

foreshore Sand on the seaward side of the berm, sloping toward the ocean, to the low-tide mark.

fracture zone Area of irregular, seismically inactive topography marking the position of a once-active transform fault.

fringing reef A reef attached to the shore of a continent or an island.

front The boundary between two air masses of different density. The density difference can be caused by differences in temperature and/or humidity.

frontal storm Precipitation and wind caused by the meeting of two air masses, associated with an extratropical cyclone. Generally, one air mass will slide over or under the other, and the resulting expansion of air will cause cooling and, consequently, rain or snow.

frustule Siliceous external cell wall of a diatom, consisting of two interlocking valves fitted together like the halves of a box.

fully developed sea The theoretical maximum height attainable by ocean waves given wind of a specific strength, duration, and fetch. Longer exposure to such wind will not increase the size of the waves.

galaxy A large rotating aggregation of stars, dust, gas, and other debris held together by gravity. There are perhaps 50 billion galaxies in the universe and 50 billion stars in each galaxy.

gas bladder In multicellular algae, an air-filled structure that assists in flotation.

gas exchange Simultaneous passage, through a semipermeable membrane, of oxygen into an animal and carbon dioxide out of it.

geostrophic gyre A gyre in balance between the Coriolis effect and gravity; literally, "turned by Earth."

gill membrane The thin boundary of living cells separating blood from water in a fish's (or other aquatic animal's) gills.

granite The relatively light crustal rock—composed mainly of oxygen, silicon, and aluminum—that forms the continents. Its density is about 2.7 g/cm^3.

greenhouse effect Trapping of heat in the atmosphere. Incoming short-wavelength solar radiation penetrates the atmosphere, but the outgoing longer-wavelength radiation is absorbed by greenhouse gases and reradiated to Earth, causing a rise in surface temperature.

greenhouse gases Gases in Earth's atmosphere that cause the greenhouse effect; these include carbon dioxide, methane, and CFCs.

groin A short, artificial projection of durable material placed at a right angle to shore in an attempt to slow longshore transport of sand from a beach. Usually deployed in repeating units.

Gulf Stream The strong western boundary current of the North Atlantic, off the east coast of the United States.

guyot A flat-topped, submerged, inactive volcano.

gyre Circuit of mid-latitude currents around the periphery of an ocean basin. Most oceanographers recognize five gyres plus the West Wind Drift.

habitat The place where an individual or population of a given species lives. Its "mailing address."

hadal zone The deepest zone of the ocean, below a depth of 5,000 meters (16,500 feet).

Hadley cell The atmospheric circulation cell nearest the equator in each hemisphere. Air in these cells rises near the equator because of strong solar heating there and falls because of cooling at about 30° latitude. See also *trade winds*.

halocline The zone of the ocean in which salinity increases rapidly with depth. See also *pycnocline*.

heat A form of energy produced by the random vibration of atoms or molecules.

heat budget An expression of the total solar energy received on Earth during some period of time and the total heat lost from Earth by reflection and radiation into space through the same period.

Henry the Navigator (1394–1460) Prince of Portugal who established a school for the study of geography, seamanship, shipbuilding, and navigation.

hermatypic Describing coral species possessing symbiotic zooxanthellae within their tissues and capable of secreting calcium carbonate at a rate suitable for reef production.

heterotroph An organism that derives nourishment from other organisms because it is unable to synthesize its own food molecules.

hierarchy Grouping of objects by degrees of complexity, grade, or class. A hierarchical system of nomenclature is based on distinctions within groups and between groups.

high-energy coast A coast exposed to large waves.

high seas That part of the ocean past the exclusive economic zone, which is considered common property to be shared by the citizens of the world. About 60% of the ocean area.

high tide The high-water position corresponding to a tidal crest.

holdfast A complex branching structure that anchors many kinds of multicellular algae to the substrate.

holoplankton Permanent members of the plankton community. Examples are diatoms and copepods. Compare *meroplankton*.

horse latitudes Zones of erratic horizontal surface air circulation near 30°N and 30°S latitudes. Over land, dry air falling from high altitudes produces deserts at these latitudes (for example, the Sahara).

hot spot A surface expression of a plume of magma rising from a stationary source of heat in the mantle.

hurricane A large tropical cyclone in the North Atlantic or eastern Pacific, whose winds exceed 118 kilometers (74 miles) per hour.

hydrogen bond Relatively weak bond formed between a partially positive hydrogen atom and a partially negative oxygen, fluorine, or nitrogen atom of an adjacent molecule.

hydrogenous sediment Sediment formed directly by precipitation from seawater. Also called *authigenic sediment*.

hydrostatic pressure The constant pressure of water around a submerged organism.

hydrothermal vent Spring of hot, mineral- and gas-rich seawater found on some oceanic ridges in zones of active seafloor spreading.

hypothesis A speculation about the natural world that may be verified or disproved by observation and experiment.

Ice Age One of several periods (lasting several thousand years each) of low temperature during the last million years. Glaciers and polar ice were derived from ocean water, lowering sea level at least 100 meters (328 feet). (See Appendix II, "Geological Time.")

interference Addition or subtraction of wave energy as waves interact. Also called *resonance*. See also *constructive interference*; *destructive interference*.

intertidal zone The marine zone between the highest high-tide point on a shoreline and the lowest low-tide point. The intertidal zone is sometimes subdivided into four separate habitats by height above tidal datum, typically numbered 1 to 4, land to sea.

intertropical convergence zone (ITCZ) The equatorial area at which the trade winds converge. The ITCZ usually lies at or near the meteorological equator. Also called the *doldrums*.

invertebrate Animal lacking a backbone.

ion An atom (or small group of atoms) that becomes electrically charged by gaining or losing one or more electrons.

island arc Curving chain of volcanic islands and seamounts almost always found paralleling the concave edge of a trench.

isostatic equilibrium Balanced support of lighter material in a heavier, displaced supporting matrix. Analogous to buoyancy in a liquid.

kelp Informal name for any species of large phaeophyte.

kingdom The largest category of biological classification. Five kingdoms are presently recognized.

krill *Euphausia superba*, a thumb-sized crustacean common in Antarctic waters.

lagoon A shallow body of seawater generally isolated from the ocean by a barrier island. Also the body of water enclosed within an atoll, or the water within a reverse estuary.

land breeze Movement of air offshore as marine air heats and rises.

La Niña An event during which normal tropical Pacific atmospheric and oceanic circulation strengthens, and the surface temperature of the eastern South Pacific drops below average values. Usually occurs at the end of an ENSO event. See *ENSO*.

latent heat of evaporation Heat added to a liquid during evaporation (or released from a gas during condensation) that produces a change in state but not a change in temperature. For pure water, 585 calories per gram at 20°C (68°F).

latent heat of fusion Heat removed from a liquid during freezing (or added to a solid during thawing) that produces a change in state but not a change in temperature. For pure water, 80 calories per gram at 0°C (32°F).

latitude lines Regularly spaced imaginary lines on Earth's surface running parallel to the equator.

law A large construct explaining events in nature that have been observed to occur with unvarying uniformity under the same conditions.

Library of Alexandria The greatest collection of writings in the ancient world, founded in the third century B.C. by Alexander the Great. Could be considered the first university.

limiting factor A physical or biological environmental factor whose absence or presence in an inappropriate amount limits the normal actions of an organism.

Linnaeus, Carolus Carl von Linné (1707–1778). Swedish "father" of modern taxonomy.

lithification Conversion of sediment into sedimentary rock by pressure or by the introduction of a mineral cement.

lithosphere The brittle, relatively cool outer layer of Earth, consisting of the oceanic and continental crust and the outermost, rigid layer of mantle.

littoral zone The band of coast alternately covered and uncovered by tidal action; the intertidal zone.

longitude lines Regularly spaced imaginary lines on Earth's surface running north and south and converging at the poles.

longshore bar A submerged or exposed line of sand lying parallel to shore and accumulated by wave action.

longshore current A current running parallel to shore in the surf zone, caused by the incomplete refraction of waves approaching the beach at an angle.

longshore drift Movement of sediments parallel to shore, driven by wave energy.

longshore trough Submerged excavation parallel to shore adjacent to an exposed sandy beach. Caused by the turbu-

lence of water returning to the ocean after each wave.

low-energy coast A coast only rarely exposed to large waves.

lower mantle The rigid portion of Earth's mantle below the asthenosphere.

low tide The low-water position corresponding to a tidal trough.

low tide terrace The smooth, hard-packed beach seaward of the beach scarp on which waves expend most of their energy. Site of the most vigorous onshore and offshore movement of sand.

lunar tide Tide caused by gravitational and inertial interaction of the moon and Earth.

macroplankton Animal plankters larger than 1 to 2 centimeters ($\frac{1}{2}$ to 1 inch). An example is the jellyfish.

Magellan, Ferdinand (c. 1480–1521) Portuguese navigator in the service of Spain who led the first expedition to circumnavigate Earth, 1519–22. He was killed in the Philippines during the expedition.

magma Molten rock capable of fluid flow. Called *lava* aboveground.

magnetometer A device that measures the amount and direction of residual magnetism in a rock sample.

Mammalia The class of mammals.

mangrove Large flowering shrub or tree that grows in dense thickets or forests along muddy or silty tropical coasts.

mantle The layer of Earth between the crust and the core, composed of silicates of iron and magnesium. The mantle has an average density of about 4.5 g/cm³ and accounts for about 68% of Earth's mass.

mariculture The farming of marine organisms, usually in estuaries, bays, or nearshore environments or in specially designed structures using circulating seawater. Compare *aquaculture*.

marine pollution The introduction by humans of substances or energy into the ocean that change the quality of the water or affect the physical and biological environment.

marine science The process (or result) of applying the scientific method to the ocean, its surroundings, and the life forms within it. Also called *oceanography* or *oceanology*.

masking pigment See *accessory pigment*.

mass extinction A catastrophic, global event in which major groups of species perish abruptly.

Maury, Matthew (1806–1873) Father of physical oceanography. Probably the first person to undertake the systematic study of the ocean as a full-time occupation, and probably the first to understand the global interlocking of currents, wind flow, and weather.

maximum sustainable yield The maximum amount of fish, crustaceans, and mollusks that can be caught without impairing future populations.

mean sea level The height of the ocean surface averaged over a few years' time.

meroplankton Temporary members of the plankton community Examples are very young fishes and barnacle larvae. Compare *holoplankton*.

metabolic rate The rate at which energy-releasing reactions proceed within an organism.

***Meteor* expedition** German Atlantic expedition begun in 1925; the first to use an echo sounder and other modern optical and electronic instrumentation.

meteorological equator Also called the *thermal equator*. The irregular imaginary line of thermal equilibrium between hemispheres. It is situated about 5° north of the geographical equator, and its position changes with the seasons, moving slightly north in northern summer.

meteorological tide A tide influenced by the weather. Arrival of a storm surge will alter the estimate of a tide's height or arrival time, as will a strong, steady onshore or offshore wind.

microtektite A small, rounded, glassy component of cosmogenous sediments, usually less than 1.5 millimeters ($\frac{1}{16}$ inch) in length. Thought to have formed from the impact of an asteroid or meteor on the crust of Earth or the moon.

Milky Way The name of our galaxy. Sometimes applied to the field of stars in our home spiral arm, which is correctly called the Orion arm.

mixed layer See *surface zone*.

mixed tide A complex tidal cycle, usually with two high tides and two low tides of unequal height per day.

mixing time The time necessary to mix a substance throughout the ocean, about 1,600 years.

molecule A group of atoms held together by chemical bonds. The smallest unit of a compound that retains the characteristics of the compound.

molluska The category of animals that includes chitons, snails, clams, and octopuses.

monsoon A pattern of wind circulation that changes with the season. Also, the rainy season in areas with monsoon wind patterns.

moraine Hills and ridges of sediments left when glaciers retreated. Some sections of primary coast have been built out by glacial moraines.

motile Able to move about.

multicellular algae Algae with bodies consisting of more than one cell. Examples are kelp and ulva.

Mysticeti The suborder of baleen whales.

nanoplankton Very small members of the plankton community. Examples are coccolithophores and silicoflagellates.

Nansen bottle A water-sampling instrument perfected early in this century by the Norwegian scientist and explorer Fridtjof Nansen.

natural system of classification A method of classifying an organism based on its ancestry or origin.

neap tide The time of smallest variation between high and low tides, occurring when Earth, moon, and sun align at right angles. Neap tides alternate with spring tides, occurring at two-week intervals.

nekton Pelagic organisms that actively swim.

neritic Of the shore or coast. Refers to continental margins and the water covering them, or to nearshore organisms.

neritic sediment Sediment of the continental shelf, consisting mainly of terrigenous material.

niche Description of an organism's functional role in a habitat. Its "job."

Niskin bottle Water-sampling device suspended from a ship or platform.

NOAA National Oceanic and Atmospheric Administration, founded within the U.S. Department of Commerce in 1970 to facilitate commerical uses of the ocean.

node The line or point of no wave action in a standing pattern. See also *amphidromic point.*

nodule Solid mass of hydrogenous sediment, most commonly manganese or ferromanganese nodules and phosphorite nodules.

nonextractive resource Any use of the ocean in place, such as transportation of

people and commodities by sea, recreation, or waste disposal.

nonrenewable resource Any resource that is present on Earth in fixed amounts and cannot be replenished.

nor'easter (northeaster) Any energetic extratropical cyclone that sweeps the eastern seaboard of North America in winter.

North Atlantic Deep Water Cold, dense water formed in the Arctic that flows onto the floor of the North Atlantic Ocean.

nutrient Any needed substance that an organism obtains from its environment except oxygen, carbon dioxide, and water.

ocean (1) The great body of saline water that covers 70.78% of the surface of Earth. (2) One of its primary subdivisions, bounded by continents, the equator, and other imaginary lines.

ocean basin Deep-ocean floor made of basaltic crust. Compare *continental margin.*

oceanic crust The outermost solid surface of Earth beneath ocean floor sediments, composed primarily of basalt.

oceanic ridge Young seabed at the active spreading center of an ocean, often unmasked by sediment, bulging above the abyssal plain. The boundary between diverging plates. Often called a mid-ocean ridge, though less than 60% of the length exists at mid-ocean.

oceanic zone The zone of open water away from shore, past the continental shelf.

oceanography The science of the ocean. See also *marine science.*

oceanus Latin form of *okeanos,* the Greek name for the "ocean river" past Gibraltar.

Odontoceti The suborder of toothed whales.

ooze Sediment of at least 30% biological origin.

orbit In ocean waves, the circular pattern of water particle movement at the air–sea interface. Orbital motion contrasts with the side-to-side or back-and-forth motion of pure transverse or longitudinal waves.

orbital wave A progressive wave in which particles of the medium move in closed circles.

Osteichthyes The class of fishes with bony skeletons.

outgassing The volcanic venting of volatile substances.

overfishing Harvesting so many fish that there is not enough breeding stock left to replenish the species.

ozone O_3, the triatomic form of oxygen. Ozone in the upper atmosphere protects living things from some of the harmful effects of the sun's ultraviolet radiation.

ozone layer A diffuse layer of ozone mixed with other gases surrounding the world at a height of about 20 to 40 kilometers (12 to 25 miles).

Pacific Ring of Fire The zone of seismic and volcanic activity that encircles the Pacific Ocean.

paleoceanography The study of the ocean's past.

paleomagnetism The "fossil," or remanent, magnetic field of a rock.

Pangaea Name given by Alfred Wegener to the original "protocontinent." The breakup of Pangaea gave rise to the Atlantic Ocean and to the continents we see today.

Panthalassa Name given by Alfred Wegener to the ocean surrounding Pangaea.

partially mixed estuary An estuary in which an influx of seawater occurs beneath a surface layer of fresh water flowing seaward. Mixing occurs along the junction.

passive margin Continental margin near an area of lithospheric plate divergence. Also called *Atlantic-type margin.*

pelagic Of the open ocean. Refers to the water above the deep-ocean basins, sediments of oceanic origin, or organisms of the open ocean.

pelagic sediment Sediments of the slope, rise, and deep-ocean floor that originate in the ocean.

pelagic zone The realm of open water. See also *benthic zone.*

perigee The point in the orbit of a satellite where it is closest to the main body; opposite of *apogee.*

perihelion The point in the orbit of a satellite where it is closest to the sun; opposite of *aphelion.*

pH scale A measure of the acidity or alkalinity of a solution. Numerically, the negative logarithm of the concentration of hydrogen ions in an aqueous solution. A pH of 7 is neutral; lower numbers indicate acidity, and higher numbers indicate alkalinity.

Phaeophyta Brown multicellular algae, including kelps.

photic zone The thin film of lighted water at the top of the world ocean. The photic zone rarely extends deeper than 200 meters (660 feet). Compare *euphotic zone*.

photosynthesis The process by which autotrophs bind light energy into the chemical bonds of food with the aid of chlorophyll and other substances. The process uses carbon dioxide and water as raw materials and yields glucose and oxygen.

physical factor An aspect of the physical environment that affects living organisms, such as light, salinity, or temperature.

physical resource Any resource that has resulted from the deposition, precipitation, or accumulation of a useful nonliving substance in the ocean or seabed. Also called a *nonliving resource*.

phytoplankton Plantlike, usually single-celled members of the plankton community.

Pinnipedia The carnivoran suborder that contains the seals, sea lions, and walruses.

piston corer A seabed-sampling device capable of punching through up to 25 meters (80 feet) of sediment and returning an intact plug of material.

planet A smaller, usually nonluminous body orbiting a star.

plankter Informal name for a member of the plankton community.

plankton Drifting or weakly swimming organisms suspended in water. Their horizontal position is to a large extent dependent on the mass flow of water rather than on their own swimming efforts.

plankton net Conical net of fine nylon or Dacron fabric used to collect plankton.

plate One of about a dozen rigid segments of Earth's lithosphere that move independently. The plate consists of continental or oceanic crust and the cool, rigid upper mantle directly below the crust.

plate tectonics The theory that Earth's lithosphere is fractured into plates, which move relative to each other and are driven by convection currents in the mantle. Most volcanic and seismic activity occurs at plate margins.

plunging wave Breaking wave in which the upper section topples forward and away from the bottom, forming an air-filled tube.

polar cell The atmospheric circulation cell centered over each pole.

polar front Boundary between the polar cell and the Ferrel cell in each hemisphere.

polar molecule A molecule with an unbalanced charge. One end of the molecule has a slight negative charge, and the other end has a slight positive charge.

Polynesia A large group of Pacific islands lying east of Melanesia and Micronesia and extending from the Hawaiian Islands south to New Zealand and east to Easter Island.

polyp One of two body forms of Cnidaria. Polyps are cup-shaped and possess rings of tentacles. Coral animals are polyps.

poorly sorted sediment A sediment in which particles of many sizes are found.

population A group of individuals of the same species occupying the same area.

population density The number of individuals per unit area.

potable water Water suitable for drinking.

precipitation Liquid or solid water that falls from the air and reaches the surface as rain, hail, or snowfall.

primary coast Coast on which terrestrial influences dominate. See also *secondary coast*.

primary consumer Initial consumer of primary producers or autotrophs; the second level in food webs.

primary producer An organism capable of using energy from light or energy-rich chemicals in the environment to produce energy-rich organic compounds. An autotroph.

primary productivity The synthesis of organic materials from inorganic substances by photosynthesis or chemosynthesis. Expressed in grams of carbon bound into carbohydrate per unit area per unit time ($gC/m^2/yr$).

Prince Henry the Navigator (1394–1460) Prince of Portugal who established a school for the study of geography, seamanship, shipbuilding, and navigation.

principle of constant proportions The principle that the proportions of major conservative elements in seawater remain nearly constant though total salinity may change with location. Also called *Forchhammer's principle*.

progressive wave A wave of moving energy in which the wave form moves in one direction along the surface (or junction) of the transmission medium (or media).

proton A positively charged particle at the center of an atom.

protosun Our sun in the protostar stage—that is, as a tightly condensed knot of material that had not yet attained fusion temperature.

pteropod Small planktonic mollusk with a calcareous shell, which contributes to biogenous sediments.

pycnocline The middle zone of the ocean, in which density increases rapidly with depth. Temperature falls and salinity rises in this zone.

radioactive decay The disintegration of unstable forms of elements, which releases subatomic particles and heat.

radiolarian One of a group of usually planktonic amoeba-like animals with a siliceous shell, which contributes to biogenous sediments.

radiometric dating The process of determining the age of rocks by observing the ratio of unstable radioactive elements to stable decay products.

random distribution Distribution of organisms within a community whereby the position of one organism is in no way influenced by the positions of other organisms or by physical variations within that community. A very rare distribution pattern.

reef A hazard to navigation. A shoal, a shallow area, or a mass of fish or other marine life.

refraction Bending of light or sound waves as they move at an angle other than 90° between media of different optical or acoustical densities. See also *wave refraction*.

renewable resource Any resource that is naturally replaced on a seasonal basis by the growth of living organisms or by other natural processes.

residence time The average length of time a dissolved substance spends in the ocean.

restoring force The dominant force trying to return water to flatness after formation of a wave.

reverse estuary An estuary along a coast in which salinity increases from the ocean to the estuary's upper reaches because of evaporation of seawater and a lack of freshwater input.

Rhodophyta Red multicellular algae.

rip current A strong, narrow surface current that flows seaward through the surf zone and is caused by the escape of excess water that has piled up in a longshore trough.

rogue wave A single wave crest much higher than usual, caused by constructive interference.

salinity A measure of the dissolved solids in seawater, usually expressed in grams per kilogram or parts per thousand by weight. Standard seawater has a salinity of 35‰ (parts per thousand) at 0°C (32°F).

salinometer An electronic device that determines salinity by measuring the electrical conductivity of a seawater sample.

salt gland Specialized tissue responsible for concentration and excretion of excess salt from blood and other body fluids.

salt wedge estuary An estuary in which rapid river flow and small tidal range cause an inclined wedge of seawater to form at the mouth.

sand Sediment particle between 0.062 and 2 millimeters in diameter.

sandbar A submerged or exposed line of sand accumulated by wave action.

sand spit An accumulation of sand and gravel deposited downcurrent from a headland. Sand spits often curl at their tips.

scattering The dispersion (or "bounce") of sound or light waves when they strike particles suspended in water or air. The amount of scatter depends on the number, size, and composition of the particles.

schooling Tendency of small fish of a single species, size, and age to mass in groups. The school moves as a unit, which confuses predators and reduces the effort spent searching for mates.

science A systematic way of asking questions about the natural world and testing the answers to those questions.

scientific method The orderly process by which theories explaining the operation of the natural world are verified or rejected.

scientific name The genus and species name of an organism.

sea breeze Onshore movement of air as inland air heats and rises.

sea cave A cave near sea level in a sea cliff cut by processes of marine erosion.

sea cliff Cliff marking the landward limit of marine erosion on an erosional coast.

seafloor spreading The theory that new ocean crust forms at spreading centers, most of which are on the ocean floor, and pushes the continents aside. Power is thought to be provided by convection currents in Earth's upper mantle.

sea island Island whose central core was connected to the mainland when sea level was lower. Rising ocean separates these high points from land, and sedimentary processes surround them with beaches. Compare *barrier island.*

seamount Circular or elliptical projection from the seafloor, more than 1 kilometer (0.6 mile) in height, with a relatively steep slope of 20° to 25°.

Seasat First satellite dedicated to oceanic research, launched in 1978.

secondary coast Coast dominated by marine processes. See also *primary coast.*

sediment Particles of organic or inorganic matter that accumulate in a loose, unconsolidated form.

seiche Pendulum-like rocking of water in an enclosed area; a form of standing wave that can be caused by meteorological or seismic forces, or that may result from normal resonances excited by tides.

seismic sea wave Tsunami caused by displacement of earth along a fault. (Earthquakes and seismic sea waves are caused by the same phenomenon.)

seismic wave A low-frequency wave generated by the forces that cause earthquakes. Some kinds of seismic waves can pass through Earth.

seismograph An instrument that detects and records earth movement associated with earthquakes and other disturbances.

semidiurnal tide A tidal cycle of two high tides and two low tides each lunar day, with the high tides of nearly equal height.

sensible heat Heat whose gain or loss is detectable by a thermometer or other sensor.

sessile Attached. Nonmotile. Unable to move about.

sewage sludge Semisolid mixture of organic matter, microorganisms, toxic metals, and synthetic organic chemicals removed from wastewater at a sewage treatment plant.

shallow-water wave A wave in water shallower than 1/20 its wavelength.

shelf break The abrupt increase in slope at the junction between continental shelf and continental slope.

shore The place where ocean meets land. On nautical charts, the limit of high tides.

siliceous ooze Ooze composed mostly of the hard remains of silica-containing organisms.

silt Sediment particle between 0.004 and 0.062 millimeter in diameter.

Sirenia The order of mammals that includes manatees, dugongs, and the extinct sea cows.

slack water A time of no tide-induced currents that occurs when the current changes direction.

sofar *Sound fixing and ranging.* An experimental U.S. Navy technique for locating survivors on life rafts, based on the fact that sound from explosive charges dropped into the layer of minimum sound velocity can be heard for great distances. See *sofar layer.*

sofar layer Layer of minimum sound velocity in which sound transmission is unusually efficient. Sounds leaving this depth tend to be refracted back into it. The sofar layer usually occurs at mid-latitude depths around 1,200 meters (4,000 feet).

solar nebula The diffuse cloud of dust and gas from which the solar system originated.

solar system The sun together with the planets and other bodies that revolve around it.

solar tide Tide caused by the gravitational and inertial interaction of the sun and Earth.

sonar *Sound navigation and ranging.*

sound A form of energy transmitted by rapid pressure changes in an elastic medium.

sounding Measurement of the depth of a body of water.

Southern Oscillation A reversal of airflow between normally low atmospheric pressure over the western Pacific and normally high pressure over the eastern Pacific. The cause of El Niño. See *El Niño.*

species diversity Number of different species in a given area.

specific heat The heat, measured in calories, required to raise 1 gram of a substance 1°C. The input of 1 calorie of heat energy raises the temperature of 1 gram of pure water by 1°C.

spilling wave A breaking wave whose crest slides down the face of the wave.

spreading center The junction between diverging plates at which new ocean floor is being made. Also called *spreading zone.*

spring tide The time of greatest variation between high and low tides, occurring when Earth, moon, and sun form a straight line. Spring tides alternate with neap tides throughout the year, occurring at two-week intervals.

standing wave A wave in which water oscillates without causing progressive wave forward movement. There is no net transmission of energy in a standing wave.

star A massive sphere of incandescent gases powered by the conversion of hydrogen to helium and other heavier elements.

state An expression of the internal form of matter. Water exists in three states: solid, liquid, and gas. A solid has a fixed volume and fixed shape; a liquid has a fixed volume but no fixed shape; and a gas has neither fixed volume nor fixed shape.

stipe Multicellular algal equivalent of a vascular plant's stem.

storm Local or regional atmospheric disturbance characterized by strong winds often accompanied by precipitation.

storm surge An unusual rise in sea level as a result of the low atmospheric pressure and strong winds associated with a tropical cyclone. Onrushing seawater precedes landfall of the tropical cyclone and causes most of the damage to life and property.

stratigraphy The branch of geology that deals with the definition and description of natural divisions of rocks. Specifically, the analysis of relationships of rock strata.

subduction The downward movement into the asthenosphere of a lithospheric plate.

subduction zone An area at which a lithospheric plate is descending into the asthenosphere. The zone is characterized by linear folds (trenches) in the ocean floor and strong deep-focus earthquakes. Also called a *Wadati–Benioff zone*.

sublittoral zone The ocean floor near shore. The inner sublittoral extends from the littoral (intertidal) zone to the depth at which wind waves have no influence; the outer sublittoral extends to the edge of the continental shelf.

submarine canyon A deep, V-shaped valley running roughly perpendicular to the shoreline and cutting across the edge of the continental shelf and slope.

succession The changes in species composition that lead to a climax community.

supralittoral zone The splash zone above the highest high tide; not technically part of the ocean bottom.

surf The confused mass of agitated water rushing shoreward during and after the breaking of a wind wave.

surface current Horizontal flow of water at the ocean's surface.

surface zone The upper layer of ocean, in which temperature and salinity are relatively constant with depth. Depending on local conditions, the surface zone may reach to 1,000 meters (3,300 feet) or be absent entirely. Also called the *mixed layer*.

surf zone The region between the breaking waves and the shore.

sverdrup (sv) A unit of volume transport named in honor of oceanographer Harald U. Sverdrup: 1 million cubic meters of water flowing past a fixed point each second.

swash Water from waves washing onto a beach.

swim bladder A gas-filled organ that assists in maintaining neutral buoyancy in some bony fishes.

Teleostei The osteichthyan order that contains the cod, tuna, halibut, perch, and other species of bony fishes.

temperature The response of a solid, liquid, or gas to the input or removal of heat energy. A measure of the atomic and molecular vibration in a substance, indicated in degrees.

terrane An isolated segment of seafloor, island arc, plateau, continental crust, or sediment transported by seafloor spreading to a position adjacent to a larger continental mass. Usually different in composition from the larger mass.

terrigenous sediment Sediment derived from the land and transported to the ocean by wind and flowing water.

territorial waters Waters extending 12 miles from shore and in which a nation has the right to jurisdiction.

thallus The body of an alga or other simple plant.

theory A general explanation of a characteristic of nature consistently supported by observation or experiment.

thermal equilibrium The condition in which the total heat coming into a system (such as a planet) is balanced by the total heat leaving the system.

thermal inertia Tendency of a substance to resist change in temperature with the gain or loss of heat energy.

thermocline The zone of the ocean in which temperature decreases rapidly with depth. See also *pycnocline*.

thermohaline circulation Water circulation produced by differences in temperature and/or salinity (and therefore density).

thermostatic property A property of water that acts to moderate changes in temperature.

tidal bore A high, often breaking wave generated by a tidal crest that advances rapidly up an estuary or river.

tidal current Mass flow of water induced by the raising or lowering of sea level owing to passage of tidal crests or troughs. See also *ebb current*; *flood current*.

tidal datum The reference level (0.0) from which tidal height is measured.

tidal range The difference in height between consecutive high and low tides.

tidal wave The crest of the wave causing tides. Another name for a tidal bore. Not a tsunami or seismic sea wave.

tide Periodic short-term change in the height of the ocean surface at a particular place, generated by long-wavelength progressive waves that are caused by the interaction of gravitational force and inertia. Movement of Earth beneath tide crests results in the rhythmic rising and falling of sea level.

tombolo Above-water bridge of sand connecting an offshore feature to the mainland.

top consumer An organism at the apex of a trophic pyramid, usually a carnivore.

TOPEX/Poseidon Joint French–U.S. satellite launched in 1992 to study the ocean.

tornado Localized, narrow, violent funnel of fast-spinning wind, usually generated when two air masses collide. Not to be confused with a cyclone. (The tornado's oceanic equivalent is a waterspout.)

trace element A minor constituent of seawater present in amounts less than 1 part per million.

trade winds Surface winds within the Hadley cells, centered at about 15° latitude, which approach from the northeast in the Northern Hemisphere and from the southeast in the Southern Hemisphere.

transform fault A plane along which rock masses slide horizontally past one another.

transform plate boundary Place where crustal plates shear laterally past one another. Crust is neither produced nor destroyed at this type of junction.

transverse current East-to-west or west-to-east current linking the eastern and western boundary currents. An example is the North Equatorial Current.

trench An arc-shaped depression in the deep-ocean floor with very steep sides and a flat sediment-filled bottom coinciding with a subduction zone. Most trenches occur in the Pacific.

trophic pyramid A model of feeding relationships between organisms. Primary producers form the base of the pyramid; consumers eating one another form the higher levels, with the top consumer at the apex.

tropical cyclone A weather system of low atmospheric pressure around which winds blow counterclockwise in the Northern Hemisphere and clockwise in the Southern Hemisphere. Originates in the tropics within a single air mass but may move into temperate waters if water temperature is high enough to sustain it. Small tropical cyclones are called *tropical depressions,* larger ones *tropical storms,* and great ones *hurricanes, typhoons,* or *willi-willies,* depending on location.

tsunami Long-wavelength shallow-water wave caused by rapid displacement of water. See also *seismic sea wave.*

turbidite A terrigenous sediment deposited by a turbidity current; typically, coarse-grained layers of nearshore origin interleaved with finer sediments.

turbidity current An underwater "avalanche" of abrasive sediments thought responsible for the deep sculpturing of submarine canyons and a means of transport for sediments accumulating on abyssal plains.

uniform distribution Distribution of organisms within a community characterized by equal space between individuals (the arrangement of trees in an orchard). The rarest natural distribution pattern.

United States Exclusive Economic Zone The region extending seaward from the coast of the United States for 200 nautical miles, within which the United States claims sovereign rights and jurisdiction over all marine resources.

United States Exploring Expedition The first U.S. oceanographic research voyage, launched in 1838.

upwelling Circulation pattern in which deep, cold, usually nutrient-laden water moves toward the surface. Upwelling can be caused by winds blowing parallel to shore or offshore.

valve In diatoms, each half of the protective silica-rich outer portion of the cell. The complete outer covering is called the *frustule.*

vertebrate A chordate with a segmented backbone.

voyaging Traveling (usually by sea) for a specific purpose.

water vapor The gaseous, invisible form of water.

wave Disturbance caused by the movement of energy through a medium.

wave crest Highest part of a progressive wave above average water level.

wave-cut platform The smooth, level terrace sometimes found on erosional coasts that marks the submerged limit of rapid marine erosion.

wave frequency The number of waves passing a fixed point per second.

wave height Vertical distance between a wave crest and the adjacent wave troughs.

wave period Time it takes for successive wave crests to pass a fixed point.

wave refraction Slowing and bending of progressive waves in shallow water.

wave shock Physical movement, often sudden, violent, and of great force, caused by the crash of a wave against an organism.

wave steepness Height-to-wavelength ratio of a wave. The theoretical maximum steepness of deep-water waves is 1:7.

wave trough The valley between wave crests below the average water level in a progressive wave.

wavelength The horizontal distance between two successive wave crests (or troughs) in a progressive wave.

weather The state of the atmosphere at a specific place and time.

Wegener, Alfred (1880–1930) German scientist who proposed the theory of continental drift in 1912.

well-mixed estuary An estuary in which slow river flow and tidal turbulence mix fresh and salt water in a regular pattern through most of its length.

well-sorted sediment A sediment in which particles are of uniform size.

westerlies Surface winds within the Ferrel cells, centered around 45° latitude, which approach from the southwest in the Northern Hemisphere and from the northwest in the Southern Hemisphere.

western boundary current Strong, warm, concentrated, fast-moving current at the western boundary of an ocean (off the east coast of a continent). Examples include the Gulf Stream and the Japan (Kuroshio) Current.

westward intensification The increase in speed of geostrophic currents as they pass along the western boundary of an ocean basin.

West Wind Drift The Antarctic Circumpolar Current, driven by powerful westerly winds north of Antarctica. The largest of all ocean currents, it continues permanently eastward without changing direction.

Wilson, John Tuzo (1908–1993) Canadian geophysicist who proposed the theory of plate tectonics in 1965.

wind The mass movement of air.

wind duration The length of time the wind blows over the ocean surface, a factor in wind wave development.

wind strength Average speed of the wind, a factor in wind wave development.

wind wave Gravity wave formed by transfer of wind energy into water. Wavelengths of from 60 to 150 meters (200 to 500 feet) are most common in the open ocean.

world ocean The great body of saline water that covers 70.78% of Earth's surface.

zone Division or province of the ocean with homogeneous characteristics.

zooplankton Animal members of the plankton community.

zooxanthellae Unicellular dinoflagellates that are symbiotic with coral and that produce the relatively high pH and some of the enzymes essential for rapid calcium carbonate deposition in coral reefs.

Credits

This page constitutes an extension of the copyright page. We have made every effort to trace the ownership of all copyrighted material and to secure permission from copyright holders. In the event of any question arising as to the use of any material, we will be pleased to make the necessary corrections in future printings. Thanks are due to the following for permission to use the material indicated.

Chapter 1

Figure 1.17 & Figure 1.18 provided by Deborah Day of Scripps Institute of Oceanography.

Chapter 2

Chapter opening vignette by Slocum, J. 1899. *Sailing Alone Around the World.* Reprint. New York: Sheridan House, 1972. **Figure 2.6** reprinted by permission of William K. Hartmann.

Chapter 3

Figure 3.22 reprinted by permission of the NOAA/National Geophysical Data Center.

Chapter 4

Figure 4.1 provided by Deborah Day of Scripps Institute of Oceanography. **Figure 4.2a** provided by Robert Profeta of Orange Coast College, Marine Science Department. **Figure 4.14** reprinted by permission of Taylor & Francis International Scientific and Educational Publishers. **Figure 4.25** reprinted by permission of Chet Raymo.

Chapter 5

Chapter opening illustration by R. Miller from History of the Earth by W. K. Hartmann and R. Miller, Workman Publishing Company, 1992. Reprinted by permission of the authors.

Chapter 7

Figure 7.11a & Figure 7.11b reprinted by permission of Alan D. Iselig.

Chapter 8

Figure 8.11 reprinted by permission of O. Brown, R. Evans, and M. Carle, University of Miami Rosenstiel School of Marine

and Atmospheric Science. **Table 8.1** from *Oceanography: A View of the Earth,* 7/e by Gross, M. G., © 1996. Reprinted by permission of Prentice-Hall, Inc., Upper Saddle River, N.J. **Figure 8.21** from William Schmitz, Woods Hole Institute.

Chapter 10

Chapter opening illustration by Arthur Rackham.

Chapter 11

Figure 11.1 from Dr. William Haxby, LDEO. **Figure 11.26** from G. H. Wheless, Old Dominion University.

Chapter 12

Figure 12.7 after Milne, 1995.

Chapter 15

Figure 15.23 from "Northern Hemisphere Millennial Temperature Reconstruction," *Geophysical Research Letters,* Vol. 26, No. 6, p. 759–762 by Michael E. Mann, Raymond S. Bradley, and Michael K. Hughes. Copyright © 1999 American Geophysical Union. Reprinted by permission of Michael E. Mann.

Appendices

Appendix III, Figures 1, 2, 3, and **4** and **Appendix IV, Figure 1,** from David Greenhood, *Down to Earth: Mapping for Everybody.* Reprinted by permission of The University of Chicago Press.

Photo Credits

Chapter 1

Chapter opening photo courtesy of NASA. **Figure 1.2b** Tom Garrison. **Figure 1.6** Painting © Herb Kawainui Kane. **Figure 1.8** Reprinted by permission of the James Ford Bell Library, University of Minnesota. **Figure 1.10** © National Maritime Museum, London. **Figure 1.11** Painting © Herb Kawainui Kane. **Figure 1.12** Reprinted by permission of the United States Naval Academy Museum. **Figures 1.14, 1.15** Courtesy of the Library of Congress. **Figure 1.16** Reprinted by permission of the Royal Geographical Society, London, archives from the journal

of Lt. Pelham Aldrich kept aboard *H.M.S. Challenger, 1872–1875.* **Figure 1.18** *Challenger Report,* Scripps Institute of Oceanography archives. **Figure 1.19a** Reprinted by permission of Hulton Deutsch Collection Limited. **Figure 1.19b** Courtesy of Alda Amundsen, SPRI. **Figure 1.21** Courtesy of the Deep Sea Drilling Program. **Box 1.1a** Courtesy of Ted Delaca, University of Alaska. **Box 1.1b** © 1998 Commander Submarine Force, U.S. Pacific Fleet. **Figure 1.22** Photo used by permission of JAMSTEC. **Figures 1.23a & b** Reprinted by permission of the Scripps Institution of Oceanography. **Figure 1.24** Courtesy of NASA.

Chapter 2

Chapter opening illustration by Rockwell Kent. **Figure 2.1a** Courtesy of Philip Richardson, Woods Hole Oceanographic Institution. **Figure 2.1b** Tom Garrison. **Figure 2.1c** © Dan Burton Photography. **Figure 2.3** Courtesy of ESO. **Figure 2.5a** Used with permission of © William K. Hartman. **Figure 2.5b & c** From Hubble Space Telescope, STScI. **Figure 2.7a** Photo by © William K. Hartman **Figure 2.7b** © Don Dixon. All rights reserved. **Figure 2.9** Reprinted by permission of Woods Hole Oceanographic Institution. **Figure 2.11** Photo by Department of Earth & Space Sciences/ UCLA **Figure 2.12, 2.13** Courtesy of NASA.

Chapter 3

Chapter opening photo © Dorian Weisel, Hawaii. **Figure 3.2** Reprinted by permission of the USGS. **Figure 3.8** © Alfred Wegener Institute for Polar and Marine Research. **Figure 3.15d** Reprinted by permission of V. Courtillot. **Figure 3.22, 3.26** Reprinted by permission of Dr. Peter Sloss, NOAA/National Geophysical Data Center, Boulder, Colorado.

Chapter 4

Chapter opening a photo courtesy of Don Walsh, International Maritime, Incorporated. **Chapter opening b** photo courtesy of Dr. Andreas Rechnitzer. **Figure 4.1** *Challenger Report,* Scripps Institute of Oceanography archives. **Figure 4.3b** Courtesy of Charles H. Langmuir and

Milene Cormier and Scripps Institute of Oceanography. **Figure 4.4a** Reprinted by permission of the U.S. Department of the Navy. **Figure 4.4c** Illustration courtesy of Karen Marks, NOAA. **Figures 4.11, 4.12** Courtesy of William Haxby, Lamont. **Figure 4.15** Reprinted by permission of the Society for Sedimentary Geology. **Figure 4.16** Reprinted by permission of the U.S. Department of the Navy. **Figure 4.18a** Courtesy of Aluminum Company of America. **Figure 4.18b** Courtesy of Lamont-Doherty Earth Observatory of Columbia University. **Fig 4.20** Photo is used by permission of Woods Hole Oceanographic Institution. **Figure 4.22** Heezen and Hollister. **Box 4.1a** Courtesy of NUYTCO Research. **Box 4.1c** Photo used by permission of JAMSTEC. **Figure 4.23** Photo is used by permission of Deborah Smith, Woods Hole Oceanographic Institution, and Joe Cann, University of Leeds. **Figure 4.26** Reprinted by permission of Dr. Peter Sloss, NOAA/ National Geophysical Data Center, Boulder, Colorado.

Chapter 5

Chapter opening illustration by R. Miller from *History of the Earth* by W. K. Hartmann and R. Miller, Workman Publishing Company, 1992. Reprinted by permission of the authors. **Figures 5.1, 5.2, 5.3** Courtesy of Charles D. Hollister. **Figure 5.4** Heezen and Hollister. **Figure 5.5a** Photo courtesy of the USGS. **Figure 5.5b** Courtesy of NASA. **Figure 5.6** Courtesy of the Department of Geology, University of Delaware. **Figure 5.7a** Courtesy of Dr. Howard Spero. **Figure 5.7b** Photo used by permission of Wim van Egmond. **Figure 5.7c** Courtesy of Roger Witmer. **Figure 5.8a** Photo used by permission of Wim van Egmond. **Figure 5.8b** Courtesy of Roger Witmer. **Figure 5.9** Reprinted by permission of Susumi Honjo. **Figure 5.10a & b** Courtesy of Unical Research. **Figure 5.12a** Courtesy of Department of Geology, University of Delaware. **Box 5.1 a & b** Courtesy of Dr. James Broda, Woods Hole Oceanographic Institution. **Figure 5.13a** Tom Garrison. **Figures 5.14, 5.15** Reprinted by permission of the Deep Sea Drilling Project, Texas A&M University. **Figure 5.17** NASA/JPL/ Malin Space Science Systems.

Chapter 6

Chapter opening photo © C. Sims, SuperStock, Inc. **Figure 6.5a** Tom Garrison. **Figure 6.5c** Photo courtesy of

William Cochlan. **Figure 6.6** Reprinted by permission of David Breiter. **Figure 6.23b** Photo courtesy of EG&C Ocean Products. **Figure 6.23c** Courtesy of WESMAR.

Chapter 7

Chapter opening a Art by George Kelvin from *Meteorology Today* by C. Donald Ahrens, Brooks/Cole, 2000. Reprinted by permission of the author. **Chapter opening b** Courtesy of Breitling, Switzerland. **Figure 7.1** © Chigmaroff/ Davison, SuperStock, Inc. **Figure 7.10** Courtesy of NASA. **Figure 7.14** Reprinted by permission of Tom Loebl, Imaging Publications. **Figure 7.15** © Bettmann Archive/ CORBIS. **Figure 7.16, 7.19** Courtesy of NASA. **Box 7.1a** Provided by the SeaWiFS Project, NASA/Goddard Space Flight Center and ORBIMAGE. **Box 7.1b** AP/Wide World Photos. **Figure 7.20** Courtesy of Jim Bell, Cornell University Astronomy Department.

Chapter 8

Chapter opening photo © Bettmann Archive/ CORBIS. **Figure 8.7b** Courtesy of NASA/JPL. **Figure 8.11** Reprinted by permission of O. Brown, R. Evans, and M. Carle, University of Miami Rosenstiel School of Marine and Atmospheric Science. **Figure 8.15b & d, 8.16a–d** Courtesy of NASA. **Figure 8.17** TOPEX/ Poseidon Team, CNES, NASA. **Figure 8.21** Used by permission of Woods Hole Oceanographic Institution. **Figure 8.22a** Tom Garrison. **Figure 8.22b** Courtesy of Philip Richardson, Woods Hole Oceanographic Institution. **Figure 8.22c** Used by permission of Webb Research.

Chapter 9

Chapter opening photo ©Ron Romanosky. **Figure 9.8** Tom Garrison. **Figure 9.10** Courtesy of NASA. **Figure 9.12** Art by Raychel Ciemma. **Figure 9.13** Reprinted by permission of the U.S. Department of the Navy. **Figures 9.16a & b** © Dennis Junor, Creation Captured. **Figure 9.17** from Erich Lessing, Art Resource, NY: S0126532 40-15-01/23 Color transp., Crane, Walter (1845–1915). Neptune's Horses. Oil on Canvas. Neue Pinakothek, Munich, Germany. Reprinted by permission of Art Resource. **Figure 9.18b** Used by permission of the University of Hawaii Press. **Figures 9.19a & b** Photos courtesy of the Thames Barrier Visitors Centre. **Figures 9.22a–c** Courtesy of the USGS. **Figure 9.23** Courtesy of the Bishop Museum, Hawaii. **Figure 9.24** Photo by Ky-

odo Photo Service. **Figure 9.26a** Tom Garrison. **Figure 9.26b** AP/Wide World Photos.

Chapter 10

Chapter opening a illustration by Arthur Rackham. **Chapter opening b** photo used by permission of BODO Arrangement, Norway. **Figures 10.19a & b** Courtesy of Scott Walking Adventures, Halifax, Nova Scotia. **Figure 10.20** Photo used by permission of F. W. Rowbotham, England. **Figure 10.21** Used by permission of Cabrillo Marine Aquarium. **Figure 10.22** Photo courtesy of Photothèque EDF.

Chapter 11

Chapter opening Tom Garrison. **Figures 11.1a & b** Courtesy of William Haxby, Lamont. **Figure 11.2** Courtesy of NASA. **Figure 11.3** Photo by Bob & Suzanne Clemenz, © Bob Clemenz Photography. **Figure 11.4** Courtesy of the USGS. **Figure 11.5** Courtesy of NASA. **Figure 11.7** Used by permission of John S. Shelton. **Figure 11.8** Photo courtesy of Dr. George P. Lafaker, USGS. **Figure 11.9** Courtesy of NASA. **Figures 11.14a & b** Photos used by permission of Gerald G. Kuhn. **Figures 11.15a–c** Courtesy of Dr. William Wallace. **Figure 11.15d** Courtesy of Dante de la Rosa. **Figure 11.16** © Carr Clifton Photography. All rights reserved. **Figure 11.19** Courtesy of Jack Mertz. **Figure 11.20** Courtesy of NASA. **Figures 11.21a & b** Photos used by permission of Gerald G. Kuhn. **Figure 11.21c** Used by permission of W. Demetrakas, O. C. Camera. **Figure 11.22** Photo courtesy of Great Barrier Reef Marine Park Authority, Queensland, Australia. **Figure 11.23** © Brian Parker/ Tom Stack & Associates. **Figure 11.26** Used by permission of Glen Wheless. **Figure 11.27** Photo used by permission of John S. Shelton. **Figure 11.28a** The Fairchild Aerial Photography Collection at Whittier College, Flight C-1670, frame 4. **Figure 11.28b** The Fairchild Aerial Photography Collection at Whittier College, Flight C-14180, frame 3:61. **Figure 11.30** AP/Wide World Photos. **Figure 11.31** Photo used by permission of John S. Shelton.

Chapter 12

Chapter opening photo used by permission of © Bruce Hall. **Figure 12.2b** Photo used by permission of Wim van Egmond. **Figure 12.3a & b** Provided by the Sea-WiFS Project, NASA/Goddard Space Flight Center and ORBIMAGE. **Figures 12.6c**

Index